Solar PV Power

Solar PV Power
Design, Manufacturing and Applications from Sand to Systems

Rabindra Satpathy
Venkateswarlu Pamuru

ACADEMIC PRESS

An imprint of Elsevier

ELSEVIER

Academic Press is an imprint of Elsevier
125 London Wall, London EC2Y 5AS, United Kingdom
525 B Street, Suite 1650, San Diego, CA 92101, United States
50 Hampshire Street, 5th Floor, Cambridge, MA 02139, United States
The Boulevard, Langford Lane, Kidlington, Oxford OX5 1GB, United Kingdom

Notices

Knowledge and best practice in this field are constantly changing. As new research and experience broaden our understanding,
changes in research methods, professional practices, or medical treatment may become necessary.

Practitioners and researchers must always rely on their own experience and knowledge in evaluating and using any information,
methods, compounds, or experiments described herein. In using such information or methods they should be mindful of their
own safety and the safety of others, including parties for whom they have a professional responsibility.

To the fullest extent of the law, neither the Publisher nor the authors, contributors, or editors, assume any liability for any injury
and/or damage to persons or property as a matter of products liability, negligence or otherwise, or from any use or operation of
any methods, products, instructions, or ideas contained in the material herein.

Library of Congress Cataloging-in-Publication Data
A catalog record for this book is available from the Library of Congress

British Library Cataloguing-in-Publication Data
A catalogue record for this book is available from the British Library

ISBN 978-0-12-817626-9

For information on all Academic Press publications
visit our website at https://www.elsevier.com/books-and-journals

Publisher: Brian Romer
Acquisitions Editor: Lisa Reading
Editorial Project Manager: Aleksandra Packowska
Production Project Manager: Prasanna Kalyanaraman
Cover Designer: Christian J. Bilbow

Typeset by SPi Global, India

Working together
to grow libraries in
developing countries

www.elsevier.com • www.bookaid.org

Dedication

Dedicated to my beloved mother, the late Smt. Charubala Satpathy,
who inspired me to be a solar engineer from my college days.
Rabindra Satpathy

Dedicated to my beloved mother, the late Smt. Pamuru Subamma.
Dr. Venkateswarlu Pamuru

Contents

About the authors

Rabindra Satpathy

He is an electrical engineering graduate from NIT, Warangal. He has more than 30 years of professional experience in the area of renewable energy, that is, in the fields of solar photovoltaic (PV) power systems, wind power generation, and others. He started his solar career in 1984 at the Odisha Renewable Energy Development Agency, installing India's first 25 Kwp solar PV diesel hybrid system, a PV wind hybrid system, and a 1.1 MW wind farm. He served as a key member of the management team at Tata BP Solar, Bangalore, from 1990 to 2002. He was instrumental in setting up India's first 5 MW project in Rajasthan and a 1 MW rooftop project for the Thyagaraj Sports Complex, New Delhi, as president of Reliance Industries Limited-Solar Group from 2007 to 2013. He was involved in establishing a solar module and cell factory as well as planning for a large manufacturing facility for polysilicon to systems.

Affiliations and expertise

Board Member of International Solar Energy Society in Freiburg, Germany.

Venkateswarlu Pamuru

He holds a Ph.D. in material science from the Indian Institute of Science, Bangalore. He worked in the Indian Space Research Organisation for 23 years as a scientist in the solar panels division. He was involved in the development of space solar panel technology with Si, GaAs, and multijunction solar cells. He worked as a project manager for the solar array for the INSAT-2C and INSAT-2D satellites and was involved in the design, manufacturing, and testing of solar panels for satellites. He was involved in the establishment of Si solar cell and module manufacturing lines. He also developed the fabrication process and produced modules at Moserbaer Photovoltaics and Reliance Industries. He is currently involved in the design and engineering of rooftop and ground-mounted solar PV power systems as well as monitoring the operational and maintenance aspects of solar PV plants.

Affiliations and expertise

Worked as General Manager, Engineering, Quanta Power Solutions India Pvt. ltd. Presently associated with Grand Solar Pvt. Ltd., Chennai, India.

Foreword

When my good colleague Rabindra Satpathy and his coauthor Venkateswartu Pamuru asked me to write a foreword for this important volume, I accepted immediately. During my 10 years as president of the International Solar Energy Society (ISES) from 2010 to 2019, I worked closely with Rabindra on the executive committee to formulate ISES programs, with particular emphasis on PV-related activities in India and around the world. ISES is the oldest continuously operating solar energy society in the world, with its roots dating all the way back to 1954 with the formation of the Association for Applied Solar Energy in Arizona. Perhaps not so coincidentally, this was the same year that Bell Telephone Laboratories patented the first commercial solar cell. The Society, which became officially known as ISES in 1970, has advocated for governments, universities, and private industry to invest in solar energy research, development, and deployment, and to report on this work through scientific conferences and technical publications. These activities, and in particular the ISES Solar Energy Journal and the ISES Solar World Congresses and more recently the ISES webinar series, have certainly had a significant influence on the remarkable growth of the solar industry from watts to gigawatts during the past half century. According to recent reports,[a] the worldwide installed capacity of PV power systems is now estimated at 650 GW (as of mid-2020) and will likely exceed 1 TW in just a few short years,[b] meeting as much as 10% of the world's electricity needs.

Solar PV Power serves as a textbook-level description of the entire solar supply chain, starting in the early chapters with detailed discussions of the basic ingredients and the manufacturing processes of silicon-based PV systems, and concluding with chapters that discuss the multitude of applications of PV systems being used today. Specifically, Chapter 1 sets the stage with a comprehensive description of the foundation of silicon-based PV systems, quartz: why this ingredient is required for polysilicon production, and its current price and availability. The chapter then continues with the basic manufacturing processes for polysilicon, covering topics such as polysilicon specifications and the process of converting quartz to solar-grade silicon. The chapter significantly highlights the Siemens manufacturing process and the equipment used in this process, but also examines alternative manufacturing processes such as fluidized bed reactors.

Chapter 2 then delves into the process of growing silicon crystals to make ingots, which are fundamental to producing silicon crystalline solar cells. The chapter summarizes processes for making ingots and ingot casts, and describes equipment required for ingot manufacturing. Topics such as quality testing and reliability are also discussed. The chapter concludes with a discussion on manufacturers and the manufacturing capacity of ingots globally as well as technology trends and an analysis of the pricing of ingots.

The conventional processes as well as new approaches for silicon wafer manufacturing are the focus of Chapter 3. The discussion of conventional processes includes first establishing specifications for wafers, the process for making wafers such as ingot slicing using wire and diamond saws, quality testing and reliability, and the equipment required for wafer manufacturing. Current silicon wafer manufacturers as well as technology trends and pricing analysis of wafers are also included in this

---

[a]REN21 Global Status Report, 2019.
[b]Solar Power Europe Global Market Outlook, 2019–2023.

discussion of conventional manufacturing approaches. Chapter 3 then brings up various topics covering new manufacturing innovations, such as the 1366 process and the Kerf loss wafer process by epitaxy. The chapter further explores new trends in these new processes, including their advantages and disadvantages, and cost reduction opportunities in direct wafer manufacturing.

From silicon wafer manufacturing, Chapter 4 goes into great detail to present the all-important topic of manufacturing crystalline silicon solar cells. Following a basic description of the physics of solar cells, there is an extensive discussion of processes in their manufacture and the many aspects related to maintaining high quality control during the manufacturing process. First, the wafers must be checked, and saw damage removed. Topics such as etching and doping, including the equipment used in these processes, are given substantial discussion. Then, processes such as etching of phosphor silicate glass, antireflection coating deposition, contact making (e.g., screen printing, photolithography and evaporation, and electroplating), and the testing and sorting of solar cells are covered. The subsection of cell manufacturing is capped off with a description of sun simulator requirements and suppliers. Chapter 4 then provides more comprehensive descriptions, including manufacturing processes, of P-type vs. N-type solar cells, the PERC/PERL cells, bifacial cells, and advanced high efficiency cells. As with previous and subsequent chapters, Chapter 4 closes with descriptions of equipment used in solar cell manufacturing, a list of silicon solar cell manufacturers, and technology trends and pricing analysis of solar cells.

Chapter 5 then covers the manufacturing of crystalline silicon solar PV models. After providing a description of the specifications of solar modules, the chapter digs into the many detailed aspects involved in the manufacturing process, such as the superstrate and substrate, encapsulant, solar ribbons, the junction box and frame, adhesives, solar cell tabbing and stringing, solar circuit layout, electrical testing, etc. Processes involving special types of PV modules, such as bifacial and glass-to-glass, are also covered. The chapter then goes into reliability and testing procedures, offers a list of manufacturers of silicon PV modules, and provides information on technology trends and pricing analysis of solar PV modules.

The remaining chapters of the book deal with the extensive types of end-use power applications offered by solar PV systems. Chapter 6 introduces a summary of the many traditional solar power applications, such as hybrid power systems, solar home systems, street lighting, water pumping, and battery charging. The chapter also presents new, innovative applications that offer possibilities of significant expansion of solar system utilization, such as floating solar PV, electric vehicle charging, and even solar-powered wheelchairs and drones.

The next chapters go into further detail on specific categories of applications. Chapter 7, for example, covers the very important topic of off-grid solar PV systems, including system components, system design, costing, and some examples. The chapter includes information about system reliability and life cycle cost analysis as well as typical examples of system performance using analysis and calculations. Approximately 1 billion people around the world, and especially in Africa and south and southeast Asia, are still without adequate access to reliable electricity services, so this chapter is key to describing how solar PV can make electrical energy accessible to everyone.

Another key application is rooftop solar (where the solar PV system is integrated into the local electrical distribution system) and building-integrated PV (BIPV). Chapter 8 provides a description of rooftop solar and BIPV, including the components involved, the design of these systems, and costing information. Field applications and examples of rooftop solar systems are presented, along with examples of system performance analysis based on calculations. Key information about the reliability and life cycle cost analysis of rooftop and BIPV systems is also provided.

Despite the ongoing importance of distributed solar applications such as described in Chapters 7 and 8, there is a growing trend to develop solar farms, serving as central power stations connected to national or regional grids. This has become the lowest-cost solution for providing bulk solar power to the grid, and recently has become the lowest-cost solution for adding power to the grid from *any* source in many cases. Chapter 9 addresses this topic in detail by providing an overall description of grid-connected solar PV systems, and the components required to make the connection, such as inverters, DC and AC cables, array junction boxes or DC string combiner boxes, module mounting structures, and other civil works. Chapter 9 also provides information on AC systems and protections and proper grounding. Design aspects, including simulation tools, costing aspects, and life cycle cost analyses, are also covered.

Finally, in Chapter 10 an overall discussion of grid integration and the performance of solar PV power systems is presented. This chapter covers topics such as requirements for AC systems, transformers, switchgears, substations, and transmission lines as well as protection requirements. The chapter presents methodologies and calculation schemes for performance analysis and presents key information on system reliability as well as operation and maintenance requirements.

Over the past half century, solar PV has evolved from a curious laboratory experiment with limited commercial use to an elegant and low-cost technology that provides, especially with storage systems, reliable electricity services almost anywhere and at almost any scale, from remote and off-grid applications to bulk energy supply into national grids. With our electricity demands increasing relative to all other energy demands, and with a growing climate crisis that places urgency on decarbonizing our energy supply, solar PV power system applications of all forms are destined to grow substantially and rapidly. The International Renewable Energy Agency's RE-MAP analysis[c] shows that possibly more than 8.4 TW of PV capacity will be installed worldwide by the middle of this century, roughly 12 times the current installed capacity. A key to this growth, however, is to educate and train a workforce that can continuously support this technology, and to inform policy makers and energy decision makers of the low cost and reliability of solar PV technologies. *Solar PV Power* serves as a very important contribution to supporting the growth of solar PV. I commend the coauthors Rabindra Satpathy and Venkateshwarlu Pamuru for this remarkable work and encourage all readers to make frequent use of this book as they develop their own solar power programs.

David Renné

Former Scientist of NREL and Immediate Past President of the International Solar Energy Society

---

[c]IRENA Global Energy Transformation, 2019 Edition: https://www.irena.org/-/media/Files/IRENA/Agency/Publication/2019/Apr/IRENA_Global_Energy_Transformation_2019.pdf.

Acknowledgments

The authors are thankful to all the people associated with this book, from its conceptualization through its completion.

Our gratitude goes to Padmabibhushan—Dr. R.L. Mashelkar, former Director General of the Council of Scientific and Industrial Research and a member of the board of directors of Reliance Industries Ltd., who supported us in writing this book, which could serve as a guide to all solar enthusiasts, engineers, scientists, young entrepreneurs, and the overall solar community at large.

Rabindra Satpathy would like to express thanks to his wife, Mamta Satpathy, and his daughter, Dr. Arushi Satpathy, for their unwavering encouragement and morale boosts during the entire journey of this book.

Satpathy would also like to convey special thanks and gratitude to his brother-in-law, Manoranjan Mohapatra (Mao), who provided the courage and stimulation needed for the completion of this book. His support was of paramount importance while working on this book in Bangalore in early 2020.

Dr. Venkateswarlu Pamuru would like to extend thanks to his sons, P. Venugopal and P. Ravikanth, for lending their expertise in drawings for various chapters as well as their constant motivation during the book writing journey. He would also like to thank his wife, P. Vijayalakshmi, for continuous support.

The completion of this book could not have been accomplished without the support of Dr. Gopal Krishnan, who provided excellent technical support on the polysilicon process, from his vast experience in the only company that makes polysilicon in India, Mettur Chemicals. Thanks also go to Prof. S. Venugopalan for his insightful input for the "Batteries, Energy Storage, and Electric Vehicle Charging". And lastly, Madan Kumar. G, for drafting and creating all the AutoCAD drawings and other images used in this book.

Finally, the authors would also like to thank all their colleagues and friends who have been with them in this endeavor and for their unforgettable help and assistance in the successful writing of the various chapters of this book.

Manufacturing of polysilicon

1.1 Introduction

There is an increasing demand for solar PV installation in the world. In 2019, 97.1 GWp of solar PV capacity has been installed globally [1]. As per GTM research, within 5 years, the installed PV capacity will reach 1 terawatt$_p$ [2]. Silicon has been the workhorse material since the inception of the solar PV industry and 90% of PV installations were done with Si solar cell-based modules. To meet this demand, very large quantities of polysilicon feedstock material are required to manufacture Si solar cells. To produce solar-grade Si, high purity quartz is required.

Quartz is available in abundance in the Earth's crust in many parts of the world. SiO_2, which is silica, is one of the major sources of quartz. It has many applications, such as in the manufacture of glass, ceramics, synthetic inorganic crystal phases, refractory minerals, and other uses. Although quartz is available in excess, it is not found in high purity in many places around the world. Quartz is a sought-after item and is used by many important industries such as semiconductors, high-temperature crucibles, and solar wafer/cell manufacturing.

Silicon is used as a raw material in the chemical and semiconductor industries. It is also used in the alloying of steel, cast iron, and aluminum. The major portion of silicon production is ferrosilicon while the remaining portion that is produced is metallurgical-grade (MG) silicon. Metallurgical-grade silicon is used in the production of solar-grade silicon for solar wafer/cell manufacturing.

Annually, millions of metric tons of ferrosilicon and silicon material are refined. Silicon is mostly used in the manufacturing of cement, mortars, ceramics, glass, and polymers. Silicon-based polymers are additionally utilized as options in contrast to hydrocarbon-based items. They can show up in numerous everyday products, for example, lubricants, adhesives, greases, skin and hair care items.

Metallurgical silicon is used in steel making and in aluminum casting as an alloying agent. It is used in making fumed silica, which is a thickening agent; silanes, which are coupling agents; and producing silicones by the chemical industries. So, there is a large demand for quartz material.

1.2 Quartz—the input for polysilicon production

There is an increasing demand for silicon metal used in the manufacture of solar cells, which is currently a growing industry. Next to oxygen, silicon is the second most abundant element available on Earth. Si occurs only in an oxidized state in the form of rock quartz or other silicate materials.

Solar PV Power. https://doi.org/10.1016/B978-0-12-817626-9.00001-0

The problem is that it never occurs naturally in a pure, elemental form. Due to the formation of silicon and oxygen compounds in nature, it is available as silicon dioxide or silicates. Silicon dioxide is found in large quantities of quartz sand or quartzite in many parts of the world.

Quartz is present in igneous rocks, but in a pure form only if excess silica exists. As magma cools, first silicates, that is, compounds with oxygen, silicon, and metal, are formed. The formation of quartz is possible with sufficient availability of Si and O_2 after silicates are formed. Quartz is the second most common mineral in the world and its average content in igneous rocks is 12%. Moreover, many rocks of the sedimentary and metamorphic types contain quartz crystals. Although purity varies significantly, there is an unlimited source of silicon.

Quartz with the highest purity and lower impurities is not available in abundance; this type of high-purity quartz is concentrated in a few locations around the world. Impurities in quartz are caused by either the fluid inclusions of other minerals or the ions of trace elements incorporated within the quartz lattice.

Spruce Pine is an idyllic mountain town in the Appalachian Mountains, 780 m above sea level, and it is the source of quartz mines of high purity. The quartz mines were shaped at Spruce Pine because of a geological event that happened 380 million years ago [3].

Because of the collision effect between North America and Africa continents, the friction resulted in the melting of rocks that were 10–15 miles under the Earth. Due to pressure at this location, there were cracks in molten rock with additional cracks in adjacent rock. This geological activity created deposits called pegmatites [4]. There was a slow cooling and crystallization of the deposits over several million years.

Due to the depth and lack of moisture present, impurities were unable to penetrate the grain structure of the material. Spruce Pine quartz is the world's primary source of the raw material needed to make the fused-quartz crucibles in which computer chip-grade polysilicon is melted for Si crystal growth.

1.2.1 Specifications/requirements of quartz

The quartz that is suitable for making metallurgical-grade Si, solar-grade Si, and electronic-grade Si should have the following requirements.

The quartz should contain SiO_2 of 98–99%. The impurity content of each of the elements of Al, Ca, and K in quartz should be <0.06%. Titanium impurity should be <50 ppm. The iron impurity should be <0.1%. The impurity content of phosphorous should be <0.001% and sodium should be <0.01%.

Boron and phosphorous are the dopants for the p- and n-type wafers, respectively; and they interfere and negatively affect the performance of solar cells. So, the boron and phosphorus contents are specified in the subparts per million (ppm) range.

The impurities of iron and transition elements will become defect centers in the cell and affect the performance of the solar cell; hence, they are specified in the ppm range. The calcium type of impurity will interfere with the molten Si and hence, it has to be removed.

The semiconductor industry imposes restricted requirements on quartz purity. Quartz should have high purity and excellent high-temperature properties so that it can resist thermal shock and have better thermal stability. There is a requirement for withstanding extreme temperature and a rapid heat transfer rate, known as rapid thermal processing (RTP). The rapid thermal process is employed to modify the properties of wafers.

There are specific requirements of impurities in quartz and the limiting values differ industry to industry. In the case of crucibles and semiconductor-based materials, the aluminum content in the refined quartz needs to be lower than 10 ppm, other metals less than 0.1 ppm, and total impurities should be less than 15 ppm, ideally.

The fineness and particle size are also criteria for quartz. The sand that is found in the river beds and sea beaches is also a form of silica, but it cannot be used for metallurgical Si production in view of the higher impurities and fineness of particles. The fineness of the particles will not provide permeability to escape for the gases generated in the furnace during the carbothermic reduction of silica.

The requirement of silicon purity grade is expected to vary for different applications. For solar cell applications, the purification level of solar-grade Si must be 99.9999%, which is referred to as six nines or 6N pure. For integrated circuits in the electronics industry, the purity is even higher, around 9–11N.

1.2.2 Quartz resources, processing, and quantity required for polysilicon production

Quartz is the main source for the mineral silica (SiO_2). Silica sand is formed of small, round particles of quartz, formed with breaking granite rocks due to weather effects combined with erosion. The quartz is obtained from the following:

Homogeneous Pegmatite: A rock equivalent generally found with a composition of 15–45% quartz and 45–85% feldspar along with small quantities of other minerals.

Fig. 1.1A–C shows images of homogeneous pegmatite.

Zoned Pegmatite: This term is utilized for pegmatite bodies with a quartz-rich focal central portion (core) and feldspar in the peripheral(minimal) zone.

Fig. 1.2 shows an image of zone pegmatite quartz.

Quartz Veins: A quartz vein is a linear body of quartz intruded in other igneous rocks. The range of width varies between 1 and 18 m with a length more than 100 m.

Fig. 1.3 shows an image of quartz veins.

Lump quartz is another commercial source of silica (SiO_2). Lump quartz originates from pegmatite and quartz veins, which are engaged with fragmentary crystallization more than quartz in silica sand. Silica sand originates from erosion and weathering of granite. It is also the most abundant feldspar-bearing rock in the Earth's crust. Fig. 1.4 shows an image of lump quartz.

Quartz float: This term is used to describe quartz debris scattered in a circular manner around a small quartz outcrop in the middle.

Fig. 1.5 shows an image of quartz.

Quartzite: The grains of quartz appear close to each other and silica cementing material is seen between the grains of quartz. Orthoquartzite demonstrates less porosity. Under a lens, quartz grains show texture interlock and metamorphism signage.

Fig. 1.6 shows an image of quartzite.

For making a 1 Wp solar pv module with an efficiency of 17%, about 5 g of solar-grade silicon is required [6]. Hence, for 1 GWp of solar PV capacity, 5000 tons of solar-grade polysilicon are required.

For making one ton of solar-grade silicon, three tons of metallurgical-grade Si are required.

For making one ton of metallurgical-grade silicon, 3.5 tons of quartz is required. In the conventional Siemens process, to make one ton of solar-grade silicon, 11.84 tons of quartz is required. In the Elkem process, to make one ton of solar-grade silicon, 5.92 tons of quartz is required.

To meet 100GWp solar PV capacity per year, a minimum of 2960KTPA and a maximum of 5920KTPA will be required, depending on the process followed, to manufacture solar-grade poly silicon.

1.2.3 Quartz purification, sources of availability and price

High-purity quartz is produced from natural quartz through intense processing. Quartz is studied under a microscope and with other advanced instruments and, we can study the presence of fluid inclusion, deformation, the size of fluid inclusions, etc.

The processing flow generally includes the following steps [7].

1.2.3.1 Crushing

In this process, mined quartz is crushed into pieces using crushing/smashing equipment. Generally, the quartz smashing plant comprises a jaw smasher, a cone crusher, an impact smasher, a vibrating feeder, a vibrating screen, and a belt conveyor. The vibrating feeder feeds materials to the jaw crusher for essential crushing. At that point, the yielding material from the jaw crusher is moved to a cone crusher for optional crushing, and afterward to effect for the third time crushing. As part of next process, the squashed quartz is moved to a vibrating screen for sieving to various sizes.

(A)

FIG. 1.1

Homogeneous pegmatite quartz [5]. (A) Example of pegmatite rock, (B) an example of pegmatite—a light blue color cleavelandite, with smoky quartz crystals available in Russia, (C) Mika pegmatite available in Tajikistan.

(Courtesy: Oleg Lopatkin, Russia.)

FIG. 1.1, CONT'D

FIG. 1.2

Zone pegmatite quartz [5].

(Courtesy for image: Oleg Lopatkin, Russia.)

FIG. 1.3

Quartz veins.

FIG. 1.4

Lump quartz.

FIG. 1.5

Quartz.

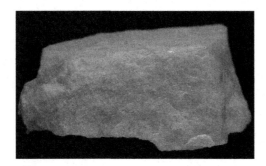

FIG. 1.6

Quartzite.

1.2.3.2 Attrition scrubbing

Scrubbing is the use of mechanical force and grinding force between the sand particles to remove the thin film iron, bonding, and muddy impurities. Further, it is used to grind the unprepared mineral aggregates on the surface of the quartz sand, and then to further purify the quartz sand by a classification operation.

1.2.3.3 Magnetic separation

Magnetic separation removes heavy minerals from silica, as they are mostly paramagnetic or ferromagnetic. These minerals are pulled in by an attractive field. Magnetic fields of moderate strength are required to remove the ferromagnetic minerals while higher strength of magnetic fields are necessary to remove paramagnetic minerals.

1.2.3.4 Froth flotation

Froth separation is another method for removing minerals. In a medium of water, the removal of minerals takes place in which quartz ore is fed in a suspended form with simultaneous agitation so that any sedimentation is avoided. The addition of a frothing agent along with air forms air bubbles. Hydrophobized mineral particles such as heavy minerals, feldspar, or mica get attached to the air bubbles and rise to the surface, forming froth. The particles remain suspended under the froth layer. The froth, which is carrying the minerals, is separated.

Apart from the physical processing methods, a suitable chemical treatment process is deployed to remove impurities in the surface to achieve the highest silica purity. Hot chlorination, acid wash, and leaching are designated processes. Mild acids such as H_2SO_4 or HCl are used for the acid wash. In the leaching process, advanced H_2SO_4 is used to remove the surface impurities properly at an elevated temperature. The impurities in microfissures and at dislocations shall be separated due to the high dissolution quartz rate. In the hot chlorination process, in an atmosphere of Cl_2 or HCl, quartz is heated at about +1000 °C.

The above-described processes help in the removal of impurities from quartz. However, there are impurities in the quartz lattice, which become part of the quartz structure.

Unimin Corporation/Sibelco has dominated the global high-purity quartz market with mining and beneficiation activities in the pegmatite deposits of North Carolina in the United States.

Table 1.1 Price and purity of different types of quartz material [8].

Type of quartz	Cost in $/MT	Percentage of SiO_2	Other elements
Clear glass sand	30	99.5%	0.5% (5000 ppm)
Semiconductor filler, LCD, and optical glass	150	99.8	0.2% (2000 ppm)
Low-grade high-purity quartz	3000	99.95%	0.05% (100 ppm)
Medium-grade high-purity quartz	500	99.99%	0.01% (100 ppm)
High-grade ultrapure quartz	5000	99.997%	0.003% (30 ppm)

Norwegian Crystallites produces high-purity quartz in its Dago plant in western Norway and Moscow-based JSC Polar Quartz is active in the production of high-purity quartz.

In China, the Donghai Pacific Quartz company serves only the huge domestic market.

There are quartz reserves in all continents of the world. The price depends on the quality of the quartz, and the cost of silica in 2012 is given in Table 1.1.

1.3 Polysilicon–basic manufacturing processes

Most of the polysilicon sold around the world is manufactured through the trichlorosilane (TCS)-based Siemens process route. To produce trichlorosilane, the required raw material is metallurgical-grade Si. The molten Si is obtained by the reduction of quartz in the presence of carbon in an electric furnace. After solidification, the molten Si will become metallurgical-grade Si, which will have 98% purity. For solar-grade Si applications, the purity of Si should be 6N that is, 99.9999%. For the electronic-grade Si application, the purity of Si ranges from 9N to 11N. The metallurgical-grade Si is made to react with HCl in a fluidized bed reactor to get trichlorosilane (TCS), which is the feedstock material for the production of polysilicon. The obtained TCS is purified by the distillation process in order to reach the required purity. The purified TCS is made to decompose into Si atoms in vapor form in the presence of hydrogen gas at high temperature in a pressurized chamber, and is then deposited on the seed rods of silicon. The deposition of Si atoms from the vapor form is called the chemical vapor deposition (CVD) technique. Thick rods of highly pure silicon will be produced in this process. The rods are separated from the reactor after deposition and broken into small pieces or chunks of polysilicon.

1.3.1 Specifications/requirements of polysilicon

The electrical requirements of polysilicon of different grades are given in Table 1.2.

1.3.2 The process from quartz to metallurgical-grade silicon

The first step in the production of polysilicon is the extraction of quartz from a silica mine. The second step is crushing and purification of the quartz material. The next step in the manufacturing of polysilicon is the production of metallurgical-grade silicon using quartz as the raw material. The quartz is loaded into a furnace along with a source of carbon, which is a mixture of coal, coke, woodchips, or charcoal. The mixture is then heated along with the quartz, and the silicon is formed by the removal

Table 1.2 Electrical requirements of polysilicon of different-grades [9].

Electrical properties	Electronic-grade	Solar-grade I	Solar-grade II	Solar-grade III
Conductivity type/dopant	N/phosphorus	N/phosphorus	N/phosphorus	N/phosphorus
Resistivity	$\geq 250\,\Omega$-cm	$\geq 100\,\Omega$-cm	$\geq 40\,\Omega$-cm	$\geq 20\,\Omega$-cm
Resistivity	$\geq 1000\,\Omega$-cm	$\geq 500\,\Omega$-cm	$\geq 200\,\Omega$-cm	$\geq 100\,\Omega$-cm
Brick lifetime	$\geq 100\,\mu s$	$\geq 100\,\mu s$	$\geq 50\,\mu s$	$\geq 30\,\mu s$
Oxygen concentration (atoms/cm^3)	$\leq 1.0 \times 10^{17}$ (atoms/cm^3)	$\leq 1.0 \times 10^{17}$ atoms/cm^3	$\leq 1.0 \times 10^{17}$ atoms/cm^3	$\leq 1.5 \times 10^{17}$ atoms/cm^3
Carbon concentration	$\leq 2.5 \times 10^{16}$ atoms/cm^3	$\leq 2.5 \times 10^{16}$ atoms/cm^3	$\leq 4.0 \times 10^{16}$ atoms/cm^3	$\leq 4.5 \times 10^{16}$ atoms/cm^3
Benefactor impurities concentration (P)	$\leq 0.3\,$ppba	$\leq 1.5\,$ppba	$\leq 3.76\,$ppba	$\leq 7.74\,$ppba
Acceptor impurities concentration	$\leq 0.3\,$ppba	$\leq 0.5\,$ppba	$\leq 1.3\,$ppba	$\leq 2.7\,$ppba
Total metal impurities: Fe, Cr, Ni, Cu, Zn	$\leq 0.03\,$ppmw	$\leq 0.05\,$ppmw	$\leq 0.1\,$ppmw	$\leq 0.2\,$ppmw

ppba–parts per billion atoms, 1 ppba is a fraction of 1 in 10^9 atoms.
ppbw–parts per billion by weight, 1 ppbw is a fraction of 1 in 10^9 units by weight, ng/g.
ppmw–parts per million by weight, 1 ppmw is a fraction of 1 in 10^6 units by weight, μg/g.

of oxygen from SiO_2. This process of the reduction of Si from SiO_2 in the presence of carbon is called carbothermic reduction, which produces liquid silicon, CO_2, and silica fumes. The silica fumes are used for other industrial processes and applications while the liquid silicon is poured out of the furnace. The liquid is further refined, then allowed to solidify. The resulting silicon material is referred to as metallurgical silicon, or metal silicon (MG-Si).

The metallic Si is produced in large industrial furnaces and it requires large amounts of electrical energy, in the order of 10–12 MWh for manufacturing one ton of metallurgical-grade Si [10].

The electric load such as 1 MW or above defines the size of a silicon furnace and it can be 30 or 40 MW.

The pot sizes of furnaces are typically up to 10 m in diameter and 3.5 m in depth. Electric energy is supplied through a three-phase alternating current by three electrodes submerged deep in the charge. The manufacturing process of a typical silicon metal plant with an arc furnace is shown in Fig. 1.7.

It is difficult to decompose SiO_2 into silicon and oxygen, and a high temperature is needed for the reaction to take place. Purified quartz, the source of SiO_2, in lumps of 10–100 mm with the appropriate purity and thermal resistance is preferred as a raw material. The quartz is combined with carbon-rich materials such as metallurgical-grade coal, coke, and wooden chips in the appropriate ratio and loaded from the top side into an electric arc furnace. The temperature of the furnace should be around 1000 °C when the raw materials are loaded into it. The coal and coke are used to provide thermal energy and they are the best reduction agents to separate oxygen from SiO_2. During heating, the coke and coal provide porosity to allow the passage of generated gases such as SiO and CO from the furnace. The wooden chips are the source of carbon and maintain good permeability for the escape of gases generated during the reaction taking place in the furnace. The reactivity of raw materials, their mixing ratio in the charge, and their porosity are extremely important factors in achieving good furnace

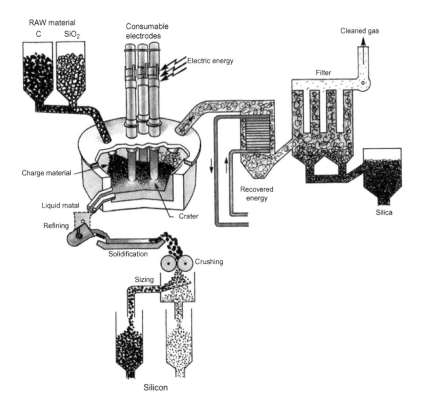

FIG. 1.7

Manufacturing process of ferrosilicon or Si metal plant with an arc furnace.

performance in terms of high material yield, lower power consumption, and good product quality. The trace impurity element content of the raw materials, including quartz, carbon source materials, and electrodes, finally decides the purity of the silicon metal produced. The electrical energy is supplied to the furnace by three large carbon electrodes, which are submerged in the raw materials of furnace. The electrodes supply a three-phase current to create the sparks between the electrode and the electrical ground of the furnace to heat the mixture to approximately 2000 °C. As the electrodes are gradually consumed, new electrodes are introduced into the furnace from the top of the electrodes.

At the bottom of the furnace, the temperature can be on the order of 2000 °C, and SiO_2 reacts with carbon and produces molten silicon. In this process, liquid silicon is produced by the removal of oxygen in the presence of carbon, leading to the formation of carbon monoxide (CO) and volatile silicon monoxide (SiO). These gases flow upward as a flux of convection. As the gases go upward in the furnace, a heat exchange with the condensed matter falling downward in the furnace takes place. Several other chemical reactions take place at this temperature. The overall major reactions are as follows [11]:

$$2\,C\,(solid) + SiO_2\,(solid) \rightarrow Si\,(liquid) + SiO\,(gas) + CO\,(gas) \tag{1.1}$$

$$SiO\,(gas) + 2\,C\,(solid) \rightarrow SiC\,(solid) + CO\,(gas) \tag{1.2}$$

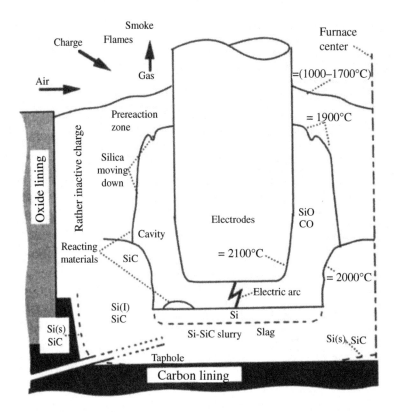

FIG. 1.8

Schematic of the polysilicon production process that is used to convert quartz into metallurgical-grade silicon.

$$\text{SiC (solid)} + \text{SiO}_2 \text{ (solid)} \rightarrow \text{Si (liquid)} + \text{SiO (gas)} + \text{CO (gas)} \tag{1.3}$$

There are several temperature zones in the furnace and different reactions take place, as shown in Fig. 1.8.

In the high- temperature zone around the electrode tip, the following reactions occur:

$$2\text{SiO}_2 \text{ (s, l)} + \text{SiC (s)} \rightarrow 3\text{SiO (g)} + \text{CO (g)} \tag{1.4}$$

$$\text{SiO}_2 \text{ (s, l)} + \text{Si (l)} \rightarrow 2\text{SiO (g)} \tag{1.5}$$

$$\text{SiO (g)} + \text{SiC (s)} \rightarrow 2\text{Si (l)} + \text{CO (g)} \tag{1.6}$$

The slowest of these three reactions are probably the SiO(g)-producing reactions (1.4), (1.5), which consume a major part of the input electrical energy.

The produced SiO and CO gases move upward in the furnace and SiO reacts with carbon and produces SiC material.

$$\text{SiO(g)} + 2\text{C(s)} \rightarrow \text{SiC(s)} + \text{CO(g)} \tag{1.7}$$

The SiC reacts with molten SiO_2 and produce molten Si, CO, and SiO.

$$SiC(s) + SiO_2(l) \rightarrow Si(l) + CO(g) + SiO(g) \tag{1.8}$$

As this process is involved with high temperatures, the furnace has to be operated as continuously as possible. Usually, the main operations of silicon furnaces are stoking, charging, and tapping. During stoking, which is nothing but the addition of charge to furnace, a thin crust on top pushes the old charge toward the electrode. The new charge is loaded in small batches in frequent intervals of time and the new charge is distributed appropriately on the top of the old charge. There are preprepared holes in the transition between the side and bottom lining of the furnace, from which the molten material is collected; and this process is called tapping.

The formed liquid silicon is taken out at regular intervals from the bottom of the furnace and used for bed casting. The exhaust gases contain the energy that is recovered through a suitable process. The silica fumes consist mainly of very fine particles of amorphous silica of less than 1 μm, which are part of the exhaust gases that are made to pass through filter cloths and recaptured in the form of microsilica. The microsilica is used in many industrial applications such as additives in the concrete and refractory industries.

The metallurgical silicon is drained from the furnace and solidified outside in the bed casting process. It is further refined, crushed into pieces, and packaged. The purity of this obtained MG-Si will be about 98% [12] and mainly depends on the purity of the raw materials used in the process. Boron, carbon, aluminum, and iron are the major impurities associated with this MG-Si.

1.3.3 Siemens process of TCS production-purification

An ultrahigh-purity feed stock polysilicon material is required for photovoltaic and semiconductor industry applications. To manufacture high-efficiency solar cells and reliable integrated chips, the impurity levels should be in the range of parts per billion atoms. The boron, phosphorous, and metallic impurities such as iron affect the performance of the solar cells. At the industry level, the semiconductor-grade or solar-grade silicon feedstock is obtained by converting metallurgical-grade silicon into a volatile silicon compound, then purifying it and decomposing it into elemental silicon. The manufacturing process may be classified into four successive steps as shown in Fig. 1.9:

  (i) Preparation/synthesis of the volatile silicon compound using metallurgical-grade Si.
 (ii) Purification of the synthesized compound such as trichlorosilane (TCS) or silane.
(iii) Decomposition of TCS or silane into elemental silicon.
 (iv) Recovery and recycling of formed byproducts.

There are four major industrial methods presently followed for the manufacturing of polysilicon feedstock material. The methods and processes are the Siemens process, the Union Carbide process, the Ethyl Corporation process, and the Wacker process.

1.3.4 Siemens process

A German company by the name of Siemens developed chemical vapor deposition technology combined with the conversion of MG- Si to trichlorosilane. Therefore, the process and reactors were named after Siemens. The Siemens reactor was created in the late 1950s and the process has been

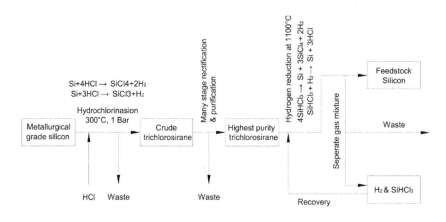

FIG. 1.9

Schematic of the Siemens process for the production of semiconductor-grade silicon [14].

the dominant production route historically for manufacturing electronic and solar-grade polysilicon materials. More than 75% of the total polysilicon produced around the world used the Siemens process for manufacturing in 2018.

There are four steps in the Siemens process as shown in Fig. 1.9 [13,14].

First, the MG-Si is reacted with hydrochloric acid and converted to trichlorosilane (TCS) in a fluidized bed reactor (FBR) through the reaction

$$Si + 3HCl \rightarrow SiHCl_3 + H_2 \tag{1.9}$$

The obtained TCS is in crude form and contains different chlorosilanes and impurities, thus requiring purification. The crude TCS undergoes fractional distillation by passing through distillation towers to achieve the required purity.

$$(SiHCl_3)_{crude} \rightarrow (SiHCl_3)_{hyperpure} \tag{1.10}$$

The hyperpure TCS is made to pass into the CVD reactor with a mix of hydrogen gas and decomposed into silicon in a reducing atmosphere at around 1000 °C via the reaction shown in Eq. (1.11). The decomposed Si atoms from the vapor are deposited on thin silicon seed rods.

$$4(SiHCl_3)_{hyperpure} + H_2 \rightarrow Si_{hyperpure} + 3SiCl_4 + 2H_2 \tag{1.11}$$

There are several byproducts (H_2, HCl, $HSiCl_3$, $SiCl_4$, and H_2SiCl_2) that are formed and recovered in step 4, and then recycled in the process. In CVD process, thick rods of highly pure silicon are produced. The rods inside the Siemens reactor are removed after the deposition process and broken into chunks. The silicon chunks, which are small in sizes, are the starting materials to be used to grow large blocks/ingots of mono- or multicrystalline materials. TCS and H_2 are feeding gases to the Siemens reactor.

1.3.5 **TCS production technology**

The first step is the conversion of metallurgical-grade Si to TCS. This is done in two routes: direct chlorination and hydrochlorination.

1.3.5.1 Direct chlorination

Traditional direct chlorination FBR was developed in the middle of the 20th century for polysilicon production. Since the 1940s or earlier, TCS and silicon tetrachloride (STC) have been manufactured using direct chlorination (DC) reactors. So, most polysilicon plants have been designed with direct chlorination technology using a fluidized bed reactor (FBR). The FBR is a chemical chamber where the reacting compounds are in the fluidized form, and multiphase reactions can take place. There is a provision for the variation of temperatures and pressures.

The process uses HCl in a fluidized bed reactor to chlorinate metallurgical silicon. HCl enters the bottom of the reactor through a vigorous blowpipe framework. The distribution of HCl over the cross-section is done by the silicon bed and blowpipe system. At a temperature between 250 and 300 °C and a pressure of 3–7 bar, the HCl gas in FBR reacts with the powered MG-Si. It requires careful chilling of the liquid bed reactor for the synthesis of TCS, and it is an exothermic reaction.

The principal reaction is:

$$Si + 3HCl \rightarrow SiHCl_3 + H_2 \tag{1.12}$$

Around 85% of TCS is yielded in this process.

Some of the contending reactions in these processes are:

$$Si + 4HCl \rightarrow SiCl_4 + 2H_2 \tag{1.13}$$

$$Si + 2HCl \rightarrow SiH_2Cl_3 + H_2 \tag{1.14}$$

Silicon tetra chloride (STC) formed as a byproduct of the reaction is used for making silanes and fumed silica, for which there exists a great demand. STC is also a byproduct during TCS decomposition in the presence of H_2 gas in the production of polysilicon material. The TCS production capacity was increased due to high polysilicon demand. Hence, larger amounts of TCS were produced along with the STC byproduct. The selectivity of TCS in the FBR was not a concern, as there was a market for the byproduct STC.

The outlet gases of the fluidized bed reactor are quenched and filtered. In the direct cooling system, the transfer fines of silicon, metal chloride, and some STC will be isolated. A compressed mixture is sent to the rectification unit containing TCS, unreacted STC, and other silanes. The H_2 gas will be reused by an oil-free reciprocating compressor to transform the fluid surface.

Excess STC supply led to the development of two prominent technologies, hydrochlorination and STC converters. Cyclones such as residue separators help to remove silicon dust, a step before nascent trichlorosilane is condensed in a direct process and fed to the rectification unit. Silicon chlorination is a proven innovation that is utilized by a several polysilicon producers.

The highly exothermic reaction proceeds rapidly at the operating conditions of the FBR. It can create localized heating and cause constraints for the high-yield production of TCS.

High temperatures and the creation of hotspots in some locations of the reactor cause the excess generation of the byproduct STC. The byproducts are settled down at the bottom of the chamber, clogging the nozzles and causing frequent shutdowns and maintenance of the FBR.

1.3.5.2 Hydrochlorination

Hydrochlorination is also called the hydrogenation or cold conversion process because STC is converted to TCS at a lower temperature than in the thermal conversion process. In the hydrogenation process, STC is used as the chlorine source for the production of TCS. STC is produced in large quantities as a coproduct in TCS production as well as in TCS decomposition and deposition of polysilicon. So, the STC and other chlorosilanes generated in these reactions are collected and recycled back to the hydrogenation FBR. All hydrogen from the process is also collected and recycled to the hydrogenation reactor. Thus, separate STC conversion reactors are avoided. Due to STC recycling, the hydrogenation process has a lower electrical energy consumption than the direct chlorination process.

The charge materials silicon tetrachloride (STC; $SiCl_4$) and hydrogen (H_2) are heated to a range of 550–650 °C and continuously fed into the fluidized bed reactor. A finely ground metallurgical-grade silicon combined with a copper catalyst is introduced into the FBR. The $SiCl_4$ and H_2 enter the reactor from the bottom side through a robust blowpipe system. Superheated STC and H_2 gases are distributed across the cross-section of the fluidized silicon bed using the blowpipe system. The FBR operates at a pressure of 20–30 bar.

The FBR is constructed with Incolloy, an alloy of steel that is a corrosion-resistant material.

The heating and cooling of the bed are done by circulating heat transfer fluids such as oil or water through internal tubes and outside coils. At a temperature of >500 °C and a pressure of >20 bar in FBR, TCS is formed with the following reaction

$$3SiCl_4 + Si + 2H_2 \rightarrow 4HSiCl_3 \tag{1.15}$$

This reaction is the summary of all the mechanisms taking place and not a true reaction of its own. The conversion rate of MG-Si into TCS mainly depends on the temperature, pressure, and ratio of H_2 and STC gases. The TCS yield may reach 20–30%. Vapor phase products such as TCS, STC, hydrogen, metal chlorides, and other chlorosilanes are produced in the reaction.

The outlet gases of the fluidized bed reactor are quenched and filtered. Using a direct cooling and condensation system, the fines of silicon, metal chloride, and some STC present in the outlet gases will be separated. The hydrogen gas will be recycled to fluid bed conversion by an oil-free reciprocating compressor. TCS, unreacted STC, and other silane compounds that are collected as condensed mixtures are made to be sent to the recycling unit. The generated products are vented at the top of the reactor into a settler where the gases are cooled and partially condensed, precipitating the solids. Hydrogen and chlorosilanes are collected and returned for recycling. Liquid and solid wastes from the bottom of the settler are collected for disposal.

The off gases from the reactors are first subjected to dry dedusting by cyclone separators, which ensures the removal of the unreacted metallurgical silicon and sends it back to the reactors. The off gases containing primarily chlorosilanes and H_2 are subjected to condensation up to 60 °C and then to a series of purification steps before final scrubbing and disposal.

Depending on the purity of the feed silicon, a slipstream from this will be taken and the material will be purged out to waste. A similar purge from the bottom of the reactor will be made to ensure the impurities do not build up inside the reactor and do not end in the chlorosilane, increasing the load on the purification stage.

After the cyclone separators, the gases are subjected to dry dedusting to remove the $AlCl_3$ by maintaining the temperature in such a way that they are removed here. The gases from the reactor are cleaned by scrubbing the effluent gas, which cools the gas and removes waste liquid and solids, called wet dedusting.

Off gases are subjected to a three-stage condensation process to condense all the chlorosilanes, which is called crude TCS. Effluent gas exiting the scrubber consists of hydrogen, hydrogen chloride, and small amounts of chlorosilanes and reaction byproducts. The TCS production of effluent gas can be directed to a dedicated effluent gas recovery system and scrubber. The HCl can be disposed of and the H_2 vented to a high point vent, as they will not meet the purity levels required in the deposition process.

The hydrogenation of STC is an endothermic reaction, so temperature across the fluid bed is easier to control. Hence, FBRs can be made with larger diameters, thus larger capacities.

The product thus collected is subjected to distillation to remove the high boiling point impurities and the chlorides of metal to a very low level for it to be used in the downstream process.

1.3.5.3 Closed loop hybrid Siemens process

The process of hydrochlorination can operate continuously for a year or more between shutdowns. In this process, impurities contained in the metallurgical Si become gaseous and leave the FBR as effluent gases and do not accumulate in the FBR. This is not the case for the process of direct chlorination, which requires cleaning the bed every 6–10 weeks. The raw material cost does not vary for both processes. In the hydrochlorination FBR process, more complicated procedures are followed for the removal of solids downstream of the reactor compared to direct chlorination FBR. These complications have been successfully addressed and both the direct chlorination and hydrochlorination processes are incorporated in modern polysilicon manufacturing plants.

The off-gas recovery in the hydrochlorination process is analogous to the process that is followed in direct chlorination, but requires the addition of a recycling hydrogen compressor. Hydrochlorination is a complex process compared to the converter and it needs the removal of HCl. The hydrochlorination FBR requires more expensive construction materials due to the operation of the reactor at higher temperatures and pressures. The construction is expensive for the quenching system and the preheat train system compared to the required systems in the direct chlorination plant. Hydrochlorination process plants bring in a scale of economy as compared to direct chlorination-based converters.

The direct chlorination and hydrochlorination processes are combined in a single reactor and this process is call the hybrid Siemens process.

A finely ground metallurgical-grade silicon is made to react with anhydrous hydrogen chloride gas in a fluidized bed reactor. The reactor produces a mixture of chlorosilanes that is primarily TCS and STC with a small amount of dichlorosilane (DCS) up to a max of 2%.

The reactions are:

Dichloro Silane reaction

$$Si + 2\,HCl \rightarrow SiH_2Cl_2 \tag{1.16}$$

Trichloro Silane reaction

$$Si + 3\,HCl \rightarrow SiHCl_3 + H_2 \tag{1.17}$$

Silicon tetra chloride reaction

$$Si + 4\,HCl \rightarrow SiCl_4 + 2\,H_2 \tag{1.18}$$

The formed byproduct STC is made to react with MG-Si along with hydrogen gas in a fluidized bed reactor. Thus, in this process, the byproduct STC is converted to TCS. This hybrid Siemens process incorporates two process steps in a single reactor. This process is cost-effective and conserves the usage of power.

The reaction is endothermic, the constituents are required to be heated by external heaters, and the resultant gas mixture is passed through for further processing.

The hydrogenation of STC as well as the chlorination of silicon are done in a one-step process as per the above reaction in the hybrid Siemens process. Excess HCl from the CVD process can be fed to the FBR; this will give the same reaction as that of the direct chlorination process.

1.3.6 Purification of TCS

The crude TCS coming out of the reactor contains multiple impurities. So, TCS must be purified before it can be used to produce high-purity polysilicon for use in the manufacture of of wafers for photovoltaic solar cells. Higher molecular weight chlorosilanes and polysilanes are produced in the hydrochlorination process. The trace quantities of moisture present in the reactor react with compounds of chlorosilanes and polysilanes to produce siloxanes. The impurities present in the metallurgical-grade silicon and hydrogen chloride react to produce metal chlorides such as aluminum, titanium, and iron chlorides as well as hydrocarbons, chlorinated hydrocarbons, and a small amount of methyl chlorosilanes.

First, TCS and the byproduct chlorosilanes formed in the synthesis of crude TCS are separated and TCS is piped to the purification system. In this process, STC, metal chlorides, higher molecular weight chlorosilanes, polysilanes, siloxanes, and hydrocarbons are removed from the TCS. In addition, the STC is purified for subsequent conversion into TCS. The TCS purification system consists of distillation and adsorption towers. The design of the distillation columns helps in removing metal chlorides, heavy impurities, and chlorosilanes. These columns/towers separate and collect STC and send it back to the hydrogenation converter for conversion to TCS. These towers also remove hydrocarbons and light-boiling compounds as well as separate carbon contained chlorosilane compounds. The separation of the compounds is based on the boiling points of the compounds. TCS boils at 31.8 °C and the boiling point of STC is 57.65 °C, so this separation is relatively easy. More problematic are compounds with boiling points close to TCS.

The distillation process helps in removing the compounds of chlorosilane along with metal chloride impurities such as iron chlorides and aluminum chlorides as well as the chlorides of phosphorous and boron. The columns of the distillation process do the following: The foremost column helps to remove impurities such as aluminum chloride, ferrous chloride, and titanium chlorides, as higher boiling residues are separated and this helps to remove all chlorides. In the next column, the next purification of chlorosilane takes place with the removal of all chlorides, making it highly pure. The removal of phosphorous and boron cannot be fully done using the distillation process only. In order to achieve final purification, the distilled chlorosilanes need to pass into another reactor filled with an absorbent such as activated silicon or alumina gel. The final material from this last reactor is considered to be of the required purity level. For the removal of suspended particles, one needs to subject chlorosilane to a basic distillation process and a stainless steel container is needed to store this pure product after distillation.

The purified TCS is fed to the deposition reactors to make polysilicon. The low-grade TCS can be sold as a byproduct or further reacted to recover HCl for recycling to the polysilicon process. The purified STC feeds the conversion reactors for conversion to the purified form of TCS. The heaviest waste goes to waste treatment for neutralization and disposal. The small amount of vent gas from the purification processes is sent to the vent gas treatment system. To produce high-purity TCS, the purification process necessitates that some of the TCS is lost with the unseparated impurities. Several of these impurities have boiling points that are very close to that of TCS and are difficult to separate.

The impurities that are not separated, and the lost TCS, are collected as low-grade TCS, which can be either sold as a byproduct or incinerated with sufficient hydrogen to produce a fumed silica byproduct and recover HCl for recycle to the polysilicon process or chlorinated to produce silicon tetrachloride for return to the polysilicon process.

Additional purification of the TCS is required to remove impurities containing phosphorus, boron, arsenic, and iron, which can negatively impact the polysilicon product quality even at extremely low levels.

1.3.7 **Why Siemens reactor-based polysilicon is the most adopted**

The impurities in the Si wafer affect the performance of the solar cells. Phosphorous and boron act as donor and acceptor atoms in the Si material. The purity of the Si wafer depends on the purity of the raw material polysilicon, which is used for the crystal growth of Si. The purity of polysilicon is very important, so the impurity level in the basic polysilicon has to be controlled.

The metallic impurities such as Fe and other transition elements create defect centers in the solar cell and become killers of solar cell efficiency. So, the impurities of phosphorous, boron, iron, and other transition elements are to be completely removed or restricted to the ppb level.

There are different methods to manufacture polysilicon materials such as chemical routes and metallurgical routes. In chemical routes, the TCS-Siemens process, the TCS-FBR process, and the silane FBR are there while upgraded metallurgical (UMG) Si is one of the methods being followed in the metallurgical route.

The polysilicon prepared in the upgraded metallurgical Si contains the impurities to a certain level and it is not possible to completely remove the impurities. The metallurgical processes have not been proven yet to eliminate phosphorous and boron impurities as compared to chemical distillation and condensation methods. UMG contains donor and acceptor impurities two or three orders of magnitude more than the polysilicon purified gas distillation and condensation method.

The FBR silane process is a chemical route process and allows low dopant concentrations as purification takes place in the gas phase. However, FBR is less efficient than the Siemens process with respect to metal, carbon, and oxygen contamination. The granular poly is less pure than the rod poly, as the seeds are prepared by grinding in the FBR process. Silicon oxide is formed on the granules due to the large surface area as soon as they are exposed to air. Hydrogen absorption takes place on the surface of the granular silicon.

Table 1.3 gives the impurity levels of polysilicon manufactured in different methods [15].

In the chemical route of the TCS-Siemens process, the purified/condensed TCS is converted to Si in the vapor phase and made to deposit on hot filaments to produce highly pure polysilicon material. So, the Siemens process is adopted for the manufacturing of high-purity polysilicon material.

Historically, the TCS-Siemens process was developed to produce highly pure polysilicon for semiconductor applications. The rejected material from the semiconductor poly was used as a raw material

Table 1.3 Impurity levels of polysilicon manufactured in different methods.

Impurity	Siemens (solar) (value range)	FBR (value range)	UMG (value range)
P (donor)	0.3–5 ppba	0.3–20 ppba	300–1000 ppba
B (acceptor)	0.1–5 ppba	0.3–20 ppba	500–2000 ppba
Total metals	20–50 ppbw	30–1000 ppbw	10–1000 ppbw
Carbon	0.25–1 ppma	0.5–10 ppma	50–200 ppma
Oxygen	0.5–5 ppmw	10–100 ppmw	(100 ppmw)
Gas inclusion		Hydrogen	

for the solar industry in the beginning. With increased demand for the solar industry, the polysilicon industry growth takes place in the range of 100 GW per year.

1.3.8 Siemens process by CVD

Most of the polysilicon sold around the world is produced through the Siemens process using CVD of trichlorosilane (TCS).

Purified TCS and H_2 are mixed in this process and introduced into a reactor that has a base plate, bell jar, and multiple silicon rods that are electrically heated to approximately 1100 °C. The Siemens solution includes silicon deposition from a mixture of diluted trichlorosilane or silane gas plus excess hydrogen on high-purity polysilicon crystal hairpin-shaped filaments. Inside an enclosed bell jar, which contains the gases, silicon growth occurs.

The power system supplies electrical current to the preheating graphite rods to heat the filaments, which are assembled as electric circuits in series, to the conductive temperature, about 400°C. Once the power is supplied directly to the filaments, they become conductive to provide the heating cycle. The filaments are heated to 1100–1175°C. TCS along with the carrier gas hydrogen is reduced to elemental silicon and deposited as a thin-layer film onto the surface of the hot seed filaments. HCl is formed as a byproduct and hydrogen doesn't participate in the reaction, but acts as a carrier gas.

This TCS-Siemens process is a batch process and polysilicon rods are removed from the reactor once they reach a predetermined diameter. At a defined deposition temperature and concentration, reactions are quite complicated, involving chlorine, hydrogen, and silicon. The simplified reactions are shown below:

$$SiHCl_3 + H_2 \rightarrow Si + 3HCl \tag{1.19}$$

$$SiHCl_3 + HCl \rightarrow SiCl_4 + H_2 \tag{1.20}$$

$$2SiHCl_3 \rightarrow SiH_2Cl_2 + SiCl_4 \tag{1.21}$$

$$SiH_2Cl_2 \rightarrow Si + 2HCl \tag{1.22}$$

As shown in the above reactions, dichlorosilane (DCS) participates in the process and contributes to the production of polysilicon. Apparently, the TCS stream has 6–9% of DCS and hence DCS is not separated from TCS by manufacturers.

The cooling of the reactor is done by circulating coolants through the bell jar and base plate; this will help to prevent Si deposition on the reactor's inner surface.

A significant amount of heat is lost to the cooled surfaces due to a large difference in temperature between the polysilicon rods and the reactor.

In the course of advancements and modifications to the Siemens CVD reactor technology, the number of rods has been increased from two to more than 100 in the largest low-pressure reactors. In addition, the atmospheric pressure has increased to 6 bar.

There are two variants of the Siemens process involving CVD. The first one is CVD of TCS and the second one is CVD with silane gas.

The schematic of the Siemens CVD reactor with TCS is shown in left side of Fig. 1.11.

The Siemens reactor is a large water-cooled chamber. It is a pressure vessel that consists of a water-cooled base plate and a bell jar, electrical connections with power supply, a gas distribution system, and water supply utilities. The base plate and bell jar are made of steel. They are hermetically sealed and can work at high pressures up to 10 bar. The base plate is provided with feed through holes to use as ports for inlet and outlet gases and for electrical connections to the filaments. The filaments assembled into inverted U-shaped hairpins of silicon slim rods are loaded into a clean reactor by mounting on the electrical feed-throughs and fastening the ends of the filaments to electrodes. The number of rods to be loaded depends upon the size of the reactor.

The electrical power supply is provided to heat the filaments assembled with poly rods to the working temperature of about 1100°C by resistive heating. After reaching the desired temperature of the conductor, the temperature on the surface of the slim rod is supported by passing electric current. The bell jar has been provided with viewports for monitoring the temperature of the hairpins containing the growing polysilicon rod using noncontact temperature gauges. Electrical power, gas flows, water flows, temperature, and pressure are measured and controlled by a computer system. A distributive control system with computer control is used to measure the electrical power, gas flows, water flows, temperature, and pressure in the chamber and controls them as per the requirements.

Both the base plate and the bell jar are provided with jackets for the circulation of cold water to cool them by removing the radiant heat from the hairpins. The base plate is placed on a firm pedestal. After assembling the slim rods, the bell jar is brought by a hoisting system to cover them. The bell jar is hermetically sealed on the base plate using suitable gaskets. The hairpins/filaments are brought to the working temperature using a sophisticated electrical power supply. A mixture of trichlorosilane and hydrogen gas is injected into the reactor through inlet nozzles in the base plate of the reactor. The gas distribution system monitors and controls gas flow and optimizes the turbulence of the gas stream in the reactor chamber. Nitrogen is used as a purge gas in the reactor. Hydrogen is used as a carrier gas to the TCS. Silicon deposition yields a large quantity of gaseous HCl, a significant quantity of STC, and smaller quantities of DCS and other chlorosilanes as byproducts. So, the Siemens-CVD reactor effluent gas stream consists primarily of TCS, STC, H_2, HCl, and DCS.

An exhaust port helps to get rid of the exhaust gases. The recovery of hydrogen and STC takes place, and these are exported back for TCS conversion through the converter. The diameter and height of standard polysilicon reactors vary between 0.8 and 1.5 m and 1–2.5 m, respectively. These reactors are capable of accommodating 9/18/24/48/52 or a higher number of hairpins. The reactor experiences high temperature and the generation of corrosive elements such as HCl, TCS, and STC during the deposition process. The reactor components are made from such material that they get cooled in the event of any heating while no impurities are imparted to the deposited silicon. The safety design aspect needs to be strictly adhered to in view of the presence of corrosive and hazardous chemicals such as TCS and hydrogen, which are generated inside the reactor due to the high temperature and pressure. In order to

FIG. 1.10

Polysilicon chunk from the Siemens process.

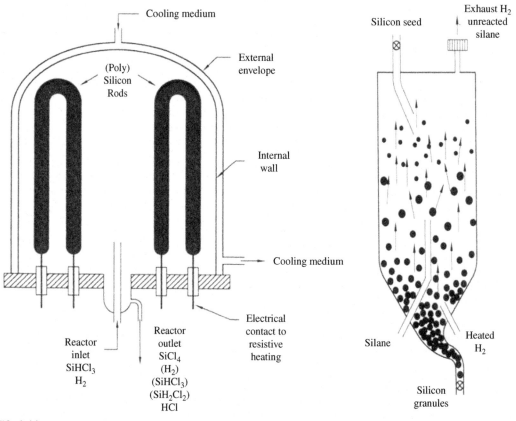

FIG. 1.11

Schematic of the TCS/Siemens process and the silane FBR process.

make the Si deposition uniform and smooth over the entire length of the hairpins, the design should be robust to provide no pores, cracks, or other defects. Fig. 1.10 shows the image of polysilicon chunk manufactured through Siemens process.

1.4 Alternate methods of polysilicon production

The Siemens process has several disadvantages. The process consumes very high amounts of energy due to substantial heat losses to the cold water-chilled walls of the metal bell jar reactor. The process is a batch-wise operation, as the slim rods are to be frequently unloaded after completion of the process. The process uses electrical contacts made of carbon, which is a source of contamination. If a power failure takes place, especially during start up, hot spot formation occurs. In this process, large amounts of byproducts are produced.

1.4.1 Upgraded metallurgical-grade process

There was a shortage of high-purity silicon feedstock from 2004 to 2008 due to the overgrowth of the photovoltaics industry. A number of projects were launched to refine silicon metal up to an adequate purity required for solar cells using the metallurgical routes.

In the solar PV industry, this type of process and product is commonly designated as upgraded metallurgical-grade (UMG) silicon. Elkem Solar and Jaco, JFE (Japan), Dow Corning (Brazil), Timminco (Canada), Silicor Materials (Canada, United States), Photosil (France), and Evonic (Norway) pursued the production of UMG silicon.

Each company mentioned above uses its own proprietary technology. Each process is normally based on several steps, each taking care of the elimination of different groups of impurities in the metallurgical-grade silicon. Separate process steps have been devised to remove impurities of the donor (P) and acceptor (B) species, transition elements such as Fe, Ni, Co, Cr, Ti, carbon (C), and less noble elements than silicon such as Al and Ca. A majority of the processes use the following steps:

Carbothermic reduction of quartz with careful selection of the raw materials to minimize the impurity content.

Ladle refining using gas (alternatively plasma) or slag. In slag treatment, molten silicon is mixed with slag, extracting impurities, especially boron. The next process is chemical leaching which is dissolution of secondary phases in acid, causing removal of impurities of phosphorous and metals.

The realized products under this category suffer from high dopant levels, which limits their application for use in the manufacturing of solar cells. The use of these products is recommended only in blends with polysilicon to achieve better results. The UMG process has lower cost and low energy consumption. The impurity contents present in UMG silicon are 2–3 orders higher than the polysilicon feedstock obtained by the Siemens process, which is a chemical process. The high donor/acceptor and metallic impurity concentrations restrict the usage of UMG silicon as a feedstock material for crystal growth.

UMG silicon is becoming a compensated material due to the presence of high levels of donor and acceptor atoms and the neutralization of the charge carriers. The high level of dopants disturbs the properties of the semiconductor and becomes a cause for the reduction of efficiency in a device.

It has been reported that the use of 100% UMG silicon as the feedstock material caused a lower yield in ingot production, driving up the cost of the ingot. This is due to the presence of wrong concentrations of dopant atoms in different locations of the ingot or brick.

It has been reported that it is possible to control the resistivity and conductivity type in the upper part of the ingot by the use of gallium as a codopant, so the lower material yield issue is technically solvable. The negative impact of using compensated material with a high boron level is the risk of light-induced degradation (LID) caused by the B–O molecular pair. UMG silicon is used to manufacture only p-type multicrystalline solar cells.

More research has to be conducted on UMG silicon for its successful use in the manufacturing of solar cells.

1.4.2 Fluidized bed reactors and FBR process

Several organizations are developing processes to produce polysilicon based on FBRs. The FBR approach to polysilicon development originated from a US sponsored project in the 1980s. The FBR approach offers the ability for a continuous production process as opposed to the batch production of the Siemens route. Further, as compared to Siemens process-based reactors, the energy consumption is less in FBR. REC Solar claims that their FBR procedure for polysilicon production consumes just around 10% of the energy needed to run a Siemens-type process.

Fluidized bed reactor is a large chamber where multiphase reactions takes place. It consists of a gas-solid mixture of reacting compounds that exhibits fluid-like properties. The bed can be represented by a single bulk density as it contains an inhomogeneous mixture of gas and solid particles.

FBRs need to adopt principles like a fluid, which will help to sink an object with higher density while enabling an object to float with lower density. The contact of the solid particles with the gas is greatly enhanced in fluidized beds, compared to packed beds. This type of behavior enables good thermal and mass transport inside the fluidized beds. In view of good heat transfer and excellent thermal uniformity, the bed has a large heat capacity while a field of homogenous temperature is maintained.

This technology uses a specially designed FBR to deposit silicon onto small silicon seeds from silane source gas. A bed of small silicon particles of about 300 μm is fluidized in hydrogen and silane gas. The wall of the reaction chamber is electrically heated to provide a bed temperature of about 600°C. At an elevated temperature and pressure, silane and hydrogen are fed through the bottom of the reactor. From the reactor top, high-purity Si particles are fed and upward gas flow helps to suspend them. Silane gas gets converted to basic Si metal at a reactor operating temperature of 750°C or higher, and this is deposited on the surface. As the seed crystals grow, their density will increase and they fall to the bottom of the reactor, where they are removed and collected continuously. Additional seed crystals are fed through the reactor top to augment the silicon granules removed.

Fluidized solid particles charged into a closed vessel may either be further grown or consumed, depending on the conditions and the reactants in the chamber. Using silane as the feed gas, 100% of it can be converted to elementary silicon with hydrogen gas as the only byproduct, according to the equation:

$$SiH_4 \rightarrow Si + 2H_2 \tag{1.23}$$

Silicon seeds in the form of fine particles are (semi) continuously loaded from the top or the middle of the reactor, whereas silane gas and hydrogen are introduced near the bottom of the reactor, as shown in the right side of Fig. 1.11. Ascending flowing gas percolates through the particle bed. When forces from the gas stream equal those of gravity, the particles begin to lift, making the bed behave as a fluid. The bed exhibits a large degree of temperature uniformity because the particles are in constant motion. Furthermore, the particles are moving freely, enabling uniform

FIG. 1.12

Granular polysilicon produced by FBR–silane process.

chemical vapor deposition. Fluidized bed reactors are designed to work within a given regime determined by parameters such as particle density and particle size distribution, bed height, gas flow, and pressure. Because, the chemical vapor deposition is responsible for the growth process of the particles, the temperature and surface area available for the deposition process have become important parameters. As the particles grow, maintaining the fluidization becomes difficult. It is therefore necessary to remove the heaviest of the grown particles, pulled downward in the bed by gravity. To keep the population or the distribution of particles in the bed in constant proportion and thus keep the bed in a steady-state condition, one needs to add a constant and continuous flow of seeds and likewise remove the largest particles to compensate the growth.

An obvious benefit of the FBR operational mode is continuous operations as both feed and exhaust gases on one hand and solid seed particles and finished granules on the other hand can be introduced and removed simultaneously. Fig. 1.12 shows the image of granular polysilicon obtained by FBR process.

1.4.3 Union carbide process

The commercial producers of FBR polysilicon use silane as the feed gas. There are at least three commercial processes to make silane. The first was developed by Union Carbide in the early 1980s and consists of the distribution of purified trichlorosilane through fixed bed columns filled with quaternary ammonium ion exchange resins acting as the catalyst. Fig.1.13 shows the steps and the process involved in production of polysilicon through Silane—Union Carbide Process.

$$2SiHCl_3 \rightarrow SiH_2Cl_2 + SiCl_4 \qquad (1.24)$$

$$3SiH_2Cl_2 \rightarrow SiH_4 + 2SiHCl_3 \qquad (1.25)$$

The reactants and products of the redistribution are separated by distillation.

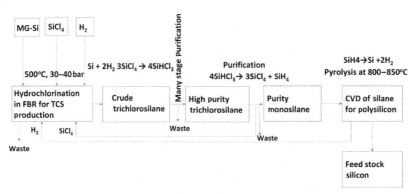

FIG. 1.13

Schematic of polysilicon production through silane-Union Carbide process [14].

Companies such as MEMC have produced granular Si using silane and the fluidized bed process for more than 10 years while Hemlock, Wacker, and others have either developed or are developing FBRs. They also employ the Siemens process to produce polysilicon.

The following are the major advantages of silane as a silicon source gas

Silane has high purity. The boiling point of silane is $-111.7°C$, so silane is easily separated from impurities by distillation. For the decomposition of silane, a low temperature is needed such as 400°C and polysilicon can be deposited as low as 600°C.

There are no byproducts of chlorosilane, as silane decomposes to silicon and hydrogen, eliminating the need for chlorosilane recovery and conversion for recycling.

Silane has been successful in producing commercial quantities of granular polysilicon at an operating cost of <$10 per kg and is commercially proven for granule deposition. A CVD deposition reactor can be designed to deposit >97% polysilicon, as it provides a high silicon yield.

The following are the major disadvantages of silane compared to TCS

- Higher energy consumption: Silane decomposes at a lower temperature, but the growth rate is much slower. Silane CVD reactors consume more than 100 kWh/kg, whereas TCS reactors consume less than 50 kWh/kg.
- Dust formation: the homogeneous decomposition of silane creates large quantities of amorphous silicon powder, a major problem in the design of CVD and fluid bed decomposition reactors.
- Higher safety risk: Silane is pyrophoric and ignites spontaneously at concentrations over 4% in air. Several explosions and fires have been reported due to leaking valves, regulator failures, and operator errors.
- Protected intellectual property: Silane CVD and FBR deposition reactor technology are not available commercially.

Some of these problems have been solved by the Union Carbide process, which replaces trichlorosilane with silane, SiH_4. That is, after the formation of trichlorosilane from metallurgical-grade silicon metal in the same manner as in the Siemens process, silane is formed by two catalytically driven reactions:

$$2HSiCl_3 \rightarrow H_2SiCl_2 + SiCl_4 \qquad (1.26)$$

$$3H_2SiCl_2 \rightarrow SiH_4 + 2HSiCl_3 \qquad (1.27)$$

Then, silane is separated from the product stream by distillation and purified before being sent to the decomposition chamber. The decomposition of silane to elementary silicon is, as in the Siemens process, obtained by pyrolytic decomposition onto heated seed rods of silicon inside a chilled metal bell-jar reactor:

$$SiH_4 \rightarrow 2H_2 + Si \qquad (1.28)$$

Thus, the Union Carbide process is also a batch process, but has a major benefit over the Siemens process in that the silane decomposition reaction may be performed at significantly lower temperatures, which means corresponding savings in energy consumption. Other benefits are that the silane decomposition process is complete so no corrosive byproducts are formed, only H_2 gas, and the process forms uniform large diameter rods free of voids. The disadvantage is that in addition to the batch-wise production, the conversion of trichlorosilane to silane involves additional process steps and thus a higher price of the volatile silicon compound as compared to the Siemens process.

1.5 Global polysilicon manufacturing scenario

The following are some of the manufacturers involved in polysilicon production. Fig 1.14 shows the production capacity in 2018 and the actual production in 2016 [16].

REC uses the FBR process with silane. MEMC also follows the FBR process.

Schmid follows the Siemens process using silane. PVinsights website publishes weekly the prices of solar PV materials. The typical price chart for polysilicon appeared in PVinsights is shown in Table 1.4 [17].

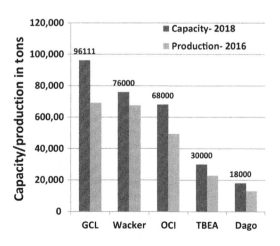

FIG. 1.14

Global polysilicon production by different manufacturers.

Table 1.4 Price of polysilicon in PVinsights website as on September 2, 2020 [16].

PV Poly Silicon Weekly Spot Price as on September 2, 2020 as per PV Insight

Item	High	Low	Average	AvgChg	AvgChg%
PV-grade PolySilicon (9N/9N+)	**12.05**	7.5	10.86	0	0
2nd-grade PolySilicon (6N–8N)	6	5.5	5.61	0	0
N Mono-grade PolySilicon in China (12N/12N+)	12.05	11.75	11.97	0	0
Mono-grade PolySilicon in China (11N/11N+)	12.05	11	11.5	0	0
PV-grade PolySilicon in China (9N/9N+)	Visit here for more Poly-Si price				
PV-grade PolySilicon Outside China (9N/9N+)	Visit here for more Poly-Si price				
Unit: USD/Kg	More		*Last Update:* 2020-09-02		

Notes: Definition of N Mono-grade: Poly silicon chunk or Chip Polysilicon with high purity can be directly produced to N-type monocrystalline ingots, mainly supplied by Wacker. Definition of PV Grade: Poly silicon chunk with high purity can be directly produced to solar PV ingots/bricks. Definition of 2nd-grade: Poly silicon chunk must be mixed with high purity polysilicon, when producing solar PV ingots/bricks. PolySilicon Price In China: The price is surveyed by RMB term with tax and then shown in USD term without 13% of VAT after April 1, 2019, 16% of VAT after May 1, 2018, and 17% of VAT before May 1, 2018.

References

[1] E. Bellini, World now has 583.5 GW of operational PV, PV Magazine, April 6, 2020, https://www.pv-magazine.com/2020/04/06/world-now-has-583-5-gw-of-operational-pv/.

[2] Global Market Outlook for Solar Power/2018–2022 www.africa-eu-renewables.org

[3] Bryan Mims The Mineral City: Exploring the Spruce Pine Mining District, https://www.ourstate.com/mineral-city/.

[4] J.J. Nortan et al. Geology and Mineral Deposits of Some Pegmatites in the Southern Black Hills South Dakota Geological Survey Professional Paper 297-E, United States Government Printing office, Washington, 1964 (https://pubs.usgs.gov/pp/0297e/report.pdf).

[5] Oleg Lopatkin web site (http://www.pegmatite.ru/smallru.htm).

[6] Policy Paper on Solar PV Manufacturing in India: Silicon Ingot & Wafer - PV Cell - PV Module, Thee Energy and Resources Institute, New Delhi, 2019, 27 pp.

[7] Mian-khalid-Habib, Silical Value -chain part-2, https://www.linkedin.com/pulse/silica-value-chain-part-2-mian-khalid-habib.

[8] J. Zhou, X. Yang, A reflection on China's high purity quartz industry and its strategic development, Material Sci & Eng. 2 (6) (2018) 300–302.

[9] GCL poly Energy Holdings Ltd. Data sheet on Poly silicon specifications of GCL poly Energy Holdings Ltd.

[10] B.S. Xakalashe, M. Tangstad, Silicon processing: from quartz to crystalline silicon solar cells, in: R.T. Jones, P. den Hoed (Eds.), Southern African Pyrometallurgy, *2011*, pp. 83–100 Trondheim, Norway.

[11] H.H.M. Ali, M.H. El-Sadek, M.B. Morsi, K.A. El-Barawy, R.M. Abou-Shahba, Production of metallurgical-grade silicon from Egyptian quartz, J. South Afr. Inst. Min. Metall. 118 (2018) 143–148.

[12] S. Bernardis, Engineering Impurity Behavior on the Micron-Scale in Metallurgical-Grade Silicon Production; 2012, (2012)Ph.D thesis. submitted to Massachusetts Institute of Technology, February 2012.

[13] H.S. Gopala Krishna Murthy Ph.D., Director, ShanGo Technologies Private Limited, Cost Reduction in Polysilicon manufacturing for Photovoltaics in private communication.

[14] V.V. Zadde, A.B. Pinov, D.S. Strebkov, E.P. Belov, N.K. Efimov, E.N. Lebedev, E.I. Korobkov, D. Blake, K. Touryan, New method of solar grade silicon production, in: 12th Workshop on Crystalline Silicon Solar Cell Materials and Processes, NREL/BK-520-32717, August, 2002, pp. 179–189.

[15] G. Bye, B. Ceccaroli, Solar-grade silicon: Technology status and industrial trends, Sol. Energy Mater. Sol. Cells 130 (2014) 634–646.

[16] M. Hutchins, The weekend read: Polysilicon and wafer manufacturer ranking, PV Magazine February 24, 2018 (https://www.pv-magazine.com/2018/02/24/the-weekend-read-from-the-top/).

[17] PVinsights Grid the world (http://pvinsights.com/indexUS.php).

Silicon crystal growth process

2.1 Specification/requirement of ingots

The solar PV industry is dominated by the wafer-based technology of crystalline silicon solar cell, which amounts to 90% of solar PV installations in the world. Multicrystalline-based solar cell technology dominated the industry through 2017. With the adoption of the diamond wire sawing technology of wafers, monocrystalline silicon technology has become prominent and has the potential for efficiency improvement. So, the mono- and multicrystalline wafers are required for the solar industry and the technology of mono- and polycrystalline silicon ingots/bricks are discussed.

Monocrystalline Si is a single crystalline material with an arrangement of Si atoms in long-range order with a particular orientation. In x-ray diffraction, the single peak is at 2θ Brag angle of 69.8° available for 100 orientation. The multi crystalline material consists of differently oriented crystals of different sizes. So, it is called poly- or multicrystalline silicon material. The crystal is not pure and it has multiple grains of single crystals of different sizes in different orientations as well as grain boundaries, dislocations etc. In x-ray diffraction, multiple peaks are observed at different 2θ angles.

The defects and the unwanted impurities in the wafer will act as charge carrier killing centers in the solar cell and will affect the efficiency. So, high-quality wafers are required for high-efficiency solar cells.

The majority of the silicon wafers used for solar cells are p-type monocrystalline and multicrystalline wafers. Czochralski (CZ) growth gives a single crystalline material while casting and directional solidification gives a multicrystalline (mc) material. As per the 20th edition of ITRPV, the share of p-type monowafers is 60% in the solar industry [1]. Directional solidification is the known process for producing multicrystalline silicon ingots, in view of the fact that there is a columnar grain growth with a planar surface [2]. The monocrystalline Si ingots are grown using the Czochralski or float zone methods to the required diameter. The multicrystalline ingots are grown by direct solidification technique or high performance (HP) mc:Si growth technique.

The following are the requirements of the ingot, which are decided based on the requirements of the wafer:

Material–Monocrystalline Si or multicrystalline Si.

Diameter of the ingot–200 mm or more for mono Si.

Crystallographic orientation– $\langle 100 \rangle$ ±3° mono c:Si.

Type of dopant-Boron-doped for p-type and phosphorous-doped for n-type.

Resistivity–0.5 to 3 Ω cm for p-type 1–7 Ω cm for n-type.

Oxygen concentration (atoms/cm^3)– $\leq$8–10 $\times$ 10^{17} for mono c:Si ($\leq$5 $\times$ 10^{17} for mc:Si).
Carbon concentration (atoms/cm^3)– $\leq$1–5 $\times$ 10^{16} for mono c:Si ($\leq$6 $\times$ 10^{17} for mc:Si).
Effective lifetime of charge carrier– $\geq$20 µs for mono c:Si ($\geq$5 µs for mc:Si).
Etch Pit density (dislocation density)– $\leq$500/cm^2 for mono c:Si ($\leq$3000/cm^2 for mc:Si).
Total bulk metal (Fe, Cu, Ni, Cr, Zn, Na) content $\leq$1 $\times$ 10^{14} atoms/cm^3 for mono c:Si. Surface should be clean and free from nicks and chips.

The final product in the mc:Si crystal growth is ready to cut Si bricks into wafers. Depending on the requirement of the size of the wafer, the ingot sizes are specified as G5, G6, G7, and G8. G5 to G8 are named based on the generation of m:Si furnace used to grow the Si crystal. G5 means generation 5 ingot capable to provide 25 bricks in 5 $\times$ 5 cutting configuration to give 156.75 $\times$ 156.75 mm or 158.75 $\times$ 158.75 mm size wafers. G6, G7, and G8 ingots provide 6 $\times$ 6, 7 $\times$ 7, and 8 $\times$ 8 bricks, respectively.

2.2 Process of making ingots (CZ and float zone method)

The basic input raw material is solar grade polysilicon, which is melted and pulled using crystal pullers to make ingots of the required diameter to accomplish crystal growth.

Several techniques have been used to convert polysilicon into single crystals of silicon; the majority of the solar PV industry uses two predominant methods. The first one is based on a pulling method known as the Czochralski (CZ) process, which is otherwise known as the Teal-Little process [3]. The second one is the zone melting method known as the floating zone (FZ) process.

In the CZ process, a single crystal ingot is grown by pulling from a melt kept in a quartz crucible. The FZ process involves passing a molten zone into and along a polysilicon rod, which helps to convert it into a single crystal ingot. In the above two processes, the input seed crystal has a dominant role in forming a single crystal with the specific orientation of the crystal structure.

As per the solar PV market estimates, 95% of single crystal silicon is manufactured using the CZ process while the balance of 5% is by the FZ process.

2.2.1 Crystal growth by Czochralski method

The solar PV industry has been using the Czochralski (CZ) method for the production of monocrystalline ingots. Ingots grown by the CZ method have been perfected over 50 years and significant progress has been made. The first process was a dislocation-free CZ Si crystal growth, which was pioneered by Dash in 1959; it uses a seeding technique [4]. The process of melt-based growth was demonstrated by Teal and Little.

The requirements of the polysilicon material are given below:

For solar-grade, the purity is 99.9999%, the growth method is Siemens CVD process, the chunk size is 2–4 inches.

Bulk type and surface resistivity: purity n-type >100 Ω-cm and purity p-type >1000 Ω-cm.

Surface morphology: Etched and nodular free.

In the CZ process, a cylindrical silicon monocrystal is pulled from a silicon melt. To begin with, the polysilicon chunks are melted along with the dopant in a quartz crucible in an atmosphere of an inert gas such as argon at a temperature of 1400°C or higher. The graphite container holds the quartz crucible. Because of the higher conductivity of heat, there is a smooth and uniform transfer of heat from the heater that surrounds the quartz crucible. The temperature of the silicon melt is uniformly maintained higher than the melting point of silicon. Dipping of the monocrystalline silicon seed crystal with the right crystal orientation is done into the melt. This is the start of the crystal formation and is aptly supported by the transfer of heat from the melt to the grown crystal. Then, the seed crystal is pulled at a slow speed of a few cm per hour from the melt. The pulling speed decides the diameter of the crystal. Homogeneous crystal growth and uniform dopant concentration are achieved by the rotation of both the crystal and the crucible in a counter direction. By increasing the pull speed continuously, a reduction in the crystal diameter to zero is achieved, which marks the end of the crystal growth. Also, this leads to no thermal stress in the ingots because of the sudden lifting out of melt. The possibility of the crystal being destroyed can be avoided.

Parts of the crystal growth puller

The crystal growth puller is used to grow monocrystalline Si ingots. The schematic view of a typical Czochralski crystal-growing puller is shown in Fig. 2.1. The following are the main parts of the CZ crystal puller.

(a) Hot zone or furnace: It consists of a quartz crucible, a graphite susceptor, a graphite heater, a power supply, a heat shield, and a crucible rotation mechanism. The furnace or hot zone is the heart of the crystal puller.

(b) Crystal-pulling mechanism: It has a seed holder and a seed rotation mechanism.

(c) Ambient temperature control: It is very important in a growth system. There should not be oxygen inside the chamber of the crystal puller. The graphite susceptor and graphite heater will react with oxygen to form CO_2. The oxygen gas should not react with molten Si. Therefore, oxygen should be removed from the chamber, which should be filled with inert gas argon. It includes an argon gas source, a flow control, and an exhaust system.

(d) Control system: A puller has a microprocessor-based control system to control the process parameters such as temperature, crystal diameter, pull rate, and rotation speed.

Fig. 2.1 depicts a schematic of typical CZ crystal growth equipment, and the process flow is shown in Table 2.1. The important steps in standard CZ silicon crystal growth sequence are as follows:

Charge weight checking: The polysilicon feedstock material comes in bags. The weight of the polysilicon is checked as per the label on the bags and the total charge is selected and weighed.

Charge preparation

The growth of a CZ silicon crystal starts with the stacking of high-purity polysilicon feedstock in the crucible, where either solar-grade or electronic-grade silicon is normally used as shown in Fig. 2.2. A charge in weight is determined by the hot zone selected in the crystal growers. The Kayex KX 360PV pullers selected can take 28″, 32″, or 36″ size crucibles. For a 36″ crucible, a 650 kg charge is required. The 650 kg polysilicon chunks or grains are loaded into the quartz crucible manually, as shown in Fig. 2.3A. This is done in such a manner so as to avoid movement during the next feedstock melting and also to minimize the contact between the crucible and the silicon melt to limit the integration of oxygen. The charge-filled crucible will be lowered into the graphite susceptor of the pullers.

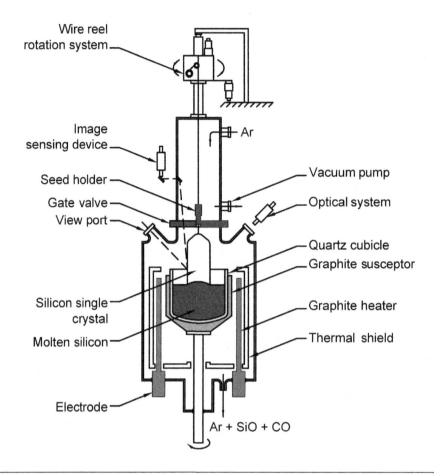

FIG. 2.1

Schematic view of CZ crystal growth puller [5].

Table 2.1 Process for monocrystalline ingot growth.

Weighing of polysilicon bags
Preparation and charging of polysilicon in a quartz crucible
Adding dopant to charge and transferring into puller
Placing the crucible with charge into a crystal grower
Heating the crucible
Dipping of seed crystal
Crystal growing starts and pulling is done using a seed cable assembly
Unloading and transportation of grown crystal in carts for ingot modification
Weighing of as-grown crystal
Cropping of tops and tails from as-grown crystal
Squaring of as-grown crystal
Surface grinding and edge grinding of squared ingot
Transportation of squared ingot to gluing station

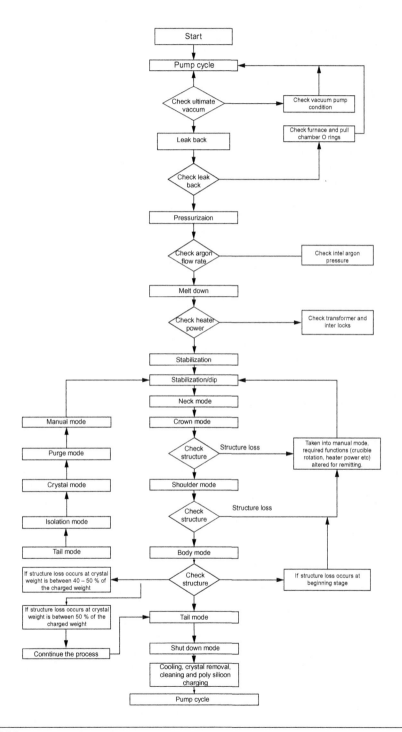

FIG. 2.2

Flow chart of CZ crystal growth process.

FIG. 2.3

A diagram of the Czochralski crystal growth technique. (A) A quartz crucible filled with polycrystalline silicon feedstock material. (B): The silicon feedstock fragments are melted together with the doping material. (C) A monocrystalline Si seed crystal is dipped into the molten silicon. (D) The seed crystal pulls a doped single crystal from the melt. (E) The schematic structure of a chamber for the crystal growing process in the Czochralski technique.

Courtesy: Micro Chemicals Gmbh.

Adding dopant to charge

Along with the charge, a trivalent material (dopant-preferential impurity) called boron is added in order to achieve p-type wafers for processing in cell manufacturing. The amount of dopant placed in a crucible with a silicon charge will determine the doping concentration in the resulting crystal.

Loading the charge into the puller: The quartz crucible, for example, with a 650 kg charge, is loaded into the crystal grower for processing. The lowering of the crucible with the charge is done using an automatic material handler/transfer machine. Fig. 2.3E shows the schematic structure of a chamber for the crystal growing process in the Czochralski technique.

Meltdown

The heating element is switched on and the crucible is heated to a temperature above the melting point of the Si material. The feedstock in the quartz crucible gets melted with the increase of the heater temperature above the melting point of silicon, as shown in Fig. 2.3B. Subsequently, it takes a few hours to achieve melt stabilization, which is the key to achieving a stable temperature of the melt and the homogeneous solute distribution. The charge is supposed to rise to a temperature above 1450°C, which is achieved uniformly.

The high temperature (above 1450°C) is maintained in an atmosphere of inert gas for the melt so that there is a complete melting and removal of bubbles to prevent any voids from being formed to create any negative crystal defects. CZ pullers are resistively heated. DC current is used to prevent the induced movement in the melt by eddy currents. In some systems, a three-phase AC current is used to set up a central rotation in the melt in order to control the growth condition. The incorporation of impurities increases with the speed rate, so the CZ rotation is lower such as 1–10 rpm.

Seeding

A small single crystal seed free from any dislocation and with a $\langle 100 \rangle$ crystallographic orientation is suspended over the crucible, touching the melt, as shown in Fig. 2.3C. Subsequently, the seed is lowered into the melt until its end becomes molten. There will be formation of a meniscus at the interface of seed end and the molten silicon. The crystal orientation of this seed will determine the orientation of the resulting pulled crystal and the wafers.

Starting of crystal growth

At this stage, the crystal pulling can start to happen. During the time period, which takes about 72 h, constant monitoring of the parameters of the puller such as vacuum levels, current indication in transformers, temperature readings, argon blanketing, argon flow rate, and other parameters is done to satisfy the health of the process. The silicon atoms from the molten silicon are likely to bond to the atoms in the seed crystal, plane by plane forming a single crystal as the seed is pulled upward.

Necking process

The formation of the neck is done with a gradual reduction of the diameter. During this process, the withdrawal of the seed from the melt takes place, as shown in Fig. 2.3D. This step is very critical. Argon gas, which is the inert gas, is allowed to flow into the pulling chamber downward throughout the crystal growth process. This helps to take out SiO and CO, which are byproducts of the reaction. Dislocations rise out of thermal shock among the cooler seed and melt and needs complete elimination before crystal growth.

A neck with a diameter of a few millimeters is grown at a 2–4 mm/min pulling rate so as to drive the dislocations to get diffused to the surface of the crystal, therefore helping to remove them. The process needs to last long enough that the crystal growth is free from any dislocation. Generally, the length of each neck is around tens of centimeters.

Crown and shoulder of the crystal

In order to achieve the required body diameter of the crystal, the next process steps to be followed are growing the crown and shoulder. With an increase of the diameter of the crystal, the growing of the shoulder and the conical portion is achieved. This increase in targeted diameter can be achieved either by a decrease in the rate of pulling, or lowering the melt temperature, or both. Growing the crown is done at a uniform pull speed and it is always less than that of the crystal body; this is done with a lower temperature of the heater. This process ensures a fixed increase in the diameter of crystal and is reliant upon both the parameters. The transition, which equals the standard diameter of the body, forms the shoulder. This transition step is required so as not to increase the diameter.

Body of the crystal

Slowly, the seed is withdrawn, causing the growth process of a single crystal by progressive freezing at the liquid-solid interface. The typical pull rate of 50–100 mm/h is used. In the process of pulling, the crucible is made to rotate in a counterclockwise direction with respect to the seed. So, there is a vertical rise of the seed cable. At last, the middle part of the body, which is cylindrical, is grown by controlling the rate of pulling and the temperature of melt. A sudden enhancement of the pull speed and a simultaneous reduction in body pull speed over a small distance will help in the vertical growth of the crystal. In order not to have a nonuniform diameter, it is recommended that the growing of the crystal body be done at a constant and uniform speed. The regulation of the speed of pull is achieved by a continuous monitoring system that inspects the meniscus and communicates with the control system. Generally, the growing of the crystal body is done at a 1 mm/min pulling rate and a typical diameter of 150–200 mm for solar applications. The variation of the speed of pulling and various parameters is done to optimize the process.

Coning and tail formation

There will be a drop in the level of melt, which is compensated during the crystal growth. Due to the drop in melt level arising out of increased heat radiation from the crucible wall, the reduction in the rate of pulling takes place at the tail end. This helps in exposing the additional crucible wall to the growing crystal. To reduce the effect of thermal shock, the reduction in crystal diameter is done gradually so as to make an end cone. This happens at the fag end of the growth process. This is done much before the full draining of the molten silicon from the crucible. The end of the process is achieved with tail growth. At this time, the diameter is reduced slowly and the end part of the ingot is formed as a conical shape. This end process needs optimization by means of an appropriate rate of cooling. It also needs to prevent high-density oxygen formation, which can lead to thermal stress as well as defects. There is a possibility of formation of slip dislocations during this last process. Once the small diameter is reached, it is advised to separate the crystal from the melt; this is free from any dislocations. At the end, the ingot is taken into a receiving chamber for its cooling down to the ambient temperature.

The next step is to unload and transfer the crystal to the next stage.

Weighing of the as-grown crystal

After transferring the as-grown crystal to the ingot modification section, the weight of the ingot is checked and recorded.

Cropping of top and tail

The top and tail of the as-grown crystal are cut using a cropping machine. Basically, this is an internal diameter saw machine that has a thin blade with a diamond coating on its inner diameter. The blade is rotated at a high rpm and it slowly moves down to cut the ingot. The crown and end cone of the as-grown ingot are cropped and the usable cylindrical portion is unloaded.

Shaping of as-grown crystal

The cylindrical body portion is fixed in a squaring saw machine. Basically, this is a band saw machine that has a saw band with a diamond coating. The sides of the round ingot are removed to make it square. Only one side of the ingot is shaped at a time.

Surface and edge grinding of the squared crystal: The surfaces and edges of the squared crystal are ground to avoid chipping and cracks in the crystal. The surface and edge grounding is done in a machine using a tool called a diamond wheel.

A schematic of the different parts of a CZ silicon ingot is shown in Fig. 2.4.

The CZ process has a short processing time and provides relatively lower thermal resistance and cost. The pull speed and temperature gradients control the diameter of the ingot. The seed/melt interface creates dislocations due to stress or thermal shock. So, the necking process has been developed by Dash.

In this CZ process, the oxygen incorporation into the silicon ingot is greater. The oxygen concentrates affect the properties of the wafer [6]. Oxygen incorporates in the range of $5-10 \times 10^{17}$ atoms/cc whereas carbon incorporates into the ingot in the range of $5-10 \times 10^{15}$ atoms/cc due to the solubility variability of oxygen in molten Si. The carbon impurity in Si affects the behavior of oxygen in the wafer. At the Si melting point, the oxygen can precipitate and can facilitate the internal gettering of impurities. A high oxygen concentration helps in the internal gettering of impurities during solar cell manufacturing process.

Active defects can be formed out of nonprecipitated oxygen and the resistivity is affected by thermal donors from oxygen. The interstitial form of oxygen in boron-doped Si severely affects the performance of the device. The B-O defect complexes cause light-induced degradation in a solar cell.

The segregation coefficient k can be defined as the ratio of the equilibrium concentration of the impurity in the solid to that in the liquid at the interface. This can be expressed as $k = \frac{C_s}{C_l}$. C_s is the concentration of the impurity in the solid and C_l is the concentration of the impurity in the liquid. The segregation coefficient, the k value of boron (0.8), determines the uniformity of distribution of the dopant along the p-type ingot. The k value for phosphorous is 0.3, which results in a lower concentration in the n-type ingot. This causes higher resistivity at the start of the CZ process. At the end of the CZ process, the ingot will have high dopant concentration resulting in lower resistivity. The relatively low value of k for the phosphorous dopant results in n-type ingots with a wide range of resistivity. The segregation coefficient k decides the uniformity of the dopant concentration in the ingot.

An inert gas ambient is used in CZ crystal growth in reduced pressure to promote the evaporation of contaminants. To reduce the oxygen dissolution in the melt, provisions are made to operate at 2–50 Torr to promote the evaporation of SiO vapors. The thermal gradient is inversely proportional to the square root of the diameter of the ingot. The stacking fault energy is 50% larger, so the tendency for the formation of dislocations is lower.

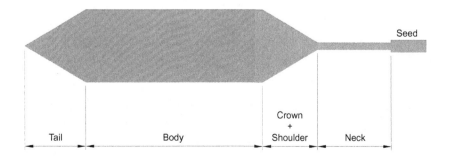

FIG. 2.4

A schematic of the different parts of a CZ silicon ingot.

FIG. 2.5

Monocrystalline silicon ingots grown by the CZ method.

Point defects, dislocations, interstitial defects, segregation coefficient of dopant, thermal gradients, turbulence in the melt, and the pull and spin rate affect the growth. For a $\langle 100 \rangle$ crystal orientation, accidental thermal or mechanical stress can result in twinning of the crystal, that is, changing the crystal orientation. Fig. 2.5 shows the grown Si ingots.

The CZ process is capable of easily producing large-diameter crystals from which large-diameter wafers can be cut. Oxygen atoms in interstitial sites improves yield strength if the concentration is up to 6.4×10^{17} atoms/cm^3. If the oxygen concentration is more, then there is a problem of the precipitation of oxygen. The precipitate attracts metallic impurities. It can also act as a sink, and these are called gettering centers. This precipitate reduces the breakdown voltage of the solar cell. Another disadvantage is the axial and radial dopant concentration in the crystal will have comparably low homogeneity. This is due to the oscillations caused in the melt during crystal growth. Hence, higher resistivity (greater than 100 Ohm cm) in CZ wafers cannot be easily done. By the use of a magnetic field, there can be retardation of the oscillations and subsequently there is improvement in the dopant to be homogenous.

Another variant of the CZ process is the continuous CZ process.

2.2.2 Crystal growth with the float zone method

To obtain the highest purity of the silicon material, the float zone process is followed. A high-quality polysilicon material manufactured through the Siemens or FBR process is required as a feedstock material. The impurities have higher solubility in molten Si compared to solid Si metal because of lower segregation coefficient of impurities. The difference in the solubility of impurities in the melt and solid makes it possible to segregate the impurities into the material to be refined. This is the principle used in the float zone process.

First, a high-purity polysilicon rod has to be made by casting or the CZ technique. Fig. 2.6 shows a schematic of monocrystalline silicon crystal growth by the float zone technique.

The polysilicon rod is suspended in a tubular zone of the FZ furnace in a vertical position. The upper part of the polysilicon rod is clamped. The feed rod has the facility to rotate. Inert gas such as argon is maintained in the chamber. A circular radio frequency (RF) coil-based heater is placed around the suspended polysilicon rod. The RF coil can be moved up and down. The RF coil heats the Si rod and a

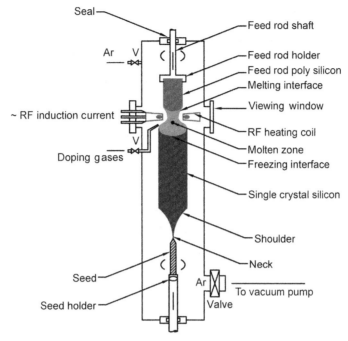

FIG. 2.6

Schematic of monocrystalline silicon crystal growth by the float zone technique.

monocrystalline silicon seed crystal of the desired orientation, that is, $\langle 100 \rangle$, is put into contact with one end of the polysilicon ingot. The seed is kept in a seed holder and can be rotated left to right. The application of the RF field to the charge ends up melting the polysilicon. Thereafter, it cools down with the passage of time and monocrystal silicon is formed with an orientation as that of the seed crystal. The full ingot experiences the movement of the melted zone and the RF coil. There is an interface between the molten Si and the solidified Si. The segregation coefficient of most impurities is lower than 1. The impurities are less soluble in the solidified silicon than in the molten silicon. So, the impurities will be carried to the newly heated zone and shifted along with the heater into the melt zone, reaching the end of the rod.

During the growth process, the melt is impregnated with impurities, which are carried by the molten zone. The concentration of impurities takes place at the end of the crystal, which can be removed. By repeating this process, any additional impurities can be removed, which reduces the impurity concentration. By adding phosphine (PH_3), arsine (AsH_3), or diborane (B_2H_6), doping happens during crystal growth.

The main advantage of the float zone technique is very high-purity crystalline silicon can be obtained. The concentration of oxygen and carbon are much lower as compared to CZ silicon, that is, below 5×10^{15} atoms/cc. The crystal growth process takes place in a vacuum or in an atmosphere of inert gas. There is no usage of a crucible or graphite container, so there is no contact of the melt with any surface. Hence, there is homogeneous dopant concentration in the final crystal. As a result of this, one can obtain a specified electrical resistivity and higher resistivity (up to 10 kohm-cm) wafers. The concentration of metallic impurities incorporated in the FZ grown crystal will be at a minimum due to the usage of high-quality feedstock material [7].

The cost of growing crystals and making wafers in the FZ method is relatively high compared to the wafers made from the Czochralski process. The surface tension of the molten Si is maintained in a vertical configuration. This avoids charge separation in the rod. The diameter of the monocrystal grown in the FZ technique has limitations imposed by the surface tension of molten Si. A maximum diameter of 150 mm with the current state of technology is possible.

2.3 Process of multicrystalline Si ingot casting

Electronic-grade Si wafers are not required for solar cells. Some of the specifications of wafers are relaxed for solar cells. High-purity crystalline Si wafers give high efficiency. Low-quality wafers also provide power but with lower efficiency due to the presence of many defects. Si casting offers a lower cost approach for large ingot growth compared to the CZ growth of monocrystalline Si ingots. In silicon casting process, polysilicon chunks are melted in a crucible separately and the molten silicon is poured into a cubic-shaped growth crucible and the molten silicon solidifies into multicrystalline ingot. If both melting and solidification are done in the same crucible, it is referred to as directional solidification. Directional solidification of silicon (DSS) is one of the most common techniques used for ingot casting of large multicrystalline silicon [8]. Due to significant advances in DSS technology, it is possible to get larger size multicrystalline ingots with improved crystalline quality at a significantly reduced cost. The size of the mc-Si ingots increased from 270 kg in 2006 to 1100 kg in 2019 due to significant advancements in DSS for manufacturing multicrystalline Si ingots. These large ingots allow for more wafers per cast, improved yields, and the potential for sizing wafers to larger dimensions. Until 2017, the mc-Si solar cell volume was expanded significantly and had a larger market share than c-Si cells. The efficiency gap between the c-Si and mc-Si cells has decreased considerably if poly PERC-based solar cells are considered. There is a considerable opportunity for DSS technology to improve wafer quality by reducing dislocation density, increasing grain size, and reducing the carbon concentration below the saturation level.

High-performance multicrystalline silicon (HP-multi-Si): During the growth of mc-Si ingots by casting or directional solidification techniques, many defects such as grain boundaries, dislocations, inclusions, oxides, and higher impurity concentration can be created in the ingot. The presence of crystal defects such as dislocations, grain boundaries, and metallic impurities like Fe are the main contributors to the lower efficiency of multicrystalline solar cells. The recombination of the minority carriers takes place at the dislocations and grain boundaries with impurities and causes a reduction in the conversion efficiency of multicrystalline Si solar cells. High-performance multicrystalline silicon (HP-multi-Si) has been developed to reduce or overcome the defects occurring in the crystal growth. In HP-multiprocess, seeds are placed in the crucible beneath the molten Si. Better control is exercised in controlling the cooling rate of the melt to ensure a uniform material with relatively small grains. HP-multi-Si produces grains of smaller size and lower dislocation density compared to mc-Si casting process. In normal casting or DSS process, thermal stress resulting from the difference in mass density of the silicon melt and the solid creates more dislocations. The generated grain boundaries mitigate the thermal stress and reduce the formation of dislocations. In the wafers of HP-multi-Si grown ingot, the recombination activity arising out of dislocations is neutralized with grain boundaries. So solar cells made with the wafers provide better performance.

In the case of mc-Si, the ingot sizes are designed to be compatible with multiple numbers of each of the standard wafer dimensions in order to maximize the geometrical yield. Multicrystalline technology can accept lower specifications of polysilicon, provided one takes care and it has purifying characteristics, which is a favorable point.

The method of making multicrystalline ingot is an easy process. The interface between the growing solid and the molten area is kept flat in a quartz crucible to extract heat from the melt. During this process, polysilicon will grow as large columns with typical heights of 25 cm. Also, there is the segregation of dangerous impurities toward the ingot top portion. To achieve a high-yield process as well as high quality, the furnace design must have proper heat control while using quality quartz crucibles.

The latest design of multicrystalline furnaces has minimal inhomogeneity while maximizing production. This ensures that the middle portion of the ingot has uniform quality and the witnessing of purification during this process achieves the required values of various impurities, excluding carbon and oxygen.

The following steps are used for the multicrystalline Si casting process, and the process flow is shown in Table 2.2 and Fig. 2.7.

Charge preparation

The process begins with the placement of the polysilicon feedstock in the crucible, which is coated with silicon nitride. These coating layers help to prevent the sticking of polysilicon to the walls of the crucible. Sticking leads to stress generation in the ingot, which gives rise to plastic deformation. This in

---

Table 2.2 Process flow for multicrystalline Si ingot.

Polysilicon is the feedstock material
Weighing of polysilicon bags
Preparation and charging of polysilicon in quartz crucible
Adding dopant
Placing the quartz crucible into crucible box
Load crucible into vacuum furnace
Heating of the crucible for melting of polysilicon feedstock
Direct solidification starts
Annealing and cooling of ingot
Unloading of ingot and crucible from the furnace
Weighing of as-grown crystal
Breaking and removal of quartz crucible mold
Transportation of grown crystal in carts for ingot modification
Cropping of ingot
Band sawing of ingot into bricks
Grinding and polishing of bricks
Transportation of squared ingot to gluing station

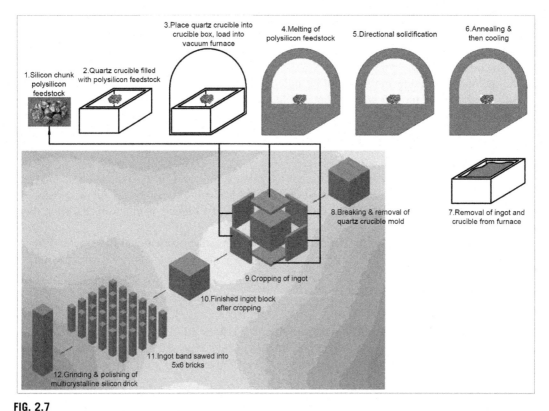

FIG. 2.7

Process of multicrystalline silicon crystal growth by DSS technique.

Courtesy: NREL.

turn develops dislocation defects. In addition, there is an added level of contamination of oxygen. A crucible is a critical material of mc-Si technology. The crucible develops cracks during the ingot cooling process, so it is treated as consumable.

Adding dopant to charge

Doping of ingots happens at a boron level of 13×10^{10} atoms/cm^3. To achieve this, the starting feedstock can have various mixtures from various sources, as typically used by the solar PV industry. As regards the resistivity range of the feedstock to be used in the solar PV industry, there are no specific requirements to dope the ingot. In the case of lowly doped silicon or virgin polysilicon feedstock, it is advised to use highly doped polysilicon powder, which is available on the market; and this can be mixed with the charge.

In the next step, the crucible is placed in the crucible box and loaded into the vacuum furnace.

Growth process

The crucible is heated with AC supply system to a temperature of 1575°C. In the presence of an inert gas in a controlled manner, the feedstock melting takes place. After complete melting of the Si and its temperature stabalization, directional solidification process is induced from the lower end to the top end by creating a vertical temperature gradient where the bottom surface of the crucible is cooled at a certain rate. Ramping down the temperature is obtained by means of the maintenance of a high gradient of vertical temperature between 500 and 1000°C/m, so as to achieve a near-planar solidification interface and ideal mixing conditions in the melt. In general, cooling is done by adopting either of these processes:

- Heat extraction in a controlled manner from the crucible bottom, which is not fixed.
- Movement of the crucible down in the midst of a vertical temperature gradient, which is described as a vertical gradient freeze.

In the presence of an inert gas such as argon, flushing of the melt surface is carried out during the whole process so that oxygen is removed by evaporation out of the melting silicon. In order to achieve the desired solidification interface, which is required for proper melt mixing, the growing rate is kept as low as 1 cm/h.

Annealing and cooling

The moment the solidification is completed, the cooling of the ingot is done carefully in order to control the stress development. The crucible is unloaded from the furnace and the ingot is removed from the crucible. The ingot is cropped on four sides to remove the red zones and edge grinding and surface finishing are done.

Typical features of DS mc-Si parameter typical values [9]:

Energy consumption— 6–7 kWh/kg.

Crystallization yield—70–80%.

Growth rate—510 mm/h.

Ingot size—800 kg in case of casting and 1100 kg in case of HP mC:Si.

Ingot base square—66 × 366 cm.

Ingot height—2025 cm.

Impurity typical value in mc-Si ingot—(ppma).

Fe—0.1.

Al—0.52.

Cu, Mn, Cr, Mg, Sr—0.1.

For getting ingot block of G5 size 860 × 860 × 350 mm^3, a charge of 580 kg crucible of size (OD) 890 × 890 × 480 mm^3 has to be used. For getting ingot block of G6 size 1010 × 1010 × 360 mm^3, a charge of 820 kg and crucible of size (OD) 1050 × 1050 × 540 mm has to be used. For getting ingot block of G7 size 1150 × 1150 × 390 mm^3, a charge of 1200 kg and crucible of size (OD) 1200 × 1200 × 540 mm has to be used. For getting ingot block of G8 size 1341 × 1341 × 390 mm^3, a charge of 1600 kg crucible of size (OD) 1375 × 1375 × 625 mm has to be used. Typical multicrystalline silicon ingots that are solidified in furnaces (Bridgman/VGF) have a shape of a square horizontal cross-section and are 1000 Kg in weight. In general, large ingots are cut with a top surface measuring

120×120 sq. mm of 6×6 bricks. The current practice is moving toward larger ingot sizes that offer higher yields while lowering wafer prices.

The multicrystalline material is cheaper compared to monosilicon material produced using the CZ process. In view of the defects present in the material such as grain boundaries and metallic impurities, such as iron, the efficiency of multicrystalline solar cells is lower than monocrystalline solar cells. In addition, the grain boundaries, which create multiple shunt paths, reduce the performance of the solar cell.

There is a diffusion of impurities such as iron into the ingot from the crucible in the course of the cooling process. This leads to the creation of highly concentrated impurities, known as the red zone. In view of the not-so-uniform distribution of impurities and the rate of cooling, the lowest-quality wafers are in the outer region of the ingot as compared to the balance of the ingot. Hence, the wafers produced from the edges are seldom used, which results in a lower yield of wafer from an ingot.

While growing multicrystalline ingots, several defects such as dislocations, inclusions, grain boundaries, and oxides can occur, in addition to the higher concentration of impurities.

In order to circumvent the above issues, there is the new development of high-performance multicrystalline silicon process. For this process, the use of seeds is combined with control of the cooling rate of the melt, which will result in a uniform material with small grains. High-performance multisilicon has a relatively lower density of dislocation as compared to standard multicrystalline silicon. The grain boundaries help to mitigate the stress and are a result of the difference in the solid and melt silicon mass density. Fewer dislocation clusters can result due to cooling, which can be controlled. There is a challenge for the neutralization of recombination activity, which arises out of dislocations. This is in addition to iron contamination and grain boundary defects.

2.4 Equipment used for ingot manufacturing

Crystal puller for the CZ technique

Fig. 2.8 shows a crystal puller for CZ growth from Linton Crystal Technologies, Kayex KX360PV.

The furnace is designed to grow Si ingot crystals of a 4 m length. The hot zone is designed to accommodate 28″, 32″, and 36″ size crucibles. A maximum of 650 kg of polysilicon feedstock can be loaded into the 36″ crucible. Ingots with a diameter of 300 mm or larger can be grown. It has an advanced control and monitoring system. The seed can be rotated 0–30 rpm, whereas the crucible rotation is 0–20 rpm.

Cropping saw

Grinding machine

Band saw or wire saw for squaring of the ingots by cutting off the edges

Edge grinding machine for smoothening of rather rough surface from sawing process

FZ puller for the FZ process

Equipment for directional solidification–casting–multicrystalline wafers:

Direct solidification furnace: Multicrystalline ingot furnace suitable for Generation (Gen7) can accommodate crucible of size 1220 mm $\times$ 1220 mm $\times$ 540 mm. The ingot can be obtained with a weight of 1150 kg. Such multicrystalline ingots can have a front surface area of up to 70×70 cm^2 and a height of up to 25 cm.

FIG. 2.8

Crystal puller for CZ growth from Linton Crystal Technologies.

Courtesy: Linton Crystal Technologies.

Cropper
Squarer
Surface grinder
Chamfer grinder
Feedstock etching station

Si_3N_4 as a coating material for crucible, argon gas, crucibles, graphite parts, diamond wire, boron dopant, and cleaning chemicals are some of the consumables are used in the Si crystal manufacturing process.

2.5 **Quality testing and reliability**

After the ingots are removed from the crystal puller, they are allowed to cool. Quality and reliability checks at each of the process steps are to be followed.

First operation: Tolerance of diameter (as it is typically grown oversized by a few mm)

Next operation: removal of shoulder and tail portions using a diamond blade cropping saw

Next operation: Body of the ingots are sawed into required lengths for slicing operation

Next operation: Grinding of the ingots to the required diameter

Ingot size is measured by tape. The brick size is measured using vernier calipers. Optical camera or visual inspection is done to check chips, scratches, microcracks on shaped and polished brick. Inspection with an infrared camera is done to check the cracks, SiC inclusions, dark spots, voids, etc., inside the bricks. The routine evaluation of ingots or boules involves measuring resistivity, evaluating their crystal perfection, and examining their mechanical properties such as size and mass. Other less routine tests include measuring the oxygen, carbon, and heavy metal impurities are measured by Fourier transform infrared (FTIR) spectroscopy technique as per the SEMI-MF-1391-1107 Standard. It is tested at brick level for 1 brick cut from top, middle, and bottom face of the ingot.

Resistivity is measured at brick level using four-point probe technique as per the standard SEMI-MF-043-00-0705. A four-point probe technique to measure the resistivity of the ingot is shown in Fig. 2.9.

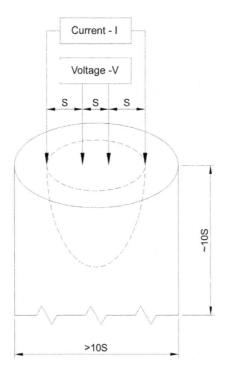

FIG. 2.9

Resistivity measurement using the four-point probe technique.

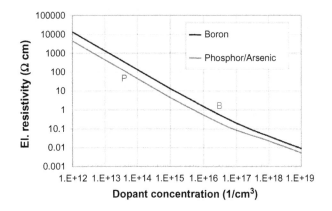

FIG. 2.10

Resistivity versus dopant density for boron and phosphorous dopants in silicon.

Courtesy: Micro Chemicals GmbH.

As shown in Fig.2.9, a current '*I*' passes through the outer probes and the voltage *V* is measured between the inner probes. The measured resistance (*V/I*) is converted to resistivity ρ (Ω-cm) using the relationship

$$\rho = \frac{V}{I} 2\pi S$$

where *S* is the probe spacing in centimeters. The calculated resistivity can be correlated with the dopant concentration using a chart such as the one shown in Fig. 2.10.

Gross crystalline imperfections are detected visually and defective crystals are cut from the boule. More subtle defects such as dislocation density can be estimated by preferential chemical etching technique as per the standard ASTM-F47-88. Chemical information can be acquired by employing wet analytical techniques or more sophisticated solid-state and surface analytical methods.

The minority carrier lifetime in the mono/multicrystalline block is measured using the microwave photoconductive decay technique and the standard is SEMI-MF-1535-0707. The bricks cut from the center, corner, and side of the ingot are tested and quantity for testing is two numbers of each type for an ingot. Crystallographic orientation measured on mono c:Si ingots/bricks using X-ray diffraction and the applicable standard is SEMI-MF-1725-1103.

2.6 **Manufacturers of Si ingots in the world**

Some of the manufacturers of mono/multicrystalline Ingots are:

GCL (CN)
LDK (CN)
China Jinglong (CN)
Yingli Solar (CN)
ReneSola (CN)

Green Energy Technology (TW)
Sornid Hi-Tech (CN)
Jinko Solar (CN)
Nexolon (KR)
Solargiga Energy Holdings
Trinasolar (CN)
Targray
Dahai New Energy (CN)
SAS (TW)
Comtec Solar
Pillar
Huantai GROUP
Crystalox
Eversol
Topoint (CN)
Photowatt
Shaanxi Hermaion Solar
CNPV
Longi Solar
Sun Power (Total energies)
Canadian Solar (CN)
Linton Crystal Technologies

2.7 **Technology trend of ingots**

New crystal growth methods are needed to meet the requirements set by solar PV ingot/wafer/cell manufacturers so as to address the high yield and performance enhancements of manufacturing. These will address issues such as: (A) large diameter, (B) controlled defect density, (C) uniform and low radial resistivity gradient, and (D) optimum initial oxygen concentration and its precipitation. Also, silicon crystal growers need to produce crystals economically while meeting high yields during manufacturing.

The major problems of ingot manufacturers are aligned with perfections of the crystal structure and the dopant distribution.

To circumvent problems with ingot growth processes, the development of alternate Ingot fabrication methods has evolved over time.

(a) Czochralski growth with applied magnetic field (MCZ)

The crystal quality of CZ silicon is strongly affected by the melt convection flow in the crucible. Originally, MCZ was intended for the growth of CZ silicon crystals that contain low oxygen concentrations and therefore have high resistivities with low radial variations. The unsteady melt convection induces growth striations resulting in temperature fluctuations at the growth interface. A magnetic field of sufficient strength can suppress the temperature fluctuations that accompany melt convection, and can dramatically reduce growth striations.

MCZ silicon growth controls the oxygen concentration over a wide range, thus providing a good-quality wafer. The MCZ method provides silicon crystals with better quality suitable to the electronics industry. The production cost of MCZ silicon may be higher than that of conventional CZ silicon because the MCZ method consumes more electrical power.

(b) Continuous Czochralski Method (CCZ)

The traditional CZ is a batch process that provides low throughput. To get more production per hour in ingot growth, it is required to increase the charge by using a large-size crucible or providing continuous charge to the crucible. In CCZ, fresh polysilicon is continuously supplied to maintain a constant melt height in the crucible. This high-purity resupply of polysilicon to the crucible dilutes the impurity build-up due to the segregation phenomena. Continuous production ensures far better utilization of raw silicon feedstock, a tighter control of dopant concentration, and competitive costs. As the wafer quality is similar to semiconductor-grade wafers, the enhanced electrical performance of each premium wafer allows solar cell manufacturers to create higher-efficiency cells with competitive wafer costs.

(c) Mono-like silicon casting

The casting of silicon with the usage of monocrystal seeding is known as cast-mono or quasimono crystalline silicon. This mono-like crystalline Si ingot uses standard mc-Si crystallization equipment such as large crucible and furnace and therefore, it is a low-cost technology. By using the common mc-Si crystallization process, mono-like or quasimonocrystalline Si ingots can be manufactured. Hence, this process involves low-cost manufacturing technology as an alternate to the CZ process.

Mono-silicon blocks do get used as seeds in this process, and are placed at the bottom of the crucible. The polysilicon feedstock with dopants constitutes the melt. In this process, the melting of seeds is partly done and followed by crystallization during which the growing crystal takes the seed orientation. Because of grain nucleation on the crucible side due to thermal gradient during cooling, only part of the ingot is monocrystalline. However, the part that is monocrystalline is of the highest quality. This can be used to produce high-efficiency solar cells due to the suitable diffusion length and minority carrier lifetime. This process has the potential to be developed so as to produce low-cost and high-quality solar PV silicon materials.

Mono-like silicon material has problems that are similar to multicrystalline silicon. These problems relate to multiplication as well as structural defect generation. In the case of mono-like Si, there is an easy propagation of dislocations, which leads to the multiplication of high dislocation density clusters. 50% of the mono-like ingots have parasitic multicrystalline Si, which is the effect of grain nucleation from the crucible wall and seed joints. Hence, this leads to various difficulties in wafering as well as texturing, which arises out of material nonuniformity.

References

[1] International Technology Roadmap for Photovoltaics (ITRPV), 11th ed., April 2020.
[2] J.M. Kim, Y.K. Kim, Growth and characterization of 240 kg multicrystalline silicon ingot by directional solidification, Sol. Eng. Mater. Sol. Cells 81 (2004) 211–224.
[3] G.K. Teal, J.B. Little, Phys. Rev. 78 (1950) 647.
[4] W.C. Dash, BGrowth of silicon crystals free from dislocations, J. Appl. Phys. 30 (4) (Apr. 1959) 459–474.

[5] F. Shimura, Semiconductor Silicon Crystal Technology, Academic, New York, 1988.

[6] F. Shimura (Ed.), Oxygen in Silicon, Academic, NewYork, 1994.

[7] A. Luedge, H. Riemann, B. Hallmann, H. Wawra, L. Jensen, T.L. Larsen, et al., High-speed growth of FZ silicon for photovoltaics, in: Proc. High Purity Silicon VII, Electrochemical, Society, Philadelphia, 2002.

[8] T.F. Ciszek, G.H. Schwuttke, K.H. Yang, Solar grade silicon by directional solidification in carbon crucibles, J. Res. Develop. 23 (3) (1979) 270–277.

[9] F. Ferrazza, Crystalline silicon: manufacture and properties, in: McEvoy's Hand Book of Photovoltaics, Elsevier Ltd, 2018.

Silicon wafer manufacturing process

3.1 Specification/requirement of wafers

The silicon feedstock material is crystallized into mono- or multicrystalline ingots or blocks. The ingots are cut into bricks with the required footprint area of the silicon wafer. The bricks are mechanically or chemically grounded and polished to improve the edge quality of the final wafers. This process will reduce the breakage rate of the wafers in the subsequent steps of cell processing. A multiwire saw or daimond wire saw is used to cut the Si brick/ingot into wafers. The wafers are separated, cleaned, and packaged.

The silicon wafer is the main material for silicon solar cells and acts as a substrate. The efficiency of the solar cell depends on the quality of the wafer used. The defects in the wafer affect the performance of the solar cell. The following are the requirements of the silicon wafer.

Material: Monocrystalline Si or multicrystalline Si.

Crystal growth method: CZ-crucible pulled for mono crystalline Si/direct solidification of silicon for multicrystalline Si.

The material and type of crystal growth are decided by the ingot.

High efficiency solar cells require high-purity monocrystalline wafers. The wafers cut from the float zone crystal-grown ingots are used for special types of solar cells such as interdigited back contact and heterojunction solar cells, which require high purity.

Orientation: ⟨100⟩: The ⟨100⟩ orientation is defined as the paralleling plane of the crystal to that of the surface of the wafer. This is applicable only for monocrystalline solar wafers. A ⟨100⟩ crystal orientation is followed for manufacturing monocrystalline silicon solar cells in the industry. Multicrystalline wafers contain different sizes of single crystals with different orientations. So, the orientation is not specified for multicrystalline wafers. This orientation is already decided by the seed crystal that is used in the crystal growth of ingot.

Si is a Face-Centered Cubic structure having more atomic density in the (111) plane. There will be a lower atomic density in the (100) plane of the Si surface, resulting in a low amount of dangling bonds. This yields higher carrier mobilities. As etching is involved in the manufacturing process of Si solar cells, the etching rate will be higher along ⟨100⟩. It provides anisotropic etching to create a pyramidal surface structure compared to the ⟨111⟩ direction. Fig. 3.1 shows the different crystal planes of silicon crystal.

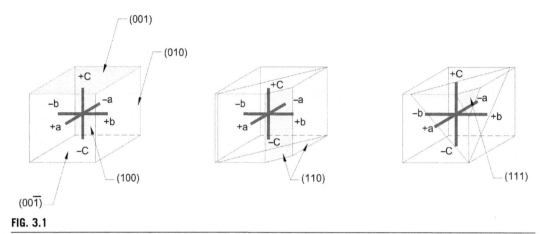

FIG. 3.1

Crystal planes in the Si crystal.

Historically, the ⟨111⟩ oriented silicon solar cells were used for space solar panels. The Young's modulus is higher for ⟨111⟩ oriented wafers, resulting in higher mechanical strength. The solar industry is satisfied with the mechanical strength offered by the ⟨100⟩ oriented wafer. Its electronic properties and ease of etching have led to its use as a standard crystal orientation for solar cells.

Size of the wafer: The size of the wafer to be cut is decided by the brick cut from the ingot. Pseudosquare or square wafers of sizes 156.75 × 156.75, 158.75 × 158.75, or 161 × 161 mm are preferred in the solar cell industry. Fig. 3.2 shows the pseudosquare wafer depicting the dimensions of length, width, and diagonal. Multicrystalline wafers will be the full square type. At present mono crystalline wafers are also available in the shape of full square and with dimensions of 182 × 182 mm and 210 × 210 mm also.

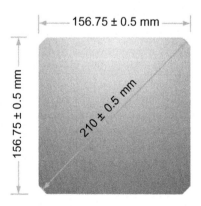

FIG. 3.2

Monocrystalline wafer with dimensions.

Table 3.1 The type and different sizes of the wafers.

Wafer type	Size (mm)	Diagonal length (mm)
M0	156×156	200
M1	156.75×156.75	205
M2	156.75×156.75	210
M3 or G1	158.75×158.75	223
M4	161.7×161.7	211
M6	166×166	223
M12	210×210	295

Different types of wafer sizes used in the industry are shown in Table 3.1. The type and size of the wafer are applicable for both mono- and multicrystalline Si wafers.

In the early days of the industry, circular wafers were used. With the evolution of crystal growth, wafer cutting, and solar cell and module technologies, the solar cell size has changed. First, wafers of $125 \, mm \times 125 \, mm$ size were used, which were later switched to the $6''$ wafer size. The $156 \, mm \times 156 \, mm$ wafer size, which is called the M0 type, dominated the industry for a long time but at present, it has disappeared from the scene.

Silicon wafers account for 30% of the cost of a solar pv module. With the increase of wafer size, there is a corresponding increase in the area exposure to solar irradiance. This leads to increased power generation and a reduction in cost. As far as the $166 \, mm$ wafer is concerned, it has reached the allowable production limit of the existing equipment and needs equipment modifications. The M6 wafer type has a surface area that is 12.2% greater than the M2 wafer type, hence providing 8.8% additional power generation. With the use of higher-wattage modules, the balance of the system cost is reduced. Based on the cost savings and the required modification of manufacturing facilities, at present the M2, M3, and M6 wafers are used in the industry. For M2 and M3 wafer usage, the existing wafer, cell, and module manufacturing lines without any upgradation can be used. The new manufacturing lines of the wafer, cell, and module can incorporate M6 wafers. At present, the G1 or M3 wafer is going to become the industry standard [1]. M12 wafers are being used to manufacture solar cells for higher wattage solar modules such as $600+W$.

Wafer Type: p-type or n-type

If trivalent atom Boron is doped in the melt, it becomes p-type and it becomes n-type, if pentavalent atom phosphorus is doped in the melt. The solar industry uses maximum p-type wafers doped with boron for mono- and multicrystalline solar cells. For high-efficiency cells such as the interdigited back contact (IBC), the heterojunction (HJT), the passivated emitter rear total (PERL), etc., the n-type wafers doped with phosphorous are used. n-Type wafers are becoming prominent to achieve higher efficiencies. To reduce light-induced degradation issue with p-type wafers, Gallium doping is being tried for making p-type wafers.

Resistivity: 0.5 to $3 \, \Omega$-cm.

The concentration of dopant decides the resistivity of the wafer. The low resistivity solar cell gives higher open circuit voltage. It influences the break down voltage of the solar cell.

Oxygen concentration: $\leq 5 \times 10^{17}$ atoms/cc

The large concentration of oxygen in a silicon solar cell creates light-induced degradation problems. The oxygen atoms pair with boron atoms in p-type solar cell and form a B-O complex.

The B-O complex gets activated during light exposure and becomes a defect center, causing degradation in the power. Hence, there is a limitation for the oxygen concentration in the wafer. The oxygen in the wafer helps to improve the mechanical properties of the wafer as well as the internal gettering of impurities in further high-temperature processes.

Carbon concentration: $\leq 1 \times 10^{16}$ atoms/cc.

The carbon impurity affects the performance of oxygen in the solar cell, and so its concentration should be limited. If the carbon concentration is high, the precipitation of oxygen atoms increases.

Effective minority carrier lifetime $\geq 10 \mu$ Sec.

If the minority carrier lifetime is more, the device efficiency will be more. For a high-efficiency cell, the wafers with higher minority carrier lifetimes are required. So, float zone-grown wafers are used for only high-efficiency solar cells.

Wafer thickness: $180 \pm 20 \mu$m standard at present.

The silicon wafer serves as a substrate for solar cells. To absorb all the light from the Si absorption coefficient point of view, the required thickness will be less than 100μm. To take care of mechanical rigidity and wafer/cell breakage issues, at present a nominal thickness of 200μm is used. If thinner wafers are used, the silicon usage for one watt of the solar cell will be reduced.

Bow $\leq 50 \mu$m.

The variation of thickness around the center point gives warpage and a bow to the wafer. A higher bow will affect the solar cell in the manufacturing process of the cell and module.

TTV $\leq 30 \mu$m.

The total thickness variation (TTV) affects the processing of the cell and module, as it requires a planar surface for the deposition of antireflection coatings of a few nm thick and the application of silver/Al paste in solar cell fabrication.

Saw damage $< 20 \mu$m.

If saw damage or saw marks are more than 20μm, microcracks are generated that weaken the mechanical strength of the wafers.

Etch Pit Density ≤ 500/cm^2.

This specification is only for monocrystalline wafers. The etch pits in the solar cell will act as defect centers and kill the efficiency of the solar cell. So, for high-quality wafers, the etch pit density should be minimum.

The surface should be free from cracks, stains, saw damage, chips, etc.

Tables 3.2 and 3.3 show the typical specifications for multi- and monocrystalline wafers, respectively.

3.2 Process of wafer making

As a material, silicon is hard, but it is also brittle. The process of converting ingots into wafers need to go through machining, chemical treatment, and polishing. Diamond is the ideal material for slicing and shaping the ingot into wafers. The wafer manufacturing process, which is cutting of ingots or bricks requires the following:

1. Best quality of cut to have superior surface quality
2. Higher yield: higher production
3. Maximum output using less process time to reduce the wafer cost
4. Best reliability process involving inspection or measurement and automation

Table 3.2 Typical specifications for multicrystalline wafers.	
P-type polycrystalline silicon wafer	
Material properties	
Growth method	Direct solidification of silicon
Conductivity type/dopant	P-type/boron
Oxygen concentration	$\leq 5 \times 10^{17}$ atoms/cm^3
Carbon concentration	$\leq 8 \times 10^6$ atoms/cm^3
Electrical properties	
Resistivity	1.0–3.0 Ω cm
Brick Lifetime	$\geq 10\,\mu$s
Geometry	
Thickness	$200 \pm 20\,\mu$m
TTV	$\leq 30\,\mu$m
Warpage	$\leq 50\,\mu$m
Length	156.75 ± 0.25 mm
Width	156.75 ± 0.25 mm
Right angle	$90° \pm 3°$
Diagonal	210 ± 0.5 mm
Microcrack	Not allowed
Saw marks	$\leq 10\,\mu$m, unlimited in the number of strips; 10–$15\,\mu$m, ≤ 5 strips/1 cm
Edge chips	Depth ≤ 0.3 mm, Length ≤ 0.5 mm, Max 2 nos. per wafer
Breakage	Not allowed
Micrograin	Single area $< 3 \times 3$ mm^2; total area $< 3 \times 3$ cm^2
Hole	Not allowed
Surface quality	No surface damage, stains, water marks, or contamination allowed

5. Thinner wafer by use of thin wires
6. Diamond wire and slurry consumption at the lowest for the reduction of environmental impact due to cutting liquids
7. Kerfloss reduction: low loss of material during cutting
8. Better mechanical strength
9. Better recyclability of the silicon material lost as kerf loss

The wafer production process starts with slicing the crystal ingot. There are three types of slicing methods that have been used in the silicon wafer industry: the internal diameter (ID) saw, the wire saw with loose abrasive slurry, and the diamond wire saw.

The following steps are followed in the manufacturing of silicon wafers meant for semiconductor applications.

- Slicing
- Edge profiling

Table 3.3 Typical specifications for monocrystalline wafers.

Material properties of monocrystalline Si wafer	
Growth method	CZ
Conductivity type/dopant	P-type/boron
Oxygen concentration	$\leq 1 \times 10^{18}$ atoms/cm^3
Carbon concentration	$\leq 5 \times 10^{16}$ atoms/cm^3
Etch pit density (dislocation density)	≤ 500 cm^{-2}
Surface orientation	$\langle 100 \rangle \pm 3°$
Electrical properties	
Resistivity	0.8–3 Ω cm
Minority carrier lifetime	≥ 20 μs
Geometry	
Cutting method	DW (diamond wire) cutting
Geometrical shape	Pseudosquare
Bevel edge shape	Round
Thickness	200 ± 20 μm
TTV	≤ 30 μm
Bow	≤ 50 μm
Warpage	≤ 50 μm
Flat to flat length	156.75 ± 0.25 mm
Diagonal length	210 mm
Saw marks/steps	≤ 15 μm

- Lapping or surface grinding
- Etching
- Edge polishing
- Polishing of surface
- Cleaning

After slicing, the wafers are cleaned and packed. There is no requirement for lapping, surface grinding, or polishing steps, as the etching step in solar cell manufacturing will take care of any saw damage that occurred during cutting the wafer.

Slicing process

When silicon is indented or scratched at load, it undergoes phase transformation rendering ductile. When the indented tip is pressed on the Si surface, it induces a high localized pressure that transfers the brittle silicon phase into the ductile phase. On unloading, the ductile phase transforms into a mixture of an amorphous and a metastable silicon phase [2].

The wafer manufacturing process begins with slicing the crystal ingot. Basically, there are three common processes utilized for wafer slicing from ingot crystals:

1. Wire saw
2. Annular saw or inside hole saw
3. Diamond wire saw

The requirements or challenges in wafer cutting process are to obtain high surface quality higher-strength thinner wafers with smaller variations in thickness with cheaper cost, low kerf loss, less damage to facilitate higher volume production for solar cells.

3.2.1 Ingot slicing using annular saw or ID saw

The internal diameter (ID) saw has a thin annular blade with a diamond-bonded region on the inside edge of the annulus. Hence, it can cut only one wafer at a time and it takes a few minutes to cut one wafer from the ingot. One wafer per each annular sawing causes a very low yield, which leads to a higher cost and more time to complete the process.

During slicing, the flexure of the blade can lead to bowing and warping in the produced wafers. Simultaneously, there is blade wearing. In view of the blade quality not being good, it creates variations in the total wafer thickness. Further, due to the variation in total thickness during cutting by a saw with bad quality, it is necessary to correct the variations in the next wafer process. The reduction in kerf losses demands controlled flexing of the thin blades by the tensioning system. Hence, in order to circumvent the large kerf loss, new sawing/slicing methods have been developed to meet the production cost requirements of wafers by the solar industry. At present, the ID sawing of wafers is not being used in the solar industry.

3.2.2 Ingot slicing using wire saw–loose abrasive slurry process

The wire saw makes several hundred parallel operations to cut several hundred wafers at a time while taking a few hours. For crystals that are more than 150 mm in diameter, the wire saws are found to be more economical than ID saws for lower kerf losses; this is achieved by the use of thin wires.

The wire saw was developed to overcome the issues in developing a tensioning system, which controls the flexing of the thin blades, to reduce the kerf loss.

Fig. 3.3 shows the schematic of the loose abrasive slurry-based wire sawing process for cutting wafers.

The main part of the system is the wire web, which consists of a delivery spool, a guidance roll, and a collector spool. A wire is taken from a delivery spool and guided by rolls/pulleys that are in place to give the wire the desired tension. From these rolls, the wire leads to the wire guidance grooves of the rolls that are coated with polymer and that feature grooves with defined spaces commonly known as pitch. The wire is wound as many times as there are grooves on the wire guidance roll. At the last groove point, the wire is again guided by a pulley to a collector spool. The wire on the wire guidance rolls forms the so-called web. Thus, a wire saw employs several segments of parallel wires to cut several wafers at one time, which helps to increase the production yield.

Free-floating abrasive particles such as SiC and diamond powders remove the material by rolling and indenting into the silicon crystal surface in the sawing channel. The particles are to be suspended in a carrier fluid in order to have good lubrication in the sawing channel. The carrier fluid considered is polyethylene glycol (PEG).

The abrasive slurry is introduced in the wire web. The wire web drags the slurry to the Si ingot. The mounted ingots are pushed into the wire web, which facilitates to cut the ingot into single wafers.

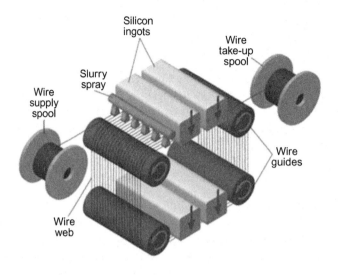

FIG. 3.3

Schematic of loose abrasive slurry-based wire sawing.

The ingot moves toward the wire web with a feed rate of 0.4–0.8 mm per minute. The wire is kept at a specified tension in the range of 18–25N and made to move at high speeds of 15 to 20 m/s. The Si ingot is pushed against the wire. The space between the Si surface and the wire is filled with slurry and abrasive particles. The wire pushes against the particles and makes them indent on the crystalline Si surface. The particles between the wire and Si surface are responsible for the removal of material from the ingot, facilitating the cut in the ingot. The particles on the side will damage the surface. Simultaneously multiple wafers are sliced at the same time when the silicon ingot is fed through the wire web.

Before 2018, the slicing of the ingot into wafers was done by slurry-based wire sawing.

The wire saw takes a few hours to make a few hundred parallel cuttings to turn one Si crystal ingot into several wafers in a single operation. For crystal ingots with a diameter of more than 150 mm, the wire sawing process is more economical than the ID sawing process. The wire saw process provides a higher yield and has lower kerf losses due to the use of thin wires. The achievement of wire sawing of crystal ingots is by free abrasive machining, which uses slurry consisting of SiC grit mixed in oil or ethylene glycol called carrier fluid. The wire helps to move the slurry to the ingot, where the grit is trapped between the tensioned wire and the ingot. This process is called rolling and indenting cutting. The speed of cutting relies upon a few factors such as the hardness and size of the grit as well as the wire speed, which determines the delivery of the grit to the cutting surface. In the event of application of high pressure, there is a distortion of the wire, which leads to warping or perhaps breaking. Hence, a cutting speed of 0.25–0.50 mm/min is recommended for SiC grit and the crystal ingot.

Table 3.4. gives the process flow and checkpoints while Fig. 3.4 gives the process flow for cutting wafers with a wire saw using the loose abrasive slurry process.

Ingot/brick weight checking

Ingots are cut into bricks suitable for the requirement of the wafer size. The ingot's weight and length are checked and labeled on the ingot in mm and kg, respectively.

Table 3.4 Wafer cutting process flow and checkpoints.

Flow of wafering process	
Start	**Check points**
Loading of ingot	Assign run number to process. Note ingot batch, weight, length
Slurry–SiC+PEG	Assign slurry batch number to process. Note batch number of SiC and PEG
Inspect and check	Adjust the specific gravity of PEG
Stabilization	Run for 15 min
Inspect and check	Check the specific gravity of the slurry and wire web
Ingot polishing	
Inspect and check	Check depth of cut and process parameters
Start ingot cutting process	
Inspect and check	Note the values for every 10 mm cut
Lifting of sliced ingot	Note process completion time and process parameters
Inspect and check	Check for wire and table movement, lifting of web, slurry flow
Unloading of sliced ingot	
Inspect and check	Correct the wire web and, if required, change the wire web
Cleaning and separation of sliced ingot	Water jet cleaning and separation of wafers
Wafer cleaning	
Wafer inspection	

Ingot preparation and gluing

The ingot is cleaned thoroughly and is pasted on a glass plate, which is in turn pasted over an aluminum plate. A two-component adhesive is used for gluing. After curing, the ingot along with the glass and aluminum plates is fixed onto the ingot mounting fixtures.

Ingot loading and stabilization

The ingot mounting fixture is fixed in the wire web in the wire saw machine. The slurry, which is a mixture of silicon carbide powder and polyethylene glycol (PEG), is pumped inside the machine through the slurry nozzles. Afterward, the machine is allowed to run to stabilize the slurry (to achieve a homogenous mixture) and the wire web. After the slurry is stabilized, the machine is opened and checked for any shifting in the wire web.

Cutting process

During the cutting process, the wire is subjected to a high speed of 12–14 m/s and the table is continuously moved down at the rate of 350 μm per minute. The wire carries the slurry flowing from the nozzles to the cutting edge and starts cutting the ingot. During the process, the slurry quality, temperature, tension arm values, etc., are monitored continuously. Once the cutting is completed, the sliced ingot is lifted from the wire web and taken to wafer precleaning.

Wafer cleaning

The process ends with cleaning the wafer using pure chemicals so that the polishing agents are taken out and it becomes free from residue.

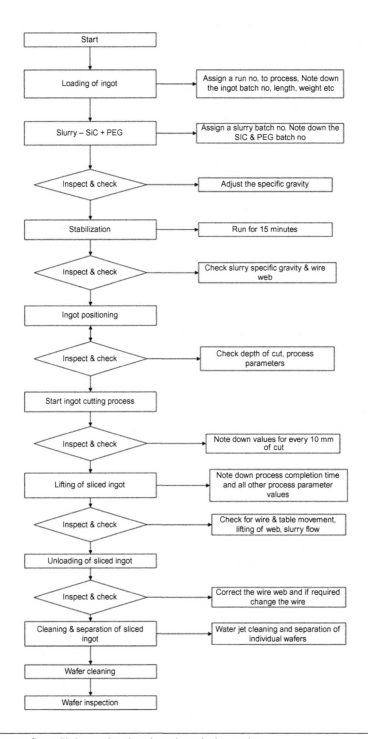

FIG. 3.4

Wafer cutting process flow with loose abrasive slurry-based wire sawing.

The freshly cut and individual wafers are cleaned one after the other in different baths. These include acidic and alkaline baths as well as pure water baths. Besides pH, the acetic acid concentration is also monitored in the acidic bath and must be in a defined range. The alkaline bath must have a constant concentration of sodium hydroxide/sodium carbonate. The exact composition of the baths and their pH is crucial for the surface structure of the wafer. Acid and alkali base concentrations are determined in the laboratory or directly at the cleaning system by the titration technique. In addition, the density is measured in the laboratory using the density meter [3].

The wafers pass through pure water baths between the different cleaning steps and at the end. The purity is monitored by measuring the conductivity to ensure that the surface is free of deposits after drying.

The abrasive grit size, slurry viscosity between the abrasives and silicon, wire speed and normal load play a dominant role in cutting process of ingot.

This process has a higher kerf loss, thus more material is used. The use of PEG slurry increases the hazardous waste and causes higher costs of waste treatment. There is more damage to the surface during sawing, requiring more postprocessing resources. So, the comparatively weaker wafers lead to reduced lifecycle of the solar cells that are made with these wafers. The yield and productivity are comparatively lower. The recovery and recyclability of silicon material from slurry is costly.

3.2.3 Ingot slicing using diamond wire saw process

During the cutting process, 50% of the expensive solar-grade Si is lost as kerf in the conventional annular saw or slurry-based wire cutting processes. So, a diamond saw process has been introduced. In diamond wire sawing (DWS) process, the steel wire is used. To the steel wire, abrasive grit (diamond) is attached with an electroplated nickel coating. Water cutting fluid is used instead of PEG slurry and all other processes are similar with LAS process.

By this technique, many wafers are cut using an automated process and with high precision.

The process of silicon ingot sawing using a diamond wire saw is depicted in Fig. 3.5. The feeding system, is designed to allow high-precision control of the wire tension during the length of the cutting process. The abrasives are electroplated diamond fixed to the wire.

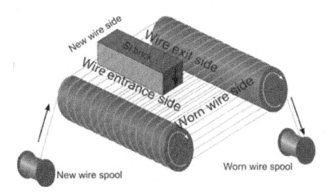

FIG. 3.5

Schematic of diamond wire sawing of Si wafers.

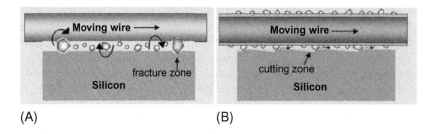

(A)   (B)

FIG. 3.6

The cutting action of the moving wire in (A) loose abrasive sawing, and (B) fixed abrasive sawing with diamond wire.

As shown in Fig. 3.6A, the slicing of brittle silicon wafers by LAS occurs through the cutting action of the hard SiC abrasive grit contained in the PEG slurry. The grit in the slurry is entrained in the cutting grooves, and the silicon wafers are sliced by a three-body material removal process occurring between the wire, the cutting grit, and the silicon ingot. With the diamond wire cutting process, as shown in Fig. 3.6B, the fixed grit indents the surface, creates microcracks, and removes material.

This method has a distinctive advantage in that it can saw several hundreds of wafers at one time using a single wire. In this process, there is reduced kerf loss, thus saving material and preserving resources for the future. As the process uses water-based cutting fluid, there is a reduction in the hazardous waste. The surface damage during the sawing process is to a smaller depth, so it requires fewer resources for postprocessing. This process yields higher productivity. The recyclability of silicon from water-based cutting fluid is cheaper [4].

Wafer cutting using a diamond wire saw enables the production of thin wafers with the use of a thin wire.

Further, there is an enhancement in the efficiency of wafer manufacturing using diamond wire sawing, which leads to a lower wafer cost of production. The generation of sawing marks by the diamond wire becomes more difficult in the subsequent etching process. In view of the higher wafer production cost due to other wafer sawing processes, the lower cost and higher production yield of wafers using a diamond wire saw has completely overtaken the solar pv manufacturing industry since 2018.

Multisilicon wafer production–process steps
Crystallization ➜ Top and bottom grinding ➜ Squaring ➜ Ductile polish grinding and chamfering ➜ Infrared inspection➜ Lifetime scanning ➜ Top and bottom soft cropping ➜ Gluing ➜ Wafering ➜ Precleaning

Cutting of multicrystalline wafers with DWS faces many challenges. Multicrystalline silicon has grain boundaries and more crystallographic defects than monocrystalline wafers such as dislocations, hard precipitates, impurities, and inclusions. These defects can significantly alter the cutting performance. The dislocation density variation in mc-Si grains was found to be correlated with variations in fracture toughness, which in turn affects the cutting characteristics.

The density of the microcracks in the saw-damaged layer more significantly affects the mechanical strength of the silicon wafers. The fracture strength of the silicon wafer depends inversely on the surface roughness of the wafer. The strength of the multicrystalline Si wafer increases with the removal of

(A)   (B)

FIG. 3.7

Image of multicrystalline wafers. (A) As diamond wire cut wafer, (B) the crystals of different sizes with different orientations are appearing.

surface damage by etching/texturization. The crystallinity has a significant effect on the mechanical strength of the multicrystalline wafer. The surface roughness and edge defects such as microcracks and grain boundaries are the probable sources for the degradation of mechanical strength.

Fig. 3.7 shows an image of multicrystalline wafers. The wafers show the multigrain structure of different crystals. Diamond wire-based sawing of multicrystalline Si wafers leaves a smooth surface that poses significant challenge in texturization of wafers. The surface with diamond wire saw cutting will appear black, if it undergoes metal catalyzed chemical etching (MCCE) treatment and so it is called black silicon. Special texturization techniques such as MCCE or adding additives to standard etchants are to be followed to remove the saw damage.

Monosilicon wafer production–process steps
Top, tail, test, and cutting → Squaring → Ductile polish grinding and chamfering → Lifetime scanning → Gluing → Wafering → Precleaning.

Fig. 3.8 shows the image of a monocrystalline wafer.

3.2.4 **Quality testing and reliability**
The following are the quality tests conducted during manufacturing of the wafer from ingot crystals.

Thickness of the wafer and its total thickness variation

Visual camera fitted to the microscope used for checking saw marks and roughness, edge defects, chips, stains, and microcracks/inclusions

The doping concentration and resistivity are measured using a four-point probe technique

The dimensions and geometry of the wafer

The bow and warpage of the wafer

The lifetime of the wafer using the microwave photoconductive decay technique

The etch pit density is measured based on the chemical etching technique

The surface roughness is measured by an atomic force microscope

The oxygen and carbon concentration are measured by Fourier transform infrared spectroscopy (FTIR)

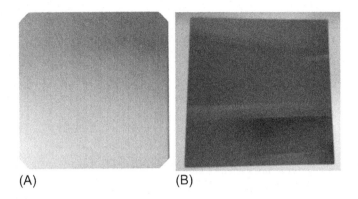

(A)   (B)

FIG. 3.8

Image of monocrystalline wafer. (A) Pseudo square type cut with slurry-based technique. (B) Full square wafer cut with diamond wire saw showing saw marks.

The wafer mechanical strength is measured by the four-point bending method and the results are statistically evaluated by Weibull analysis

The following factors such as surface roughness, crack defects at the edges, the amount of grain boundaries, and the saw-damage layer thickness are going to affect the fracture strength of a processed silicon wafer. When Si ingots are cut into thin wafers, a multiwire sawing process is used, which creates a highly stressed and damaged layer.

Without the damaged layer, the fracture strength is inversely proportional to the surface roughness. The surface roughness profile is the second most detrimental factor affecting the mechanical strength of silicon wafers. The effect of crystallinity features on the mechanical strength of the silicon wafer.

3.2.5 Equipment used for silicon wafer manufacturing

The various equipment used for wafer manufacturing are given below:

Precision weighing scale for ingot weight measurement
TTV and thickness measurement setup
Viscometer to measure the viscosity of the slurry
Hydrometer to measure the specific gravity of the slurry
Wire tension meter to measure the wire tension
Brick gluing station
Wire saw
Wafer precleaning setup
Wafer separator setup
Wafer transport system
Wafer inspection and sorting machine

3.2.6 Manufacturers of silicon wafers

Around the world, the current silicon wafer making capacity is aggregated to several GW.

Some of the manufacturers of silicon wafers of 2018 are [5] shown in Table 3.5.

Table 3.5 Top 10 solar wafer manufacturers in 2018.		
1	GCL Poly Energy	20,060 MW
2	Xi'an Longi Silicon Materials Corp	22,500 MW
3	Inner Mongolia Zhonghuan Photovoltaic Material	9000 MW
4	Jinko Solar	8000 MW
5	LDK Solar	4800 MW
6	Sornid Hi-Tech	4200 MW
7	Green Energy Technology Inc.	2625 MW
8	Yingli Green Energy	2452 MW
9	Renesola	2900 MW
10	JA Solar	4500 MW

Source: IHS market. PV Magazine International, February 24, 2018.

3.2.7 Technology trend and pricing analysis of wafers

The polysilicon material requirement has been brought down from 16 g/Wp to less than 4 g/Wp, as there has been an increase in efficiencies along with the production of thin wafers. Since 2018, the polycrystalline material consumption for wafer production has been reduced to less than 4 g/Wp. After 2018, the diamond wire saw method has been the only one being used, which has encouraged thin wafer production. Nowadays, solar cell manufacturing lines are automated, which is forcing the move to using thin wafers. In view of about a 50% loss of silicon material during the sawing process, new approaches to make direct wafers are gaining momentum in order to be cost competitive. The new process are the 1366 process, the kerf loss wafer process by epitaxy, and a few new approaches, which are described in the next section.

3.3 Direct wafer process (DWP)

Silicon wafers are being manufactured using the conventional process of casting and sawing. At first, the casting of highly pure silicon is made into large crystalline ingots/blocks, then the ingots/blocks are cut to various sizes. This is followed by polishing and subsequent slicing by a wire/diamond wire saw into thin wafers of the required typical size (156.75 × 156.75 mm) with a $200 \pm 20\,\mu m$ thickness.

While doing the sawing, almost 50% of the silicon material is lost as dust or kerf. Hence, this conventional sawing method leads to a high cost of wafer production in terms of capital as well as operating cost. Also, this sawing process takes more time with high energy consumption and involves several steps. It also results in quality defects that occur during wafer processing.

In recent times, direct wafer process (DWP) technology [6] has evolved to circumvent the old wafer process, enabling the production of wafers from molten silicon every 15 s. The polysilicon material filled in a crucible made up of graphite or Silicon carbide or Silica and heated to molten state. An interposer sheet made up of ceramic SiN is used as a mould to deposit Si material. The ceramic sheet is a free standing, very thin, flexible, porous, and able to withstand chemical and thermal environment of molten silicon. The thin mould sheet is fixed to the bottom of a vacuum plenum. The mould sheet is made to contact with the surface of molten silicon material. The mould assembly remains in contact with the Si melt for a time on the order of 1 s. The amount of contact time between the mould and the melt will depend on the required thickness of the Si wafer to be fabricated, the thickness of the mould sheet, and the temperatures of molten Si and mould sheet. The molten silicon from the crucible freezes on to the mould sheet. The vacuum will cause the formed Si sheet to be held against the mould sheet. The mould assembly system is lifted out of the melt carrying the Si sheet. The vacuum is released and the formed Si wafer can be separated from the mould sheet. Thus the Si wafer is fabricated in direct wafer process technology [7].

- DWP needs only one-third of the consumable cost as compared to standard wafer manufacturing.
- This process uses more than 90% of the silicon, as compared to more than a 50% waste of material in the sawing process.
- At present using DWP wafers, the average solar cell efficiency has reached 20.5% with a road map to achieve higher efficiency. DWP consumes only 35% of the energy as compared to standard sawing methods.
- This process helps to achieve a variety of wafer features that are not achievable by a conventional wafer manufacturing process such as sawing.
- In view of the fact that the wafer cost is 40% of the total solar pv module cost, DWP, which eliminates more than 50% of the silicon loss, helps to achieve a lower cost of the wafer in bulk production (several GW level). Therefore, the final solar pv module cost is lower and the solar cells are of good quality.
- With DWP technique the purity and quality of the wafer can be improved by minimizing dislocations and improving grain structure as well as improving the nucleation behavior for a better crystal structure.
- DWP is moving toward higher solar cell efficiency due to a better microstructure and improved consistency of the wafer. If the DWP is extended to tandem solar cells, the efficiency could surpass 30%.
- The DWP has a distinct ability to produce a doping gradient, which is grown in with a much higher concentration of dopant on the back side of the wafer. This helps to collect carriers in an efficient manner, which leads to a high voltage and fill factor due to good conductivity through the bulk of the wafer.
- DWP helps to grow wafers with uniform microstructures because there is control over the growth process environment for each wafer. The occurrence of fresh nucleation does not allow any propagation of dislocations due to the shorter process time of wafers in DWP.
- DWP helps to grow thinner wafers with thicker borders. This process enables control of the thickness of the wafer locally. One can manufacture a thinner wafer due to thicker reinforcement, which is required at the edge of the wafer.
- DWP during 2013 saw the first solar cell efficiency of 16% on a 156 × 156 mm wafer. After 4 years, this improved to 20%.
- During May 2019, the first solar pv modules using wafers from DWP (thin wafers with thick borders) were made. This enabled a reduction in silicon usage to less than 1.5 g/Wp as compared to around 4 g/Wp for a standard wafer-based solar pv module

3.4 Kerf loss wafer process by epitaxy

Kerf loss mono crystalline silicon wafers are produced directly from the gas phase, using vapor-phase epitaxy (VPE). Epitaxy is a method to grow or deposit monocrystalline films in which new crystalline layers are formed with a well-defined orientation with respect to the crystalline surface of a substrate. VPE is a modification of chemical vapor deposition (CVD) process. With this process of wafer production, the most cost-intensive steps of silicon PV technology such as polysilicon production, crystal growth of ingots, and the machining and wafering of the ingots are eliminated. This enables a substantial reduction in production cost and usage of silicon compared to conventional wafer manufacturing process. Trichlorosilane with hydrogen as carrier gas is used in CVD chamber. The process uses restructured porous silicon substrate on which a release layer is coated. This closed surface is the foundation and the seed layer for the monoepitaxial growth of the silicon wafer. In CVD chamber, TCS is made to vaporize, decompose, and Si atoms are deposited on top of the release layer of the substrate as a thick material. The next process is to separate the grown Si wafer from the Si substrate. The Si substrate is reused after cleaning the release layer. To get good quality wafer with wafer-to-wafer film thickness uniformity, the substrate is heated and rotated and sufficient quantity of TCS is provided in the CVD reactor.

In this process, a monocrystalline seed wafer is cloned. There is a closed seed wafer loop and nearly no kerf loss that allows for a low production cost. Wafer thickness of standard 180 μm or thinner, can be produced and there is no problem to produce 80 μm thick wafers.

3.5 What is the new trend in wafer manufacturing?

There have been advancements in wafer technology and the landscape has been completely changed during recent years. The introduction of the diamond wire saw is an improvement in terms of the stability and cost reduction in wafer manufacturing. The slurry-based wafer technology completely shifted to DWS in 2018 for mono- as well as multicrystalline Si wafers. With the introduction of passivated emitter rear contact (PERC) technology in solar cell production, p-type wafers dominated the market. The dominance of casted multicrystalline Si has gone and its share shrunk to less than 40% in 2019. The P-type high-performance multicrystalline Si is supporting the market, and the market share of mono-like casting wafers is 1%.

The market share of monocrystalline wafers, including the p- and n-type, was 60% in 2019. The n-type wafer utilization has been increased with the introduction of HJT and TopCon-based solar cells. The thickness of the wafer is reduced with usage of DWS in cutting process of the wafers. Electroplated diamond wire is the dominant material for cutting.

With the maturity of DWS, the reduction in kerf loss, the increased productivity, the increased quality of the wafer, and the reduction in wafer price have affected the kerfless wafer manufacturers. With the advancements in wafer technology, the consumption of polysilicon per wafer has been reduced to 16 g per wafer in 2019.

Direct wafer technology is not expected to yield significant cost advantages compared to ingot growth and wafer cutting technology due to advancements in the wafer cutting process using DWS. For the monocrystalline Si wafer cutting process, the electroplated diamond wire technique is ruling the industry. A resin-bonded diamond wire is being tried for cutting wafers.

References

[1] C. Lin, Transitioning to Larger Wafers, PV Magazine International 2020. 13th January 2020.

[2] A. Bidiville, K. Wasmer, J. Michler, P.M. Nasch, M. Van der Meer, C. Ballif, Mechanisms of wafer sawing and impact on wafer properties, Prog. Photovolt. Res. Appl. 18 (2010) 563–572.

[3] L. Candreia, Meyer Burger AG and Their Solutions for the Solar Industry, https://www.meyerburger.com/de/.

[4] A. Kumar, S.N. Melkote, Diamond wire sawing of solar silicon wafers: A sustainable manufacturing alternative to loose abrasive slurry sawing, Science Direct, Proc. Manufact. 21 (2018) 549–566.

[5] M. Hutchins, The Weekend Read: Polysilicon and Wafer Manufacturer Ranking, PV Magazine International, 2018. February 24, 2018.

[6] 1366 Direct wafer technology https://1366tech.com/technology-2/#directwafer.

[7] R. Jonczyk, E.M. Sachs, Patent on "Making semiconductor bodies from Molten material using a free-standing interposer sheet". Patent number: US 2014/0113156A1, published on 24 April 2014.

Making of crystalline silicon solar cells

4.1 Introduction

The global demand for solar photovoltaic (PV) modules is continuously growing. A record of ~97 GWp was added during 2019, resulting in the total installed capacity worldwide exceeded ~580 GWp by the end of 2019 [1]. With forecasts of 100 GWp per year, the annual growth rate may continue in the foreseeable future. The electricity tariff rates due to solar PV systems have already surpassed the grid tariff rates in some countries.

Crystalline silicon material has been a workhorse since the inception of solar cell technology. The silicon technology has matured in the microelectronics industry and most of the processes are adopted for solar cell manufacturing. Si material is abundant and nontoxic while the material is stable under different processing conditions. Si has an optimum band gap of 1.12 eV corresponding to a light absorption cut of wavelength about 1160 nm. The band gap suits the terrestrial solar energy spectrum to efficiently convert light energy to electrical energy. There are two types of silicon material in the form of wafers used for solar cells. One is a monocrystalline silicon wafer and the other one is the multicrystalline silicon type. Of the solar PV installations in the world, 90% are with solar modules of crystalline silicon technology. So, it is relevant to discuss the manufacturing technology of crystalline silicon solar cells.

4.2 Basics of solar cell/physics of solar cell

Solar cells are fundamentally quite simple devices. In most cases, they are made from semiconductor materials that can absorb light energy and produce electron-hole pairs. A solar cell is a well-designed semiconductor p-n junction that can separate and collect the electron and holes in a specific direction, as depicted in Fig. 4.1.

The solar cell is basically a p-n junction-based device. The base material is p-type mono- or multicrystalline silicon material. By doping phosphorus atoms on top of the surface, the n-type region is created, which forms a p-n junction with the wafer. The front side of the solar cell is covered with an antireflection coating on top of the n-type region. The antireflection coating is covered with a metallic grid as a front contact to collect the charge carriers. The solar cell structure shown in Fig. 4.1 is the n/p configuration type. In summary, the solar cell is a p-n junction device having contacts on both sides with an antireflection coating on the front side.

Solar PV Power. https://doi.org/10.1016/B978-0-12-817626-9.00004-6

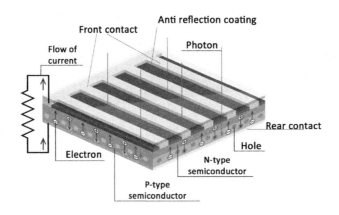

FIG. 4.1

Structure and working principle of a solar cell.

Solar cells work on the principle of photovoltaic effect. When photons from solar irradiance are incident on a semiconductor material, the light energy gets absorbed and creates a differentiated potential of two semiconductor materials, which is called the photovoltaic effect. Due to the photovoltaic effect, electrons are dislodged from a semiconductor material when it absorbs the photon of light energy.

The light spectrum consists of the distribution of energy in different wavelengths. Each wavelength of the spectrum corresponds to a photon and the energy of a photon, E, can be expressed as

$$E = h\nu = \frac{hc}{\lambda} = \frac{1.24}{\lambda} \tag{4.1}$$

where h is Plank's constant, ν is the frequency of light, c is the velocity of light, and λ is the wavelength of a photon in μm. The product of hc gives 1.24. With the above formula, the energy of a photon of any wavelength can be estimated.

When sunlight strikes a solar cell on its front side, it reaches the solar cell and gets absorbed in different depths of the cell. In the event of an incident photon of energy, the $h\nu$ is higher than the semiconductor band gap, which helps in the absorption of energy to liberate one electron-hole pair. The amount of light transmission to the semiconductor is enhanced by the deposition of an antireflective layer under the grid lines. By bringing in n-type and p-type semiconductors in close contact, a p-n junction forms. This junction formation is achieved through diffusion, deposition, or the implantation of specific dopants.

In the form of a metallic layer at the back of the solar cell, and grid like metallization on front side will form the electrical contacts to the p-n junction.

To understand the details of how a solar cell works as well as how to characterize its performance and how to improve it, each step of the electricity generation process needs to be learned. There are three basic processes involved in electricity generation from a solar cell. They are (a) light absorption and generation of charge carriers, (b) recombination of charge carriers, and (c) separation of photo-generated charge carriers and transport to the respective contacts.

4.2.1 Light absorption and generation of charge carriers

The most common solar cells are made of semiconductor materials that have an intermediate property between the conductor and insulators. Their energy band has a certain gap between the valence band and the conduction band. This structure is different from the conductor and insulator. The conductor has no gap between the bands that allows the electron to move freely. The insulator has a very large band gap, and it is very hard to allow electron transport in such materials. A semiconductor, however, has a band gap between the insulator and the conductor.

Si is a semiconductor material with an atomic number of 14. Its electronic configuration is $1s^2\, 2s^2\, 2p^6\, 3s^2\, 3p^2$. The valency of Si is 4. It is covalently bonded with its neighboring atoms. It has a band gap of 1.166 eV at 0 K. When the phosphorus P atom is doped into Si, it occupies the position of Si atoms and gives an extra electron. If more P atoms are doped, there will be more electrons. As the doping contributes to negative electrons, hence this is called an n-type semiconductor.

If a boron atom, which has three valence electrons, is doped into the Si element, it occupies the position of the Si atom, but there is a shortage of electrons to satisfy the valency of Si. The shortage of electron is a positive charge and it is called a hole. If more boron atoms are doped, there will be more holes. As the doping contributes to a positive charge, it is called a p-type semiconductor.

If the p-type and n-type semiconductors are joined, the excess electrons on the n-type will flow to the p-side and excess holes will flow to the n-side until thermal equilibrium and the p-n junction is formed. There is a built-in field in the junction due to positive ions on the n-type side and negative ions on the p-type semiconductor side. There is also a band bending at the interface of p-n junction.

The p-type region can be considered the base and the n-type can be the emitter for n/p configured solar cell. The p-type semiconductor is doped with 1×10^{17} atoms/cc and contributes 10^{17} holes/cc. Usually, the n-type region is doped with 10^{19}–10^{20} atoms/cc and contributes to electrons in the same range.

Sun irradiance spectrum: The surface of the sun is about 6000 K. It is like a black body and is known as the photosphere. The radiation intensity of the sunlight incident on a plane perpendicular to the rays at sun-earth mean distance is called the solar constant. It is about 1.353 kW/m². The air mass (AM) is a value that indicates the influence of sunshine absorbed by air during penetration through the atmosphere. The equation is given by:

$$Air\, Mass = \frac{1}{\cos\theta} \tag{4.2}$$

θ = light incidence angle (when it is 0, it signifies that the sun is a right angle).

The sun spectrum with a different air mass is shown in Fig. 4.2 [2].

If photons are incident on the solar cell, they will be absorbed in different depths of the Si solar cell depending on the energy of the photon and the light absorption coefficient of the Si material. The high energy photons that have lower wavelengths get absorbed on the surface within a range of a few micrometers. The long wavelength photons, which have low energy, are not able to be absorbed and be transmitted. The threshold energy a photon for getting absorbed is 1.1 eV, which is the band gap of the Si material.

The generation of electron takes place with the absorption of photon energy that is used to excite an electron from energy level E_1 to the higher level E_2, as shown in Fig. 4.3. The presence of differentiated energy levels such as E_1 and E_2 only helps in the absorption of photons and is defined as $h\nu = E_2 - E_1$.

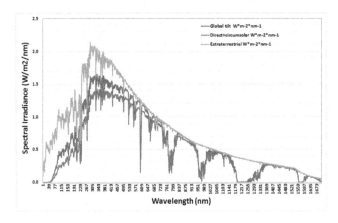

FIG. 4.2

Irradiance spectrum of black body, AM0, and AM1.5.

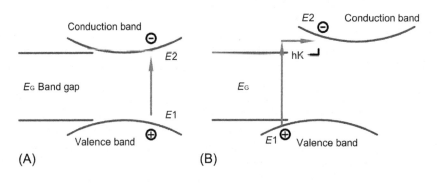

FIG. 4.3

(A) Semiconductor with a direct band gap, and (B) semiconductor with an indirect band gap.

The electrons populate energy levels below the valence band edge, E_V, and above the conduction band edge, E_C. There is no population of electrons between these two energy bands, as there are no energy states. The band gap $Eg = E_C - E_V$, which is the difference in energy. Any photon with less energy than Eg that reaches a semiconductor will not be absorbed and transmitted without any interaction with the bulk semiconductor.

The absorbed energy causes the generation of electron-hole pairs in different regions of the solar cell. The junction region is called the space charge region. In the n-type region, the hole is the minority charge carrier, whereas the electron is the minority charge carrier in the p-type region. With increased generation of electron-hole pairs, the population of minority charge carriers will increase in the n- and p-type regions of the solar cell.

The generated carriers have to be moved to the respective contacts for collection, and they contribute to the function of the device. In the p-type base region, electrons are minority charge carriers; they have to cross to the n- region for collection at the negative contact. Similarly, the holes are minority charge carriers in the emitter region, and they have to cross the junction and move toward the positive contact of the cell for collection.

In the event that the photon energy $h\nu$ is higher than the band gap energy, the photon will be absorbed. An electron from the valence band E_1 will be excited into the conduction band E_2, leaving a hole in the valence band and forming an electron-hole pair, as shown in Fig. 4.3.

In a real semiconductor, the valence and conduction bands are not flat, but vary depending on the so-called k-vector that describes the crystal momentum of the semiconductor. Electrons can be dislodged from the valence band to the conduction band in the event of the occurrence of the valence band at maximum while the conduction band is at minimum; this will happen without any conservation of crystal momentum. This is known as direct band gap material. In indirect band gap material, a change in the crystal momentum cannot happen without the excitation of electrons. In the case of indirect band gap material, the absorption coefficient is lower as compared to the direct band gap material. Hence, the absorber is thin for direct band gap materials. In the event of an electron being displaced from E_1 to E_2, a hole is created at E_1. This is like a positive charge in notion. As depicted in Fig. 4.3, an electron-hole pair is the result of the absorption of a photon. The chemical energy of the electron-hole pair is due to the conversion of the radiative energy of the photon.

4.2.2 Recombination of charge carriers and lifetime

The generated electron-hole pair tends to move back to its equilibrium due to recombination, in which case an electron drops from the conduction band to the valence band; this eventually eliminates the valence-band hole.

There are several recombination mechanisms important to the operation of solar cells, including recombination through traps (defects) in the forbidden gap, commonly referred to as the Shockley-Read-Hall recombination; band-to-band radiative recombination; and Auger recombination. Fig. 4.4 illustrates different recombination mechanisms in a solar cell.

(a) Shockley-read-hall recombination (SHR)

The Shockely-Read-Hall recombination is an avoidable recombination, comes from the impurity (defects) of the material. The defect in a semiconductor will act as recombination center in a solar cell. The impurity and defect centers in a semiconductor give rise to allowable energy levels in the forbidden gap. These defects effectively trap an electron or hole in the forbidden region while the hole or electron attains the same energy state before the electron recombines in the trap. This is a two-step process. The electron released from the conduction band is captured by a defect/trap level. From the trap level, it is moved to the valence band, annihilating a hole. A high defect density leads to a low lifetime of the carriers. The recombination trap center can be imagined as a target presented to the traveling carrier with a certain velocity. For a given time, if the velocity is high it will have a higher chance to be captured by the trap. Also, the size of the trap affects the recombination possibility. With a larger size, it is easier to trap the carrier and lower the lifetime. SHR recombination happens more commonly in bulk crystalline solar cells with a thick

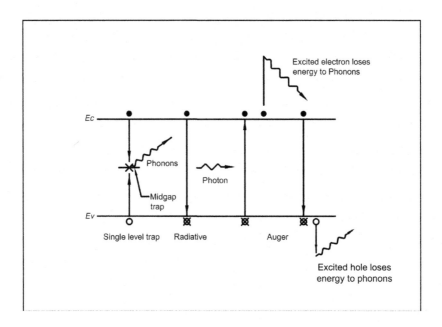

FIG. 4.4

Recombination mechanisms in a solar cell.

substrate. The charge carriers are more likely to recombine in a thick substrate containing defects during their diffusion. So, for a thin-film solar cell with a very thin layer, SHR is not the dominant recombination.

The front and rear surface of the semiconductor will have defects such as dangling bonds in the crystalline structure. These defects give rise to allowed energy levels in the forbidden gap. SRH recombination occurs effectively at the surfaces. The recombination of the charge carriers per unit area depends on the surface recombination velocities of the holes and electrons.

(b) Radiative recombination

This is the reversal of the light absorption process. Radiative recombination presents a phenomenon that an electron from the conduction band (higher energy state) recombines with the hole directly in the valence band (lower energy state) and releases the energy difference between the states in the form of a photon. This is also the basic working principle of the light-emitting diode (LED). The phenomenon very often happens in the direct band gap semiconductor materials such as GaAs. In the solar cells that are made from silicon with an indirect band gap, this type of recombination is very weak.

(c) Auger recombination

The recombination process is the same as that of radiative recombination. However, instead of releasing the energy through a photon, the Auger recombination gives the energy to another carrier. This electron (or hole) relaxes the excess energy and momentum to the photon and is called thermal relaxation. In heavily doped materials, Auger recombination is very important.

To express how fast the recombination happens, there is a critical parameter called the recombination rate. The minority carrier diffusion length and the lifetime are two important parameters for recombination. The minority carrier lifetime, τ_n or τ_p, is defined as the excited state average time that the carrier spends after electron-hole pair generation and before any recombination happens. A short carrier lifetime is the diffusion length in the base and this is less than the base thickness. In this case, the carriers created are about one diffusion length in base and are not likely to be collected. Thus, lifetime is a key property for choosing the material for a solar cell.

4.2.3 Separation of the light-generated charge carriers

Usually, the electron-hole pairs will recombine, that is, the electron will fall back to the initial energy level E1, as illustrated in Fig. 4.4. The energy will be released either as a photon in the case of radiative recombination or transferred to other electrons, holes, or lattice vibrations in the case of nonradiative recombination.

The electron-hole pair is generated following photon absorption; normally, holes and electrons do combine. Holes and electrons inside a semiconductor are separated using electric field created by the p- and n-layers. Typically, an electric circuit is driven using separated electrons. The holes get recombined with electrons after the electrons are passed using the electric circuit. The p-n junction separates the p- and n-type layers with a space charge region and the built in electric field enables the charge carriers to flow out from one region (say p-type/n-type) to other region (n-type/p-type). Solar cell is a minority carrier device. Electron is a minority carrier in p-type region. It has reach to reach n-type region and the n-contact to contribute to external current. So, it reaches junction by diffusion process and from there by drift mechanism, it will reach the n-region in the front side. If it encounters defects during its travel before reaching the n-contact, it will be annihilated.

Before any recombination of holes and electrons, it is essential to design the solar cell such that they can reach the n or p regions. This specifies that the time required by the charge carriers to reach the n or p regions is less than their lifetime, which determines the thickness of the absorber.

Charge carrier separation

The generated electron and hole in the semiconductor now can move like free particles with effective masses m_n^* and m_p^*, respectively. Electrons and holes need to move in opposite directions so that we can obtain the required current to separate the holes and electrons.

There are two mechanisms of carrier separation: drift and diffusion.

Drift

Drift is defined as the response of a charged particle to an applied electric field. In the application of an applied electric field across a uniformly doped semiconductor, holes, which are positively charged and are in the valence band, move in the same direction of the applied electric field. The electrons, which are negatively charged and are in the conduction band, move in the opposite direction of the applied electric field. The electrons will collide with the lattice atom or an impurity atom. This will reduce the excess velocity of the electron that is being picked up by the electric field. The average velocity increase of electrons between collisions caused by the electric field is called drift velocity, v_d. Electrons and holes tend to accelerate boundlessly as there is no impedance to their motion. However, due to various other objects inside the semiconductor crystals, the carriers do collide and get scattered. The other objects are crystal defects, dopant ions, atom components of the crystal, and also

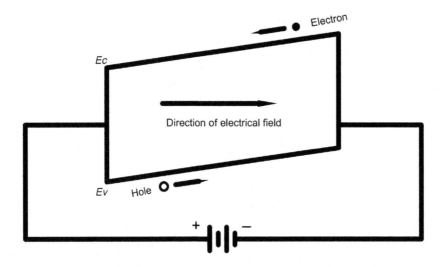

FIG. 4.5

Illustration of the concept of drift in a semiconductor.

other holes and electrons. This results in the movement of carriers at a constant velocity, v_d, which is called drift velocity. Fig. 4.5 illustrates the concept of drift in a semiconductor. It is directly proportional to the applied electric field, E.

$$|v_d| = |\mu E| = |\mu \nabla \Phi| \tag{4.3}$$

μ is defined as the carrier mobility and Φ is the electric potential at the junction interface. The carrier mobility is not directly dependent on the electric field strength unless the electric field is strong. This is not the situation in general in solar cells. The carrier mobility is decided by intrinsic lattice scattering for a low level of impurity and by ionized impurity scattering for high levels.

The current densities for holes and electrons due to drift can be written as

$$J_e - drift = qnv_d = q\mu_e nE \tag{4.4}$$

$$J_h - drift = qpv_d = q\mu_h pE \tag{4.5}$$

where μ_h and μ_e are the mobilities of holes and electrons, respectively. n and p are number of n-type and p-type dopants, respectively, in a semiconductor.

Diffusion

Diffusion is defined as the random thermal motion of electrons and holes in semiconductors. This is the movement due to diffusion from a high concentration to a lower concentration region. This is like how the ink gets evenly spread inside the water without any external forces or like the even distribution of air inside a balloon. This process is defined as *diffusion* and the hole diffusion current density ($J_h diff$) and electron diffusion current density ($J_e diff$) are expressed as:

$$J_h diff = -qD_h \nabla p = -qD_h \frac{dp}{dx} \tag{4.6}$$

$$J_e diff = qD_e \nabla n = qD_e \frac{dn}{dx} \tag{4.7}$$

$$D_e = \frac{kT}{q}\mu_e \text{ and } D_h = \frac{kT}{q}\mu_h \tag{4.8}$$

where D_e and D_h are diffusion coefficients of electron and hole, respectively. The total current density of electrons (J_e) and holes (J_h) are

$$J_e = q\mu_e nE + qD_e \frac{dn}{dx} \tag{4.9}$$

$$J_h = q\mu_h pE - qD_h \frac{dp}{dx} \tag{4.10}$$

Solar cell *I-V* characteristics

Solar cell *I-V* characterization is one of the critical parts from which we can obtain the solar cell efficiency and other electrical properties. From Eq. (4.11), we can get the total current, *I* for a solar cell of area *A*:

$$I = A[J_n + J_p] \tag{4.11}$$

The solar cell equation is:

$$I = I_{sc} - I_{o1}\left(e^{\frac{qv}{1KT}} - 1\right) - I_{o2}\left(e^{\frac{qV}{2KT}} - 1\right) \tag{4.12}$$

I_{o1} is the dark saturation current due to recombination in the quasineutral regions while I_{o2} is the dark saturation current due to recombination in the space-charge region. The 1 and 2 in the exponential functions of equation 4.12 are the ideality factors of the diodes 1 and 2. I_{sc} is the short-circuit current, which is the maximum current the cell generates under illumination. The value of I_{sc} depends on the solar cell design, material properties, and associated operating conditions. For low energy bandgap material, the I_{sc} is higher. For maximizing the I_{sc}, the light absorption should be maximum with minimization of the reflectance of light and with minimum grid shadowing on front surface. Under open circuit conditions, no current can flow and the voltage is at its maximum, called the open circuit voltage (V_{oc}). A solar cell is defined as an ideal current source (I_{sc}) in parallel with two diodes, in which case Diode 1 represents the recombination current in the quasineutral region, I_{o1}, while Diode 2 represents recombination current in the depletion region, I_{o2} (Fig. 4.6).

The *I-V* curve of a solar cell is plotted in Fig. 4.7. There are four parameters to characterize the solar cell performance: open circuit voltage (V_{oc}), short circuit current (I_{sc}), power at maximum power point (P_{max}), and fill factor (FF). I_{sc} is directly proportional to intensity of the radiation incident on the solar cells. More intensity of radiation means more number of photons will be absorbed in the solar cell and

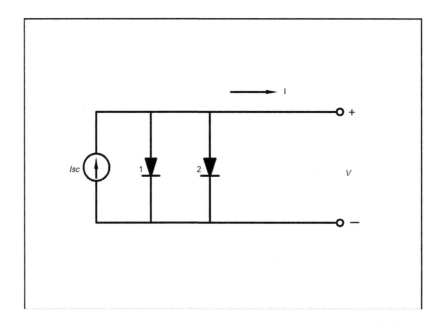

FIG. 4.6

Simple solar cell circuit model.

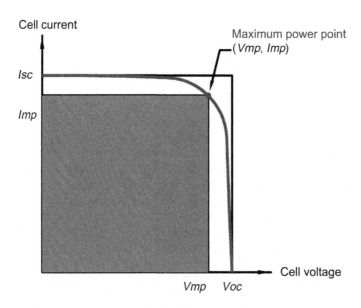

FIG. 4.7

Current-voltage characteristics of a solar cell.

generates more number of charge carriers causing increase in current. I_{sc} slightly increases with the increase of the temperature of the solar cell. This is due to increased absorption of low energy photons due to reduction of the band gap with the increase of temperature. The current will rise in the order of 15–25 µA/cm^2 with the rise of 1°C temperature.

Fig. 4.7 shows the solar cell electrical characteristic parameters. The maximum power point, P_{mp}, is particularly interesting, and illustrates the power conversion efficiency and fill factor. By referring to Fig. 4.7, the maximum power point defines a rectangle whose area, given by $P_{mp} = V_{mp} \times I_{mp}$, is the largest rectangle within any point in the *I-V* curve.

So, we shall define the following:

$$Fill\,Factor = \frac{V_{mp} \times I_{mp}}{V_{oc} \times I_{sc}} \tag{4.13}$$

$$P_{mp} = FF x V_{oc} \times I_{sc} \tag{4.14}$$

$$Efficiency = \frac{Power\,output\,per\,unit\,area}{Power\,input\,per\,unit\,area} \times 100 = \frac{\frac{Pmp\,in\,mW}{Cell\,Area\,in\,Cm2}}{Input\,solar\,power\left(\frac{100mW}{Cm2}\right)} \times 100 = \frac{Pmp}{cell\,area} \tag{4.15}$$

Power input is the intensity of solar irradiance incident that is determined by the properties of the light spectrum incident upon the solar cell.

From Eq. (4.14), it is evident that a high-efficiency solar cell will have a higher short circuit current, I_{sc}; a higher open-circuit voltage, V_{oc}; and a unity or close to fill factor of one. I_{sc} is proportional to the internal current collection efficiency and the light-generated current. The internal current collection efficiency is related to the recombination velocity and carrier lifetime. In an ideal case, the recombination velocity tends to be 0 and the life time is infinite. For higher light-generated current, the solar cell should be with least grid shadow, should have minimum reflectance and should absorb the solar spectrum as much as possible. V_{OC} is the open circuit voltage (for simplicity, Diode 2 that has less effect for a good solar cell is ignored), which can be written as:

$$V_{oc} = \frac{KT}{q} \ln\left(\frac{I_{sc}}{I_o} + 1\right) \tag{4.16}$$

From Eq. (4.16), it is clear that V_{oc} depends on the saturation current and the light-generated current. In contrast, I_{sc} has a small variation, and the key effect is the saturation current.

Hence, by reducing I_{o1}, one can improve the open circuit voltage. I_{o1} can be reduced by minimizing the recombination velocity.

Increasing the V_{oc} can improve the fill factor. To sum up, there are two ways to optimize solar cell performance:

Minimization of recombination rates throughout the device.

Maximization of the light-generated current.

Resistance effect

The *I-V* characterization above neglects the resistance in a real solar cell. A modified model with the series resistance and shunt resistance is shown in Fig. 4.8.

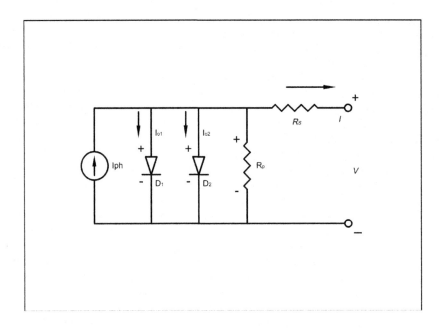

FIG. 4.8

Modified circuit model of a solar cell.

So, Eq. (4.12) becomes like this:

$$I = I_{sc} - I_{o1}\left(e^{\frac{qv}{1KT}} - 1\right) - I_{o2}\left(e^{\frac{qV}{2KT}} - 1\right) - \left(\frac{V + IR_s}{R_{sh}}\right) \tag{4.17}$$

The shunt resistance has no effect on the short circuit current while it reduces the open circuit voltage. The series resistance, however, does not change the open circuit voltage but reduces the short circuit current. Series resistance comes from the metal contact of the grid and the transverse flow of the current in the emitter to the front contact. This can be minimized with an innovative design of the solar cell. Shunt resistance is typically due to manufacturing defects rather than the design of the solar cell. Power losses are a result of low R_{sh}, which provides an alternate current path for the photon-generated current. The effects of Rs and R_{sh} are illustrated in Fig. 4.9.

Loss mechanisms

The loss mechanism (for single band gap solar cells) is the inability to convert photons with energies lower than the band gap to electricity and also thermalization of photon energies higher than the band gap. About 50% of the solar incident energy is lost in the conversion process due to these two mechanisms. Hence, the maximum energy conversion efficiency of a single-junction solar cell is always below the thermodynamic limit.

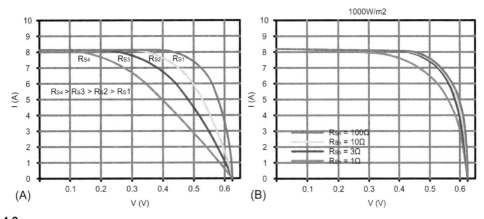

FIG. 4.9

Effect of series and shunt resistances on the fill factor: (A) Rs effect and (B) R_{sh} effect.

The performance of a solar cell is reduced by different loss mechanisms such as optical loss, thermalization loss, and electrical loss.

The optical loss arises due to shadowing of the front grid lines, the surface reflection, and the non-absorbance of long wavelength radiation. The longer wavelength photons with energy lower than the band gap energy of Si, that is, 1.1 eV, cannot be absorbed and transmitted through Si material. This is due to insufficient energy to create electron-hole pairs. The loss of low energy photons is 28%. Anti-reflection coating on the front side, the front surface texturization, the minimization of the front contact area, and the back surface reflector have been implemented to reduce the optical loss.

The thermalization loss is the nonutilization of the complete energy of the photon for electron–hole pair generation. To generate electron-hole pairs, 1.1 eV energy is required in the photon. If a photon has 2 eV of energy, only one electron-hole pair will be generated with the utilization of 1.1 eV energy, and the remaining 0.9 eV will be lost as heat dissipation.

Electrical loss arises due to the inefficient collection of the generated charge carriers. Charge carrier recombination losses occur in the bulk, front, and rear surfaces. This reduces the open circuit voltage of the cell. The internal resistance of the bulk semiconductor and the contact resistance contribute to electric loss and affect the fill factor. Surface passivation and the reduction of contact resistance minimize the electrical loss.

4.3 Specifications of the solar cell

The solar cells should have:

- Very high energy conversion efficiency
- Very high short circuit current density
- Low dark saturation current
- High open circuit voltage

- High fill factor
- Low series resistance
- High shunt resistance
- Low temperature coefficients of P_{max} and V_{oc}
- Better performance at low irradiance condition
- Ability to withstand extreme low and very hot temperatures
- The contacts are solderable and should provide good adhesion strength with the interconnector after soldering
- The light-induced degradation in power should be minimum
- The antireflection coating should have good adherence with the cell
- Easy to manufacture industrially on a large scale
- It should have good mechanical strength such that it is able to withstand handling loads
- It should withstand a high-humidity environment
- It should not show potentially induced degradation

To meet the above requirements

- The Si wafer should have good quality, that is, with less dislocation density
- Higher carrier lifetime
- Higher diffusion length
- Lower surface recombination velocity at the contact interface
- Should have a mechanism to reduce recombination losses and dark current
- Processes and contact materials are chosen such that they have low resistivity and form ohmic contacts
- Textured surface and broad band antireflection coatings for minimum reflection loss
- For maximum utilization of photons, a reflector in the rear side
- A mechanism to maximize the generated carrier collection efficiency
- A mechanism to reduce the recombination of carriers at the contact and semiconductor interface
- Low temperature operating processes

4.4 Manufacturing process of solar cell

Solar cell manufacturing involves the formation of the p-n junction and applying the electrical contacts onto the device. The as-cut silicon wafer is chemically etched to remove the saw damage caused during cutting of the wafer. Next, the wafer is texturized to absorb more light and reduce the reflectance of the surface. The opposite polarity dopant atoms of the wafer type are doped by diffusion or some other technique, and the junction is formed. During phosphorus doping, phosphosilicate glass is formed on the rear side and edges, and this will be etched out. A dielectric material such as SiN_4 is deposited to serve as an antireflection coating to the solar cell. Silver and aluminum pastes are screen printed on the front and rear side, dried, and sintered in a thermal furnace to form contacts on both sides. The final cell is tested for electrical parameters.

Table 4.1 shows the main steps of a simple process for conventional BSF solar cell fabrication based on screenprinting. With more or less minor modifications, this process is currently used by many manufacturers and dominated the industry for many years. The main virtues of this PV technology are easy

Table 4.1 Process flow of p-type BSF technology-based solar cell.	
Process for P-type BSF	**Process checks/equipment**
Load wafers	
Wafer sorting and inspection	Check for cracks, geometry, surface defects, edge chips by optical camera
Surface damage removal	
Texturing on front side	Reflectance
P doper + diffusion	Resistivity check by four-point probe
Edge isolation + phosphosilicate glass removal	Shunt resistance
Surface treatment with ozone and high radiation to make potential induced degradation (PID) resistant	
SiN-ARC deposition by PECVD	Color, thickness, refractive index, reflectance by ellipsometry
Metallization printing on front side	Print quality check by surface profiler
Print drying	
Wafer flipping	
Contact pad printing rear side	
Pad drying	
BSF printing rear	
Contact firing	Solder joint pull strength test
Testing and sorting	I-V parameters, EL check by cell tester

automation, reliability, good usage of materials, and high yield. The process flow and quality checks for each process are shown in Table 4.1.

The solar cell manufacturing process steps are briefly described in the following text.

4.4.1 Wafer sorting inspection

The industry uses monocrystalline and multicrystalline wafers to manufacture solar cells. Monocrystalline wafers are Cz-Si grown, cut from the ingot, and trimmed to a pseudosquare shape. At present full square mono and multicrystalline wafers are available in sizes 158.75 × 158.75, 166.75 × 166.75, and 210 × 210 mm. Multicrystalline wafers are a square shape. The wafer dimensions are between 156.75 × 156.75 mm to 210 × 210 mm and between 180 and 200 µm thickness.

During the sawing process of wafers from ingots or cast bricks, the wafer surface gets damaged. During the free abrasive slurry-based wire sawing process, the wafer surface is contaminated with organic and inorganic contaminants. The sawing operation is believed to create microcracks in the wafer.

The wafer quality is the key aspect in achieving the required efficiency of a solar cell. As part of the incoming material inspection, the sample wafers will be checked for the following parameters:

- Wafer thickness
- Size and geometry
- Total thickness variation
- Doping concentration in the wafer and its distribution

- Wafer bow
- Oxygen ion concentration
- Resistivity of the wafer
- Etch pit density and crystal defects
- Visual inspection for surface defects, contamination, cracks, holes, edge chips, etc.

The wafer cartridges are loaded into the wafer handling machine. The robot arm picks up the wafers and places them on the line. The wafers are inspected for cracks, microcracks, edge chips, etc., using a built-in camera, and good wafers are sorted out for further processing.

4.4.2 Saw damage removal and texturization

The sawing process to cut ingot/bricks into silicon wafers by a slurry-based wire saw or the diamond wire method leaves a damaged crystal lattice surface with an irregular crystal structure. If the surface damage is not completely removed during cell manufacturing, it will hamper the performance of the device and could lead to reliability issues. The saw marks and damage to the wafer surface cause micro-cracks, propagation of surface cracks, or high levels of mechanical stress, all of which could result in macroscopic cracks within the cell. The latent stresses around saw marks can facilitate the formation of defects such as stacking faults and dislocations during subsequent thermal treatments. So, it is necessary to etch a few microns on each side of the wafer before taking up the wafer for further processing of the solar cell.

The optical losses from the front surface are one of the most important factors limiting the efficiency of the solar cell. So, as part of the light-trapping technique, solar cells need a textured front surface to reduce reflectance. The textured surface minimizes the reflection of light from the solar cell surface and promotes the internal bouncing of photons. The probability of light absorption improves due to the reflection of light from one angled surface to another surface. The textured surface promotes short wavelength absorption due to the increased front surface area. To achieve a high-quality solar cell, the wire saw damage has to be effectively removed, the front surface has to be textured, and a safe wet processing system has to be used to clean the wafers. The saw damage removal and texturization of the wafer are done in a single step by the wet chemical etching process.

In the wet chemical etching process of the Si wafer, an oxidizer such as HNO_3 oxidizes the surface of the Si material to SiO_2. The formed oxide has to be dissolved in an acid medium such as hydrofluoric acid and the deionized water removes the dissolved oxide material from the Si surface. The etch rate depends on the concentration of chemicals used and the temperature and stirring condition of the chemical bath. Two types of chemical etching solutions are used for this purpose.

4.4.2.1 Alkaline etching

Alkaline etchants such as potassium hydroxide (KOH) or sodium hydroxide (NaOH) with water are used for etching monocrystalline silicon wafers. Etching of Si involves the oxidation of the silicon followed by the dissolution of the oxidized product. For commercial applications, alkaline etching is mainly used for saw damage removal and surface texturing of monocrystalline wafers. Alkaline etchants such as KOH, NaOH mixed with tetramethyl ammonium hydroxide (TMAH), or isopropyl alcohol (IPA) are used in the solar industry to form random upright pyramids in monocrystalline wafers. The alkaline etchant provides the anisotropic etching along the (100) plane of the silicon crystal, which exposes the (111) plane. The etchants etch the (100) planes much faster than the (111) planes as the (111) plane contains a very high atomic packing density. Therefore, when the slow-etching planes like the (111) orientation are exposed, they intersect at the surface to form square-based upright pyramids of random size. The sides of the upright square pyramids are formed by (111) planes and the base is the (100) plane; they are distributed randomly on the surface. The pyramid size must be optimized because very small pyramids lead to high reflection while very large ones can hinder the formation of the contacts. To ensure complete texturing coverage and adequate pyramid size, the concentration, the temperature, the agitation of the solution, and the time duration of wafer present in the chemical bath must be controlled.

The role of the IPA is to aid in controlling the alkaline etch rate and selectivity to the (111) crystal plane, which also helps in the formation of the random pyramid structure. Balancing the etch rate with pyramid formation is important to ensure that enough silicon is removed to optimize saw damage removal while maintaining the appropriate height, base width, and period of the pyramids to minimize front surface reflectance without increasing the surface recombination. The characteristic size of these textures is typically 3–10 μm. The decreasing height of the pyramids and uniform distribution of their sizes are advantageous in decreasing the reflectance.

The main difference between saw damage removal and texturization is the rate of etching, which depends on the concentration of etchants, the process temperature, the agitation of the solution, and the time duration. For texturization, the etch rate needs to be low, that is, 2 μm/min or lower. Fig. 4.10 shows the textured surface with randomized square-based upright pyramids.

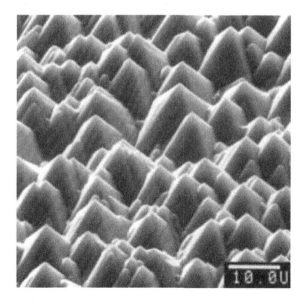

FIG. 4.10

Textured surface of a monocrystalline silicon wafer.

4.4.2.2 Acidic etching

Multicrystalline silicon wafers contain many silicon grains with different crystal orientations. The alkaline etchants, which are anisotropic in nature, are not suitable for etching multicrystalline surfaces. Due to the anisotropic nature of the alkaline etchants, different crystalline planes in multicrystalline wafers would get etched at different rates. This will cause nonuniform thicknesses across the surface. The grain boundaries in multicrystalline wafers would undergo defect etching, creating dislocations that can propagate through the wafer if alkaline etchants are used. So, acidic texturing, which is an isotropic wet etching process, is preferred for texturing multicrystalline wafers.

The process relies on the preferential etching of residual defects present in the wafer to create light-trapping features.

In acidic etching, the HNO_3 is the oxidant and HF is the acid, which dissolves the oxidized product. The acetic acid or water is used to prevent HNO_3 dissociation and facilitate wetting of the Si wafer surface, thereby improving etch quality. The texturization equipment for a multicrystalline silicon solar cell is shown in Fig. 4.11 and the process flow is shown in Table 4.2.

The saw damaged regions, which are highly stressed, are etched by an etching solution (HF+HNO_3) at a high rate compared to the other areas of the surface. This results in a textured surface. A thin porous silicon is formed at the surface during acidic texturing. The wafers are rinsed in deionized water and etched by dilute KOH to remove the formed porous silicon from the surface. After rinsing with deionized water, the remnants of KOH present on the wafer are neutralized by rinsing with dilute HCl. The critical process parameters are the temperature of the etch bath (HF and HNO_3) as well as the composition of the etchants. So, it requires the precise control of process parameters and agitation of the bath to get reproducible results in acidic texturing with multicrystalline wafers.

The acetic acid acts as a wetting agent during the etching process. The surface wetting improves the ability of HNO_3 to remain in contact with the silicon surface by uniformly getting distributed. The

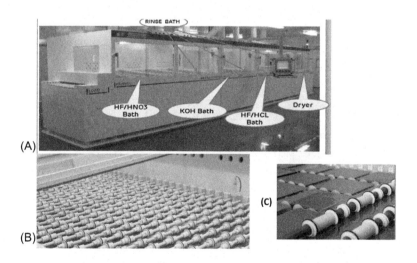

FIG. 4.11

Acid texturing machine for multicrystalline solar cells. (A) Machine for surface damage removal and texturization, (B) rollers for wafer travel, and (C) wafers on roller.

Table 4.2 Process flow of acid texturization.

Load P-type wafers
$HF + HNO_3 +$ deionized water
Rinse in deionized water
KOH to remove porous silicon layer
Rinse in deionized water
Rinse in HCl to remove remnants of KOH
Rinse in deionized water
Cleaning with HF to remove any leftover residues
Rinse in deionized water
Unloading of P-type wafers

formed texutured surface morphology highly depends on the concentration and mixing ratio of the chemicals in the texturing bath. The mixing ratio of $HF:HNO_3$ will decide the polished or rough surface of the silicon wafer. If the ratio is low, the surface can become increasingly polished, resulting in a higher reflectance and thus a lower J_{SC}. Higher ratios will result in a rough texture with low reflectance; however, defect etching may result in lower V_{OC}.

Agitation of the acid bath is essential to facilitate the movement of the solution around the bath. This can assist in reducing thermal gradients between the bottom and top of the solution–which can result in nonuniform etching across the wafer–and also facilitates ventilation of hydrogen gas pockets and replenishing of the texturing solution. Fig. 4.12 shows the etched surface of the multicrystalline wafer.

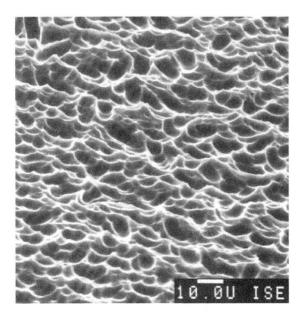

FIG. 4.12

SEM image of acid-textured multicrystalline Si surface.

The light-trapping effect to reduce the optical losses on the multi crystalline surface obtained by acid texturing is inferior compared to alkaline texturing the surface of monocrystalline silicon wafers.

4.4.2.3 Equipment used for etching and the suppliers

The saw damage-induced defects negatively affect the electrical properties of solar cells and reduce the mechanical strength of the wafers. The saw damage removal and surface texturization processes are carried out in a single integrated set up.

The following are some of the suppliers of wet chemical benches to carry out saw damage removal and texturization of the silicon wafers:

RENA Technologies GmbH, Germany
RCT Solutions, Germany
Schmid Group, Germany
Singulus Technologies AG, Germany
SC New Technology Corporation, China
Beijing Naura Microelectronics Equipment, China

The alkaline etching of monocrystalline wafers is done in a batch process, whereas the etching of multicrystalline wafers is done in an inline process, as the etching rate of the acidic process is much higher than the alkaline one.

The equipments are developed based on the following: the footprint of the equipment, the chemistry of etching, the lowest consumption of chemicals, a higher etch rate, a wafer breakage ratio of 0.01% or less, and higher throughput.

The tools are built with multiple baths, some for active etching and some for rinsing and cleaning.

4.4.3 Doping of silicon material

The incorporation of specific impurities in selective locations of a semiconductor to alter its electrical properties is referred to as doping. There are two main methods of doping silicon material in the manufacturing of solar cells. One is thermal diffusion and the other is ion implantation. The process of thermal diffusion involves two steps. In the first step, the dopant source is introduced on the surface of the material, and in the next step, the dopant atoms are driven inside the surface of the material by the thermal method.

Creating an emitter in the bulk silicon is the most critical step in manufacturing an industrial crystalline Si solar cell as it forms the p-n junction. For example, the front surface of the p-type Si wafer is counterdoped with a relatively high concentration of phosphorus atoms to form the p-n junction in a solar cell. This n-type region, which is doped with phosphorus atoms, is typically referred to as the emitter, and together with the p-type base, forms the p-n junction. The function of a diffused region is [3]:

(i) Collection of photogenerated carriers by charge separation
(ii) Acting as an electron-selective contact or an electron transport region
(iii) Providing electrical contact with metal fingers that extract the current from the cell
(iv) Enabling the lateral transport of carriers between the metal fingers
(v) Minimizing the recombination of carriers at the front surface

This is achieved by making the concentration of electrons much higher than that of holes, that is, by creating a large asymmetry between the conductivities for electrons and holes.

A high concentration of phosphorus (typically in the 1×10^{20} cm^{-3} range) is required to achieve a sufficiently low contact resistance between the metal and the semiconductor. This high concentration of doping promotes electron transport by quantum-mechanical tunneling across a thin potential barrier that arises as a consequence of the different work functions of the two materials [4].

The optimization of the emitter profile for individual solar cell performance is dictated by the compromise between the competing requirements of minimizing carrier recombination losses and minimizing resistive losses associated with extracting carriers out of the cell. Diffused regions such as emitters are typically described by their sheet resistance and/or doping profile. Sheet resistance describes the resistance of the diffused layer as seen by carriers moving laterally between the metal fingers. The distribution of phosphorous dopant atoms in the emitter is described by its profile with its notable features of surface concentration, shape, and depth. Diffusions are typically described as "light" or "heavy" based on their high or low sheet resistance values, respectively. It should be noted, however, that the average sheet resistance is an integral of the active dopants across the junction depth, and as such, emitters with very different profiles can result in the same value.

Earlier, the values of sheet resistance were in the range of 30–60 Ω/□ (ohms/square) with a junction depth $> 0.4\,\mu$m with a high phosphorus dopant concentration. At present, the sheet resistance values are in the range of 90–110 Ω/□ with a 0.3 μm junction depth and low phosphorus dopant due to improvement in the front side silver paste. Higher sheet resistance allows capturing more light in the UV and blue spectra, and reduces the recombination current that in turn increases V_{oc}.

Metrology techniques for sheet resistance include a four-point probe, surface photovoltage probing, and sheet resistance imaging. In general, diffusion engineers typically rely on two profile techniques: secondary ion mass spectrometry (SIMS) to track how dopant atoms are being introduced into the substrate via processing and either a spreading resistance probe (SRP) or electrochemical capacitance-voltage (ECV) to monitor the electrically active dopants.

4.4.3.1 Quartz tube diffusion for P-type wafer

Thermal diffusion is a two-step process that includes a deposition step in which the dopant source is supplied onto the surface of the wafer and a drive-in step in which the source dopants are diffused into the wafer by thermal energy to create the required concentration gradient.

The different sources for the dopant atoms are in solid, liquid, or gaseous sources. Some examples of dopant materials and their sources for Si are antimony (Sb)–Sb_2O_3 (s); arsenic (As)–As_2O_3 (s) AsH_3 (g); phosphorus (P)–$POCl_3$ (l), P_2O_5 (s), PH_3 (g); and boron (B)–BBr_3 (l), B_2O_3 (s), BCl_3 (g).

There are five heating zones in the diffusion tube. There is a loading zone area from where the wafers are loaded into the tube. The central loading zone is the area between the loading and center zones, and the center zone lies in the center of the tube. The center gas zone is between the center zone and the gas zone. The gas zone is the area from where the gas moves out in the exhaust. $POCl_3$ is the dominant phosphorous dopant source used for emitter formation in industrial solar cells. In a $POCl_3$ batch diffusion process, Si wafers sitting vertically in a slotted quartz boat are loaded into a heated horizontal quartz tube furnace. Fig. 4.13 shows the P-doper machine.

Emitter formation is the highest temperature process step for conventional industrial solar cells. As a result, diffusion is susceptible to the incorporation of unintended impurities or contaminants into the silicon wafer. The high temperatures used to drive phosphorus atoms into the wafer can also serve to

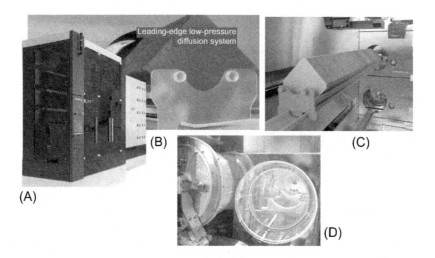

FIG. 4.13

Thermal diffusion chamber for phosphorus doping. (A) Low pressure diffusion chamber, (B) wafers loaded in the boat, (C) boat loaded with wafers entering the vacuum tube, and (D) vacuum tube closure.

Courtesy: Centrothem.

drive in contaminants that are present in the wafer. Contaminants such as Fe and Ni metals can create recombination centers, which reduce the minority carrier lifetime and, as a result, degrade the overall solar cell efficiency. Impurities can also be introduced into the starting material during ingot growth. Fe is the most common contaminant for c-Si PV. Multicrystalline Si wafers are prone to significant amounts of in-grown interstitial iron, Fe, incorporated during their low-cost crystal-growth process. Si wafers, including monocrystalline wafers, are also susceptible to extrinsic Fe contamination during cell processing. Fe is a very common and prevalent element that is very difficult to completely eliminate from a production line. Reducing the concentration of impurities in Si wafers, a process referred to as gettering, can occur via several mechanisms, including P gettering.

Industrially, the most widespread method for emitter formation in Si solar cells is P diffusion from a liquid phosphorus oxychloride (POCl$_3$) source in a closed quartz tube. Once the samples are loaded in the tube, the furnace needs to be heated to the desired temperature before any processing can occur. A carrier nitrogen gas is passed through a bubbler containing liquid POCl$_3$. The carrier gas thus feeds liquid POCl$_3$ into the heated quartz tube. Oxygen is simultaneously introduced into the tube and reacts with the POCl$_3$ to form phosphorus pentoxide (P$_2$O$_5$) on the sample surfaces, described by [5]:

$$4POCl_3 \text{ (g)} + 3O_2 \text{ (g)} \rightarrow 2P_2O_5 \text{ (l)} + 6Cl_2 \text{ (g)} \tag{4.18}$$

During predeposition, Cl$_2$ is continuously vented from the system. The supply of POCl$_3$ is eventually removed to control the thickness of the growing phosphosilicate glass. At the surface, P$_2$O$_5$ reduces elemental phosphorus during the driving step. The diffusion of P from the surfaces takes place upon the reduction of P$_2$O$_5$ by Si according to:

$$2P_2O_5 \text{ (l)} + 5Si \text{ (s)} \rightarrow 5SiO_2 \text{(s)} + 4P \text{(s)} \tag{4.19}$$

Due to the high-throughput requirements in industrial production, however, many solar cell manufacturers currently perform $POCl_3$ diffusion back to back so that double the amount of samples can be processed simultaneously. This is one of the major advantages of $POCl_3$ diffusion: the large processing capabilities. Another benefit of this technology is the independent control of the predeposition and the drive in, so that the surface source can more easily be made finite. Thus, this offers greater control of the surface concentration of P diffused emitters as well as an additional parameter of freedom during the process.

Phosphorus diffusion offers the additional advantage of impurity gettering. Briefly, gettering is the process where the impurities of transition metals such as iron, nickel, chromium, etc., diffuse from the bulk of the wafer toward the location of phosphorus diffusion. Once they are in the highly doped n +surface region, these impurities are no longer harmful to device operation. This means that relatively impure, and thus cheaper, wafers can be used, and making the complete cell process will become more cost-effective. Phosphorus gettering has become instrumental to enable the popular usage of multicrystalline silicon solar cells.

Centrotherm developed a low-pressure diffusion system suitable for both p-type and n-type c-Si solar cell manufacturing. This is a batch-type production system. It has five independently operated stacked quartz tubes.

A low-pressure diffusion system is used for processing high-efficiency solar cells, including p- and n-type with the use of $POCl_3$ and BBr_3 diffusion. Centrotherm's LP diffusion system can provide higher sheet resistance and better uniformities while also being suitable for advanced cell concepts such the TopCon type.

With the LP diffusion process, $POCl_3$ and BBr_3 diffusion can be done on both sides of the wafer.

4.4.3.2 Phosphoric acid spray

Phosphorous doping can be achieved using a phosphoric acid (H_3PO_4) spray combined with an inline furnace. A dilute H_3PO_4 solution containing some surfactants is sprayed onto the wafer surface. The wafers are then subjected to a low temperature (less than 50°C) dehydration step in order to convert the H_3PO_4 into P_2O_5 and H_2O. The P_2O_5 serves as the dopant source and when subjected to a high temperature, it is reduced to free phosphorous atoms. To perform the P drive-in, wafers transition onto a conveyor belt and run through an infra red heated inline furnace. The doping strength can be modulated via temperature settings and belt speed, although some further tuning capability can be demonstrated with lower spray flow rates. Processing with phosphoric acid spray and an inline furnace has a limited profile tuning capability compared to $POCl_3$ diffusion.

4.4.3.3 Ion implantation

Ion implantation is a relatively newer doping technique that operates close to room temperature. It is a physical process of doping, not based on a chemical reaction. Because ion implantation takes place close to room temperature, it is compatible with conventional lithographic processes, so small regions can be doped. Also, because the temperature is low, lateral diffusion is negligible. In this process, dopant atoms are ionized, separated, accelerated with electric and magnetic fields, and made to impinge on the wafer surface. These atoms penetrate some depth into the material and get embedded into the wafer. The common source material for phosphorus dopants are arsene (AsH_3) and phosphine (PH_3) gases while solid sources such as As and P are also used; BF_3 is used for a boron dopant. Electron bombardment is used to create ions. A mass analyzer, which is a 90 degree magnet, is used to separate the ions.

Based on the mass of the ions, the magnet separates them. The selected desired ions are then accelerated and made to strike the wafer surface. Using electric field coils, beam scanning or rastering is done to deflect the ion beams. The accelerating field modifies the energy of the ion, which decides the penetration depth in the wafer. The maximum dopant concentration will be at a certain depth below the surface, which is called the range. In thermal diffusion, the maximum concentration of the dopant is at the surface and the concentration decreases with depth. The range depends on the ion type and the energy. There are two stopping mechanisms: the nucleus of the wafer atoms and the interaction of the positive ions with the electrons. In ion implantation, the beam density, ion energy, and orientation of the wafer matter. As the wafer surface is bombarded by high-energy ions, knocking the Si atoms from their position, this causes local surface structural damage. To rectify the surface damage requires a high-temperature annealing treatment. There are two ways to perform thermal annealing. The first one is with a tube furnace, which is a low-temperature annealing (600–1000°C) process used to minimize lateral diffusion. The second one is a rapid thermal annealing process, which requires higher temperatures but for shorter times. Ion implantation can be used to create shallow junctions by having a small range. It can also be used to dope small regions.

4.4.4 Etching of phosphosilicate glass (PSG)

After diffusion, an amorphous glass of phosphosilicate is formed at the surface that is usually etched off in diluted HF because it can hinder subsequent processing steps.

The wafers will be continuously transferred by a roller transfer system with a constant speed through the system and placed on parallel lanes. During transport, the wafers will be processed. The wafers pass a predefined sequence of several bordering process tanks that are filled with chemicals. The HF is used to etch out the phosphosilicate glass formed during the diffusion process. The different chemicals are supplied by dosing units. The process will include the removal of the phosphosilicate glass layer (PSG) that was generated by the diffusion step and in addition, a junction isolation etch process will be performed. Fig. 4.14 shows the chemical bath in the PSG etching equipment. A surface clean is the final step in order to provide a clean surface for the subsequent plasma enhanced chemical vapor deposition (PECVD) process.

The surfaces of the solar cells are treated to make them resistant against the potential induced degradation (PID) effect, which occurs in modules where the power of the solar modules decreases after a certain time in the field. This process before PECVD will prevent module power degradation due to PID in the field. During the treatment, the wafer surface, which is cleaned and dried after PSG removal and the surface cleaning process, will be activated by high energy radiation in the presence of an ozone atmosphere. It serves as pretreatment before the PECVD deposition processes. The process is suitable for all common cell concepts and can increase the cell efficiency in the best case by up to 0.2%.

4.4.5 Edge isolation

When the wafers are undergoing the diffusion process, the entire surface of the wafer including the edges is exposed to the dopant source. So, during the diffusion process, phosphorus atoms are doped on all sides of the wafer, including the edges, creating an n-type region on the edges also. The *n*-type region at the wafer edges would interconnect the front and rear side contacts of the solar cell. This creates a current path from the front junction to the rear side of the device, causing the shunting of

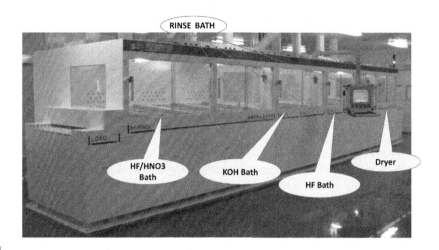

FIG. 4.14

PSG etching equipment.

the solar cell junction. The recombining carriers in the shunting path do not contribute to the power output. Therefore, immediate to the diffusion process to electrically isolate the front and rear contacts, the unwanted diffusion around the edges of the solar cell has to be removed. For this, an edge isolation step is employed. The edge isolation process can be done in three ways: dry etching, wet chemical etching, and laser grooving.

4.4.5.1 Plasma etching

Plasma etching is also called dry etching and it is a low-temperature process. In this process, the wafers are coin stacked on top of one another in such a way that only the edges of the wafers remain exposed and the front and back surfaces are protected. The coin-stacked wafers are then loaded into a plasma chamber. The plasma is obtained by exciting a fluorine-based compound such as CF4 or SF6 and O_2 gasses by a radio frequency field. The plasma containing highly reactive ions and electrons quickly etches the exposed silicon wafer edges. Thus, plasma etching effectively removes the diffused silicon at the edges. This is a batch process, but a large number of wafers can be processed. The quality of the edge isolation depends on the energy of the reactive ions in the plasma and the time duration of the etching process. If the time duration for the etching process is short, it results in insufficient isolation between the top and bottom contacts and therefore gives lower shunt resistance. In contrast, extended etch process timing may result in excessive damage to the edges and the diffused junction. This type of damage enhances recombination in this region, affecting the diode ideality factor of the solar cell.

4.4.5.2 Wet chemical etching

This is an inline process. The wafers are loaded on the rollers of a conveyor belt carried through a chemical bath containing a mixture of hydrofluoric acid and nitric acid. The chemicals in the bath touch the rear side and edges of the wafer and etch out the unwanted diffusion regions. In this process, the samples are etched on one side only, without affecting the front surface in a single-side etch tool. In this process, the complete removal of the rear side diffusion and edges is ensured. This process enables high throughput and low wafer breakage risk due to the inline nature of the single side etch tools. For this reason, in the production of industrial silicon solar cells, the wet chemical etching process is followed for edge isolation. However, one risk with this process is the potential for the etchant to splash and come in contact with the front side of the wafers. So, the etching process has to be controlled and monitored carefully to avoid splashing the etchant onto the front side while the wafers are being moved on the rollers. The removal process of the phosphorus silicate glass and edge isolation can be combined as a single step in a single chemical etch tool. This is being followed in the industry.

4.4.5.3 Laser grooving

Laser cutting of the wafer edges is an alternative process being used in industrial solar cell manufacturing to isolate the front and rear contacts. In this process, on the front surface of the wafer close to the edges, a laser groove is formed by the laser ablation technique. The groove will be few a micrometers deep and effectively isolate the current path to the edge. In this method, some area on the front side is consumed for trench formation. Edge isolation by laser grooving is performed after cofiring of the front and rear contact metallization.

4.4.6 Antireflection coating

A silicon substrate has a reflectance of 30% in the wavelength range of 0.3–1.1 μm. To minimize the reflection losses of light, an antireflection coating material has to be deposited on the solar cells. The industrial solar cells are deposited with an antireflection coating layer on the front surface to reduce light reflection. A well-designed antireflection coating on the front surface of silicon reduces the reflectivity from 30% to less than 5%. Fig. 4.15 shows the principle of interference of light between two media.

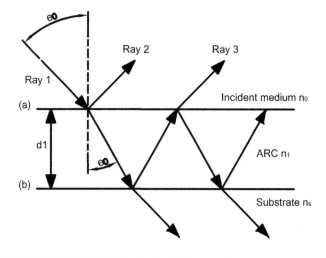

FIG. 4.15

Reflection at the interface between two media. Ray 2 and Ray 3 interfere destructively.

Let us consider a Si solar cell coated with a thin dielectric layer of refractive index n_1 while n_0 and n_s are the refractive indices of the incident medium and the silicon substrate, respectively. Ray 1 is traveling from the medium of air incident on the surface of the solar cell, which is coated with a thin dielectric layer of refractive index n_1. It is reflected at the top surface of ARC and emerges as Ray 2. The intensity of light transmitted and reflected between the two media depends upon the change in the refractive index of the media and the angle of incidence of the light at the interference. The intensity of the reflected light is given by the Fresnel equation of reflection and the reflection coefficient R is [6]:

$$R = \left(\frac{no - ns}{no + ns}\right)^2 \tag{4.20}$$

An interlayer could help to reduce the light reflection, and the optimum value is given by the geometric mean of the two surrounding indices [7]:

$$n1 = \sqrt{no \times ns} \tag{4.21}$$

The thickness t of the antireflecting coating is

$$t = \frac{\lambda o}{4n1} \tag{4.22}$$

For example, if the refractive index (rf) of silicon ns is 3.50 and the rf of air is 1.0, then the theoretical refractive index of the best ARC n_1 is 1.871.

The antireflecting materials used and their refractive indices are listed in Table 4.3. But for incident light with a wavelength different from λ_0, reflectivity will increase. So, in order to increase the output of the solar cell, the distribution of the solar spectrum and the relative spectral response of silicon solar cell should be taken into account and a reasonable wavelength will be chosen. The peak energy of the solar spectrum occurs at 0.5 μm, whereas the peak of the relative spectral response of the silicon cell is in the range of 0.8–0.9 μm wavelength. Therefore, the wavelength range of the best antireflection is 0.5–0.7 μm.

Dielectric films are most commonly used as antireflection layers on the front side of the c-Si solar cells. These thin-film coating layers not only act as antireflection layers to minimize the reflection losses, but also as passivation layers for the Si interface. Previously, SiOx, TiO$_2$, and Ta$_2$O$_5$ materials were used as single-layer antireflection coating materials for Si solar cells. As these materials suffered due to inherent problems such as defect generation during exposure to UV light, degradation of the passivation property after a damp heat test, and the intrinsic absorption of light, SiN$_4$ became the choice material for antireflection coating applications for industrial solar cells. The PC1D simulation as shown

Table 4.3 Antireflection coating materials and their refractive indices [8].

Material	Refractive index
MgF$_2$	1.4
SiO$_2$	1.5
Al$_2$O$_3$	1.8–1.9
Si$_3$N$_4$	~1.9
ZnS	2.3–2.4
TiO$_2$	~2.3
Ta$_2$O$_5$	2.1–2.3

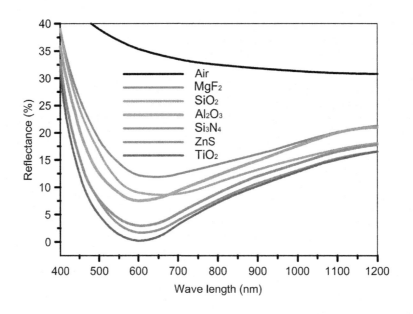

FIG. 4.16

Reflection at different wavelengths for different materials.

in Fig. 4.16 shows that the Si₃Ni₄ material appears to be the best choice for the antireflection coating material for a Si solar cell of standard technology [9].

For the industrial production of crystalline Si solar cells of standard Al-BSF technology, hydrogenated amorphous silicon nitride films are employed as antireflection coatings.

The following are the advantages of SiN₄ thin film with the performance of the following functions:

(a) It acts as an antireflection coating for the solar cell to minimize the reflection losses on the front surface

(b) It reduces the surface recombination of minority carriers by passivating the dangling bonds in the semiconductor-metal interface

(c) It acts as field effect passivation due to the presence of a positive fixed charge at the semiconductor/dielectric interface that deflects minority carriers from the surface

(d) It doesn't have parasitic absorption of incident solar radiation on the surface due to its higher band gap

(e) It releases hydrogen present in the films during the contact firing step and passivates the defects and grain boundaries present in the bulk silicon material to improve the solar cell performance

(f) It facilitates the easy penetration of screenprinted silver paste to the emitter surface during firing of contacts, so that better contact is formed with the underlying silicon material

(g) The deposition technology of PECVD is matured and adapted from the microelectronics industry

(h) It is easy to process the deposition of SiN₄ to get very high throughput in the production of industrial solar cells

(i) If the SiN₄ is properly optimized and the surface of the semiconductor is pretreated properly, the layer protects the solar cell from the effect of potential induced degradation

(j) Silicon nitride can be used as passivation layers to rear side of solar cell at contact and semiconductor interface. It acts as protecting cap layer for passivating aluminum oxide film in PERC-based solar cells.

The emitter saturation current density (J_{0e}) is reduced due to passivation. This significantly improves the open-circuit voltage (V_{OC}) of the solar cell. By utilizing a nonabsorbing thin film with an appropriate refractive index and thickness, a reduction in reflection can be achieved. The desired refractive index is dependent on the refractive indices of the Si wafer and the incident medium, which is typically 1.5 for a solar cell encapsulated within a PV module. The purpose of controlling the thickness of the film (typically 75 nm in practice) is to provide destructive interference at one specific wavelength of 600 nm in providing a minimum in the reflectance vs. wavelength curve that reduces the total reflectance integrated overall wavelengths of interest. Sputtering and plasma enhanced chemical vapor deposition (PECVD) techniques are used to deposit silicon nitride thin films on the solar cell.

4.4.6.1 Sputtering technique

Sputtering is classified under the physical vapor deposition technique. In this process, high-energy particles are made to bombard the target or source material to eject the atoms from the target material. The ejected atoms are made to fall onto a substrate getting deposited as a thin film layer. The schematic description of a sputtering system is shown in Fig. 4.17. The sputtering system consists of a high-vacuum chamber, a gas inlet, a pump connection, a sputter target, and the required gases. The target from which the atoms are to be ejected is negatively charged and the substrate on which the thin film has to be deposited is positively charged. Plasma is generated by applying direct current (DC), alternating current (AC), or radiofrequency (RF) and by exciting the gases such as argon in a vacuum chamber. The ions in the plasma are accelerated to a sufficiently high kinetic energy and made to bombard the target. The kinetic energy imparted to the bombarding particles by plasma energy is much higher than conventional thermal energies. This results in the removal of material from the

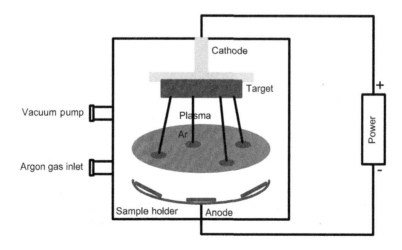

FIG. 4.17

Schematic of the sputtering process.

sputter target. The substrate is maintained at a negative charge and the source atoms are accelerated toward the substrate, resulting in thin-film deposition. For electrically conductive materials, DC discharge is preferred while an RF plasma is suitable for all materials, including dielectric target materials. To increase the sputter yield, often a so-called magnetron sputtering process is used that uses magnetic fields to confine the plasma to the sputter yield.

The sputtering technique is used to deposit silicon nitride thin films as antireflection coatings of crystalline silicon solar cells. Sputtering is also used to deposit indium tin oxide films as transparent conducting layers on silicon wafer-based heterojunction solar cells.

4.4.6.2 PECVD of silicon nitride

Plasma enhanced chemical vapor deposition (PECVD) is used for the deposition of dielectric thin films. This process has been developed for the semiconductor industry and is extensively used in microelectronics applications. This process has been adapted to the solar industry and it is a key deposition technique used in the manufacture of industrial silicon solar cells. For the deposition of thin-film layers such as hydrogenated silicon nitride (SiN_x) and aluminum oxide (AlO_x) in the manufacturing of solar cells, PECVD reactors are employed. PECVD reactors are also used to deposit intrinsic and doped amorphous silicon (a-Si) thin films for silicon wafer-based heterojunction solar cells. So, the PECVD technology is widely used in photovoltaic solar cell manufacturing [10].

Depending on the type of plasma, PECVD reactors are classified as either direct or remote systems. In a direct PECVD reactor, silicon wafers are placed between two parallel electrodes in direct contact with the electrodes and the excited plasma. In a remote PECVD reactor, the plasma is introduced onto the surface of the wafer and the substrates are not placed in direct contact with the plasma-forming electrode. A main disadvantage of the direct PECVD reactor is that there can be near surface damage due to ion bombardment, which increases the recombination rate in the affected region. However, SiN_x:H deposited by direct PECVD is typically found to be denser with a lower pin-hole density. Both types are currently used in solar cell manufacturing. Schematics of a direct and a remote PECVD system are shown in Fig. 4.18.

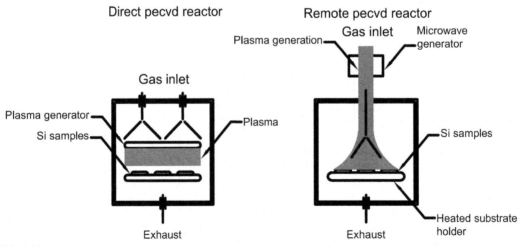

FIG. 4.18

Schematics of a direct and a remote PECVD reactor.

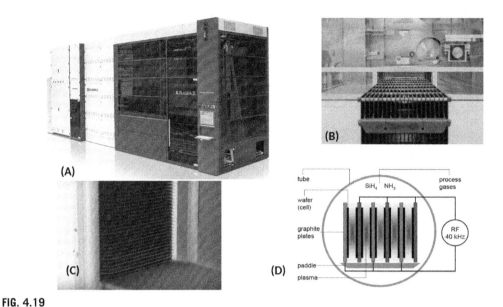

FIG. 4.19

PECVD equipment. (A) PECVD chamber with 10 quartz tubes, (B) graphite boat for loading the wafers, (C) ARC deposited wafers in the boat, and (D) schematic of direct plasma PECVD reactor.

Courtesy: Centrotherm.

A typical deposition process occurs on a heated substrate, typically in the temperature range of 350–450°C. The most commonly used precursors for the deposition of SiN_x:H are silane (SiH_4) and ammonia (NH_3) typically mixed with inert gasses such as argon (Ar) or nitrogen (N_2). The rate of the gas flows and operating temperature controls the properties such as thickness, refractive index, and hydrogen content in the growing thin film. In most cases, the SiN_x:H film is optimized to provide optimal surface and bulk passivation after the high-temperature firing step. Fig. 4.19 shows the PECVD equipment of Centrotherm. It is a direct plasma-based PECVD reactor. It has 10 quartz tube reactor chambers fully automated boat transport. Water cooling system prevents thermal interference between different tubes. This can be used to deposit silicon nitride, silicon oxide, silicon oxynitride, aluminum oxide films. The following reaction takes place:

$$SiH_4 + 2NH_3 + N_2 \rightarrow Si_3N_4 + 9H_2. \tag{4.23}$$

4.4.7 Contact making

In standard solar cell technology, the screen-printing methodology with a combination of annealing is used to make contacts to the device. This is a key process in solar cell manufacturing.

Screenprinting is a thick-film application process as opposed to the deposition/evaporation of thin films followed in the microelectronics industry. It is used to deposit a metallization material such as a

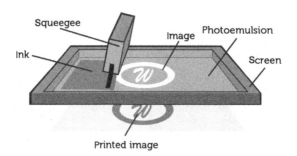

FIG. 4.20

Screen mesh with printed image.

paste containing silver powder to the front surface of the wafer in the desired pattern of fingers and bus bars. In this process, the wafer is subjected to considerable pressure. This can pose a problem for thin wafers and their breakage.

Screens and pastes are the essential elements of this technology.

Screens: Screens are tight fabrics of synthetic or stainless steel wires stretched on an aluminum frame, as shown in Fig. 4.20. The screen is covered with a photosensitive emulsion, which is treated with photographic techniques in such a way that it is removed from the regions where printing is desired.

For printing fine and thick layers, as is needed for the front contact of a solar cell, the wires must be very thin and closely spaced. On the other hand, the opening of the reticule must be several times larger than the largest particle contained in the paste to be printed. Screens for solar cell production typically feature 200 wires per inch, a wire diameter around 10 μm, and a mesh opening around 30 μm, corresponding to a nearly 50% open surface, that is, not intercepted by wires, and a total thickness (woven wires plus emulsion) of around 100 μm.

There are three types of screens being used for contact printing in the fabrication of standard solar cells of Al-BSF technology. The back side bus bars print using Ag/Al (silver/aluminum) paste. On the back side of the solar cell, the full area is printed with Al (aluminum) paste to create a back surface field. The front side metallic grid consisting of bus bars and thin grid lines is printed with silver paste. The images of the three types of screen meshes are shown in Fig. 4.21.

Pastes: The pastes are the vehicles that carry the active material to the wafer surface. Their composition is formulated to optimize the behavior during printing. A paste for the metallic contacts of the solar cell is composed of the following:

Organic solvents that provide the paste with the fluidity required for printing.

Organic binders that hold together the active powder before its thermal activation.

Conducting material, which is a powder of silver composed of crystallites of tenths of microns; for the p-contact, aluminum is also present. This amounts to 60–80% in weight of the paste and glass frit, present in the paste amounts to 5–10% in weight. The paste also contains powder of different oxides (lead, bismuth, silicon etc.). These powder of oxides will have low melting point and high reactivity at the process temperature, and enables movement of the silver grains and etches the silicon surface to allow intimate contact.

The paste composition is extremely important for the success of the metallization and is critically linked to the heat treatment.

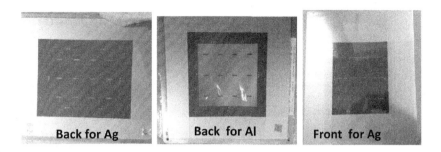

FIG. 4.21

Patterns in screen mesh for front and back metallization.

Printing: Fig. 4.22 illustrates the process of printing a paste through the patterned emulsion on a screen. The screen and the wafer are not in contact but are a distance apart, which is called the snap-off. After dispensing the paste, pressure is applied to the squeegee, which can be made of metal or rubber, and this puts the screen in contact with the wafer.

The squeegee is then moved from one side of the screen to the opposite one, dragging and pressing the paste in front of it. When an opening is reached, the paste fills it and sticks to the wafer, remaining

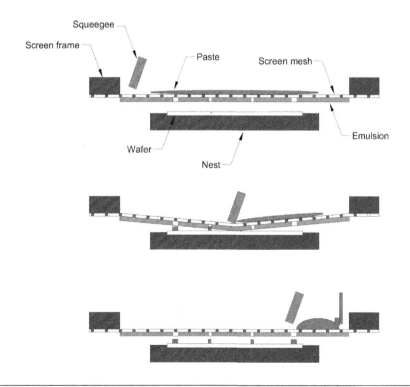

FIG. 4.22

Process of printing paste using screen mesh.

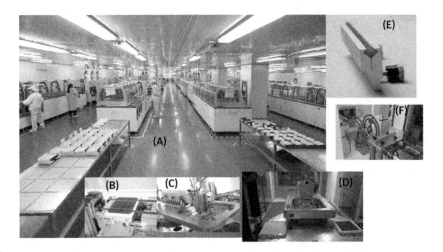

FIG. 4.23

Screen printing line over view. (A) Screen printing line, (B) solar cell inspection system, (C) printer head, (D) printing machine, (E) squeegee, and (F) flipper and buffer.

Screen Printing Line Image Courtesy: Chaudary, Risen Energy.

there after the squeegee has passed and the screen has elastically retired. Fig. 4.20 shows the printed image. Fig. 4.23 shows the overall view of the screen printing line.

The amount of printed paste depends on the thickness of the screen material, the emulsion, and the open area of the fabric. It also depends on the printed line width. The viscous properties are of utmost relevance: when printing, the paste must be fluid enough to fill without voids all the volume allowed by the fabric and the emulsion, but after being printed it must not spread over the surface. The critical parameters of this process are the pressure applied on the screen, the snap-off distance, and the velocity of the squeegee.

Drying: Solvents are evaporated at 100–200°C right after printing so that the wafer can be manipulated without the printed pattern being damaged.

Firing: Firing of the pastes is usually done as a three-step process in an infrared belt furnace. In the first step, when heating, the organic compounds that bind the powders in the paste together are burnt in the air. In the next step, the highest temperature between 600 and 800°C is reached and maintained for a few minutes. Higher temperatures are needed for penetration of the AR coating and contacting the semiconductor surface. The crystal orientation and paste composition must be considered too. In the last step, the wafer is cooled. Fig. 4.24 shows the contact firing furnace of Centrotherm. It is fast firing furnace and a solution for burning out and sintering of solar cell metal contacts. The firing zone is equipped with short wave infrared light elements. A volatile organic compound handling system consisting of a condenser and electrostatic filter sends out the solvents released during the drying process.

The phenomena that take place during firing are very complex and not completely understood. The oxides forming the glass frit melt, enabling silver grains to sinter and form a continuous conductor so that the layer can present low sheet resistance.

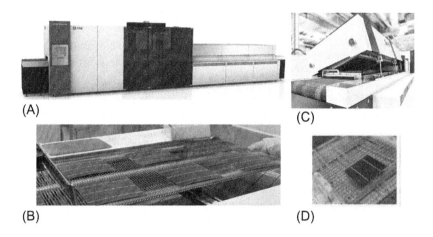

FIG. 4.24

Contact firing furnace, solar cell front and back side on the belt. (A) Firing furnace, (B) contact formed solar cells, (C) firing process chamber, and (D) solar cell on the belt.

Courtesy: Centrotherm.

Neither the silver melting point nor the silicon-silver eutectic temperature is reached, with sintering consisting of the intimate contact of solid silver crystallites. At the same time, the reactive molten glass etches some silicon, and silver grains are allowed to form intimate contact with the substrate. The amount of etched silicon is on the order of 100 nm. When a layer of SiN is present, the glass frit is able to etch through it. In fact, the quality of the contact improves because of a better homogeneity.

The contact resistance of the printed contacts is much higher than that of an evaporated contact to n-Si of the same doping. It seems that although enough silver grains make good contact with silicon, not all of them are connected to the grains in the outer layer; many remain isolated by the glass.

When the paste, in the case of back metallization, contains aluminum as well as silver, the formed and recrystallized Al-Si eutectic ensures a good contact. With dielectric layers, the contact appears to be localized as well, and some beneficial role is attributed to the metal atoms in the frit.

Limitations and trends in screenprinting of contacts: High contact resistance and the etching action of the glass frit require the front emitters to be highly doped and not very thin if screenprinting is used. Only improved paste formulation and processing can overcome this limitation. Narrow but thick fingers with good sheet conductance are also needed. Well-defined lines must be much wider than the pitch of the woven fabric; 32 μm lines seem achievable. Incrementing the amount of transferred paste implies increasing the thickness of the emulsion or the pitch-to-diameter ratio of the wires, which are both limited. Besides, screens become deformed with usage and a continuous deterioration of printed patterns is observed.

Front contact print and dry: The requirements for the front metallization are low contact resistance to silicon, low bulk resistivity, low line width with high aspect ratio, good mechanical adhesion, solderability, and compatibility with the encapsulating materials. Resistivity, price, and availability considerations make silver the ideal choice as the contact metal. Copper offers similar advantages, but it does not qualify for screenprinting because subsequent heat treatments are needed, during which its high diffusivity will produce contamination of the silicon wafer.

The paste is a viscous liquid due to the solvents it contains; these are evaporated in an inline furnace at 100-200°C. The dried paste is apt for subsequent processing.

Back contact print and dry: The same operations are performed on the backside of the cell, except that the paste contains both silver and aluminum and the printed pattern is different. Aluminum is required because silver does not form ohmic contacts to p-Si, but cannot be used alone because it cannot be soldered. The low eutectic temperature of the Al-Si system means that some silicon will be dissolved and then recrystallize upon cooling in a p-type layer.

Although a continuous contact will in principle give better electrical performance (lower resistance), most commercial wafers feature a back contact with a mesh structure. Apart from paste-saving considerations, this is preferred to a continuous layer because the different expansion coefficients would produce warping of the cell during the subsequent thermal step.

The front side paste is applied on the AR coating, which is an insulating layer, and the back paste is deposited on the rear side of the wafer. Upon firing, the active component of the front paste must penetrate the ARC coating to contact the n-emitter without shorting it. Too mild a heat treatment will render high contact resistance, but too high a firing temperature will motivate the silver to reach through the emitter and contact the base. In extreme situations, this renders the cell useless by short-circuiting it. In more benign cases, small shunts appear as low shunt resistance or dark current components with a high ideality factor that reduces the fill factor and the open-circuit voltage. The back paste, in its turn, must completely perforate the parasitic back emitter to reach the base during the firing. In order to comply with these stringent requirements, the composition of the pastes and the thermal profile of this critical step must be adjusted very carefully.

4.4.8 Testing and sorting of solar cells

The illuminated I-V curve of the finished cells is measured under a sun simulator with an artificial light source with a spectral content similar to the sun spectrum at a controlled temperature of 25°C. The solar cell is mounted on a temperature-controlled fixture and the spring-loaded contact system will contact the front and back sides of the solar cell. The sun simulator might be either LED-based or xenon arc-based. The spectral composition and the variation in the intensity should be as per the testing standards. A reference solar cell with similar structure and spectral response should have been evaluated and calibrated with a standard testing facility. The intensity of the cell tester is calibrated based on the reference cell current values. The temperature is controlled based on the V_{oc} value of the solar cell. If the temperature of the cell is deviating 25°C, appropriate correction will be applied. The contact probe system will be placed on the contact bus bars of the solar cell and aligned properly. The light from sun simulator is flashed on the solar cell and with the usage of power supply, the solar cell is sweeped from I_{sc} to V_{oc} and the current is measured at different voltages. With the software, the data acquisition system provides all the electrical parameters and I–V curve of the solar cell. The electrical parameters such as I_{sc}, V_{oc}, V_{mp}, I_{mp}, P_{mp}, and fill factor, series resistance, Shuch resistance, dark current and the intensity of radiation, and the temperature of the cell are acquired by a data acquisition system.

Along with I-V characteristics, the solar cells are tested for reverse current at $-12\,V$ and shunt resistance. The reverse current at $-12\,V$ should not be more than 1A and the shunt resistance of a $6''$ cell should not be lower than $60\,\Omega$ [11]. This is required to filter out hotspot-prone solar cells. All the solar cells are tested with EL tester to filter out any cell with microcracks.

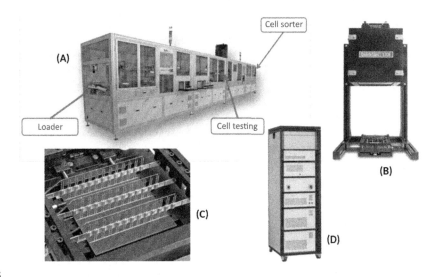

FIG. 4.25

Solar cell tester and sorter. (A) Solar cell tester and sorter, (B) sun simulator, (C) cell contact connection system, and (D) power supply and data acquisition system.

Solar cells with defects such as cracks and edge chips are then rejected, and the rest are classified according to their output. The manufacturer establishes a number of classes, typically, to the cell current at a fixed voltage near the maximum power point. Modules will subsequently be built with cells of the same class, thus guaranteeing minimal mismatch losses. If, for instance, cell currents within a class must be equal within 3%, the accuracy and stability of the system must be better than that. Automatic testing systems that meet the very demanding requirements of high-throughput processing are available. Fig. 4.25 shows the solar cell tester and sorter combined in a single unit. Fig. 4.26 shows the I–V curve and EL images of the solar cell.

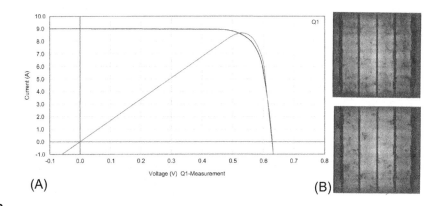

FIG. 4.26

Testing of solar cells. (A) I–V curve obtained from solar cell tester and (B) EL image of solar cell.

4.5 P-type vs N-type solar cells

In recent times, there has been tremendous interest among solar cell manufacturers and researchers to increase the efficiency of the solar cell and reduce the levelized cost of energy of solar PV systems. The 2020 edition of the International Technology Roadmap for photovoltaics (ITRPV) predicted the market share in 2020 for n-type monocrystalline wafers as 10%. It also predicts that there will be a clear shift from p-type to n-type substrates and by 2029, the market share of n-type silicon will be 40% [12]. Choosing n-type solar cells in module technology is because of certain significant advantages of n-type silicon over p-type silicon. For manufacturing Si solar cells, the silicon wafer is the basic raw material, which acts as a substrate as well as an absorber for the solar cell. If boron is doped during the crystal growth of Si, it results in a p-type wafer as the type of charge carrier is positive based, whereas the n-type wafer is negative charge carrier based as the material is doped with group V elements such as phosphorus and arsenic. Since the inception of Si solar cells and their fabrication in Bell Laboratories, n-type cells have been used for satellite power applications [13]. The space environment contains energetic charged particles such as electrons and protons at different belts. When they impinge on the solar cell and move into the Si material depending on the energy, they cause lattice damage to the emitter and base regions of the solar cell. The phosphorus-doped n-type-based solar cell was prone to more lattice damage. Therefore, the p/n cells exhibited more degradation in electrical parameters compared to the n/p type [14] and so, the type of solar cell was switched from the n-type to the p-type.

In a solar cell, the overall spectral response gives the I_{sc} and the contribution for I_{sc} is from the emitter as well as the base regions. The emitter region contributes current from high-energy photons, whereas the base or absorber region contributes current from higher wavelengths of the visible spectrum; this amounts to more than 70% of I_{sc} [15].

The minority carrier mobility, diffusion length, and ease of fabrication are some of the criteria for the selection of the type of wafer. Historically, p-type wafers have dominated over n-type wafers in the solar industry due to the following reasons.

In p-type wafers, the minority carrier is the electron and the mobility of the electron in the base region of the wafer is about three times higher than that of holes in n-type wafers. So, p-type wafers have longer diffusion lengths due to the high mobility of the minority charge carrier. The boron has a segregation coefficient of 0.8 in silicon, so the dopants are distributed uniformly in the ingot. So, boron-doped p-type wafers provide uniform resistivity, which results in a small variation of dopant concentration in the area of the wafer.

The process of formation of the p-n junction, that is, the incorporation of phosphorus atoms onto the p-type wafer surface by thermal diffusion, is simple compared to boron diffusion with n-type wafer. There is a proven technology for thermal diffusion of phosphorus on p-type wafers compared to the relative novelty and higher temperature of boron diffusion.

During the thermal diffusion of phosphorus, a gettering action takes place and the impurity atoms will be segregated to the edge; this doesn't affect the performance of the solar cell. This is more beneficial to multicrystalline wafers and low-quality wafers and also can be used to get a reasonable efficiency.

There is a disadvantage of the p-type wafer for solar cells with light-induced degradation in electrical parameters. This effect is more pronounced with monocrystalline wafers compared to multicrystalline wafers. The boron, which is a dopant in p-type Si, interacts with oxygen in the wafer and forms boron-oxygen complexes. These get activated due to light, causing defective centers and reducing the diffusion length of charge carriers and hence degradation in the electrical parameters.

The following are the advantages of n-type Si over p-type Si.

(a) The minority carrier lifetime in the n-type wafer is higher than the p-type Si wafer at equal doping and defect concentration levels.

Earlier, the comparison between carrier lifetimes of minority charge carriers in the p- and n-type wafers was done considering the equal resistivity of wafers, so the lifetime of the minority carrier in the p-type wafer showed a higher value. It is felt that the comparison of the minority carrier lifetime between p-type and n-type silicon has to be done at equal dopant and defect concentration levels. If the minority carrier lifetime (τ) is measured in practical p- and n-type silicon wafers with comparable dopant and defect concentrations, n-type silicon usually shows a much higher τ than p-type silicon [16]. This is due to a higher Auger limit to the lifetime for p-type crystalline silicon than for n-type crystalline silicon at higher doping levels. The carrier lifetime killers in silicon are transition metals, which generally have much larger carrier capture cross-sections for electrons than for holes, causing a lower diffusion length of electrons in p-type wafers [17]. Due to the above reasons, the carrier lifetime of n-type silicon is usually much higher than that of p-type silicon wafers with comparable dopant and defect concentrations.

(b) The n-type wafer-based solar cell is free from light-induced degradation (LID)

This is due to the absence of boron oxygen-related defects as the n-type wafer is doped with phosphorus atoms. Even the solar cell made with a higher oxygen concentration will not exhibit LID issues. It means the presence of oxygen in n-type silicon will not affect the minority carrier lifetime. So, the Czochralski method grown n-type wafers can be used for solar cells.

(c) It has higher minority carrier diffusion lengths

The n-type wafer exhibits resistance against common impurities. The interstitial transition element impurities such as iron have larger capture cross-sections to electrons. As the impurities possess a positive charge state, they can effectively capture the electrons in silicon material. As the minority carriers in n-type silicon are holes, so the capturing of holes by defects is not effective. Therefore, the n-type wafer offers higher minority carrier diffusion length as compared to p-type c-Si substrates with similar impurity concentrations.

(d) In the n-type solar cells, the use of a phosphorus-doped back surface field (BSF) or PERC with suitable surface passivation with dielectric material results in a higher diffusion length and better rear internal reflection.

(e) There is a potential for improvement of efficiency with solar cells of different structures due the higher diffusion length, higher bulk lifetime, etc., in n-type wafers.

(f) The solar cells made with n-type wafers such as HJT and PERT/PERC configurations can be easily adaptable to bifacial structures.

(g) The process of making bifacial designed solar cells using PERC/HJT based technologies also generates opportunities to produce cells with higher efficiencies.

Though n-type wafers have advantages with the factors mentioned above, the photovoltaic industry prefers p-type silicon material as the choice for solar cells. The cost of the n-type wafer is higher compared to the p-type wafer. All the wafers cut from the ingot cannot be used, as there is a strong variation of wafer resistivity along the ingot due to the low segregation coefficient of phosphorus in silicon. This issue can be circumvented by going for charge carrier-selective contacts. The base resistance does not significantly influence the cell performance in the one-dimensional current flow of solar cells with charge carrier-selective contacts. It was demonstrated for the first time that efficiencies greater than 25% can be achieved for base resistances between 1 and 10Ω-cm with n-type wafers [18].

4.6 PERC/PERL cells and the manufacturing process

BSF technology with mono- and multicrystalline wafers dominated the solar industry for more than a decade. This BSF-based cell type suffers from some inherent limitations and the efficiency has reached the saturation point with this technology. The metallic film of aluminum, which is applied on the rear side of the cell to serve as the BSF, cannot support the efforts in reducing rear recombination velocities of, say, less than 200 cm/s. Only 60–70% of the infrared (IR) light reaching the rear aluminum is reflected back. To mitigate the limitations of BSF technology, the researchers adapted PERC as a promising technology for industrial solar cell production. The standard back surface field (BSF) cell architecture has been extended to get passivated emitter and rear cell (PERC). The electrical and optical losses that occur on the rear side of the BSF-based solar cell can be greatly reduced by applying an additional dielectric passivation layer on the rear side—and that is PERC. The main aspect of PERC technology is incorporating a rear side passivation scheme to standard BSF cell technology. This concept deals only with the optimization of the rear surface, especially aiming at reducing recombination losses on the cell's dark side, and it is fully independent from the front side. That means that any progress on the front side, such as employing innovative metallization concepts and improved junction properties, compliment the effort to improve cell performance with PERC architecture. The concept is also relatively simple to implement in production—only 2–3 additional steps need to be added to the standard processing sequence.

The concept of a PERC solar cell is not new, but a concept that dates back more than three decades. Leading PV scientist Martin Green [19] first proposed this concept in 1983 and made solar modules for some niche applications, such as solar racing cars. But this technology was not able to enter the mainstream due to the higher cost of the manufacturing process.

The optimization of the rear surface is aimed at getting rid of the inherent limitations of the metallic film of aluminum in BSF solar cells. The new layer reduces the recombination losses on the cell's back side, causing a reduction of electrical and optical losses. The PERC solar cell is a simple technology highly compatible with the existing PV production lines, requiring only minimal changes to the existing solar cell processing lines. As a result, a small increase in production cost is required, making it preferred for many crystalline silicon solar cell manufacturers. Fig. 4.27 shows the structure of the PERC solar cell.

The solar cell is made with p-type mono- or multicrystalline Si wafers. As shown in Fig. 4.27, there is no change in the structure with respect to the front side of the solar cell, and accordingly, there is no change in the process to achieve front side configuration. The front side is textured to maximize the absorption of the solar energy. A phosphorus-based emitter is formed by diffusion on the front side to form a junction with the p-type wafer. The junction is a shallow type with a depth of around 0.3 μm to allow the absorption of UV and blue light to enhance the blue response of the solar cell. The junction will assist in separation of the charge carriers. The emitter is highly doped, so that suitable ohmic contact can be made with the solar cell on the front side. The diffusion process to form the emitter will getter the unwanted impurities such as iron and transition elements so that the defect centers can be reduced in the solar cell.

An antireflection coating of hydrogenated amorphous silicon nitride is deposited on the front side of the solar cells. It minimizes the reflection losses from the silicon surface and enhances the harvest of maximum energy into the solar cell. The silicon nitride layer has a positive charge. It passivates the dangling bonds in the emitter side and reduces the front side surface recombination velocity [20].

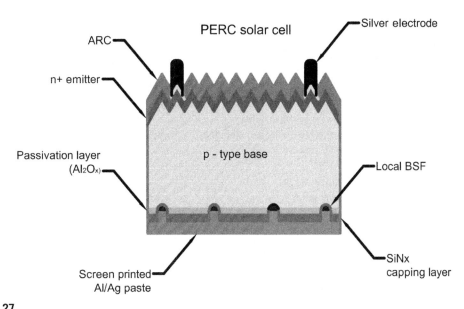

FIG. 4.27

Structure of the p-type PERC solar cell.

The ARC is hydrogenated and releases hydrogen during contact firing, passivating the grain boundaries of multicrystalline solar cells.

Silver paste makes the ohmic contact with the highly doped n+region. The front side will have five bus bars and 100 grid lines each 30 μm wide to reduce the series resistance of the emitter region. Thus, the front side of the PERC cell is similar to that of the BSF cell.

On the rear side, an aluminum oxide layer is deposited to passivate the dangling bonds of the rear surface. This layer will have a negative charge and passivate the surface while also acting as a reflector to long wavelengths that are traversing the cell. As the Al_2O_x layer is very thin, to protect it, a capping layer of SiN_4 is deposited over the Al_2O_x layer. With the laser ablation technique, the stack of dielectric layers on the rear side is opened and a specially made Al paste is screen printed on the rear side and fired to form a contact.

The key point of the PERC cell is the deposition of the rear passivation film. In general, passivation layer stacks such as Al_2O_3/SiN_x or SiO_2/SiN_x are applied as the rear side passivation scheme for high-efficiency PERC cells [20]. On one hand, by saturating the dangling bonds on the silicon rear surface, Al_2O_3 shows outstanding chemical passivation. On the other hand, a high fixed negative charge density ($\sim 10^{13}$ cm^{-2}) enables Al_2O_3 as an excellent field effect passivation Scheme [21]. As the thickness of Al_2O_3 is in the range of a few nanometers, it gets damaged during the metallization step of the process. So, the Al_2O_3 layer is capped by a thicker SiN_x layer deposited by PECVD that protects the rear passivation film from metallization. The SiN_x as a cap layer compensates for the rear passivation stack thickness and improves the internal reflection on the back of the cell, causing enhancement in the long wave response of the solar cell. Thus, the enhancement of passivation and a long wavelength response on the rear side improve the short-circuit current. In general, Al_2O_3 can be deposited by atomic layer

Table 4.4 P-type PERC solar cell process.
P-Type PERC
Surface damage removal
Texturing on front side
$POCl_3$ diffusion
Laser doping for selective emitter
PSG removal, edge isolation + rear polish
Anti-PID SiO_2 process
Rear side AlOx
Rear side SiN PECVD
Front side SiN PECVD
Laser opening of dieclectric stack on rear side for contact
Metallization on rear side
Metallization on front side
Contact firing
LID regeneration
Testing and sorting

deposition (ALD) [22] or plasma enhanced chemical vapor deposition (PECVD) [22]. On the Al_2O_3 layer, the SiN_x layer will be deposited by PECVD. The laser ablation technique is used to open the dielectric stack of the Al_2O_3/SiN_x layer at locations of bus bar area and silver paste is applied and contact is formed between the Ag and bulk silicon. Ag/Al paste is also used for rear side contact metallization. The P-type PERC process is shown in Table 4.4.

Basics of passivation

Nothing is perfect—even a carefully grown silicon ingot cannot escape from inheriting some crystallographic defects. On top, a very basic step of sawing wafers from the ingot itself causes disruptions of a crystal lattice at both wafer surfaces. These interruptions in the periodical arrangement of silicon atoms result in dangling bonds, working as recombination centers. The impurities and defects that are present in the bulk and on the surface of the silicon wafer adversely affect the performance of a solar cell. Passivation is a process in which these defects are made inactive to reduce the surface recombination of charge carriers, safeguarding the cell efficiency.

Passivation can be attained by two complementary methods—strongly reducing the charge carriers of one polarity reaching the surface and reducing the interface states by saturating the dangling bonds. The latter can be accomplished again in two ways. One is to simply saturate the dangling bonds on the surface by providing suitable conditions to grow a surface layer that allow sufficient time and energy for atoms to reach the optimal energy levels to saturate these dangling bonds. Alternatively, one can deposit a hydrogen-rich dielectric film that releases hydrogen in subsequent firing steps. This hydrogen occupies vacant sites of dangling bonds, thereby pacifying them. This method is called chemical passivation. There exists another mechanism called field effect passivation that involves creating an electric field close to the surface that can repel the charge carriers of similar polarity. It can be achieved by means of descending dopant density from a high surface concentration. Alternately, applying a dielectric layer with high fixed charges also creates an electric field gradient near the surface, which provides field effect passivation.

Silicon nitride has become the most reliable passivation material for the sunny side of a standard cell. That is because of its superior antireflective properties as well as its ability to provide good field effect passivation together with a moderate reduction of the interface states. As for the rear surface, the current state of the art is to apply metallic aluminum that alloys with silicon during a firing step to form a p-type region acting as the BSF. This Al-BSF is devised just to mimic the field effect passivation, which means that when designing the standard solar cell structure, the chemical passivation aspect of the cell's dark side was completely neglected. The PERC concept is all about adding full-scale rear surface passivation attributes to standard cells.

4.7 Bifacial solar cells and the manufacturing process

Increasing the yield of PV power plants is a major challenge, as power tariff rates are coming down and grid parity has been achieved in different countries. As scientists and engineers continue to improve their solar panel designs to increase efficiency, the yield of a PV power plant can be further increased by using modules with bifacial solar cells. The standard monofacial solar cells capture light only from one side and convert it into electrical energy. The bifacial solar cell is made such that it absorbs light energy from both sides. The solar cell's rear side generates charge carriers during its exposure to diffused or reflected light radiation. The bifacial solar cell with metallic contacts on both sides generates more power compared to the standard monofacial cell under the same conditions of operation. Depending on cell design, the intensity of the light falling on the rear side due to site albedo, and mounting conditions, the power gain in the bifacial module varies between 5% and 30% [23].

The concept of a bifacial solar cell is not new, and it has been known for many decades. Research on bifacial PV cells dates back to the dawn of the solar industry, according to Andrés Cuevas' often cited article, "The Early History of Bifacial Solar Cells" [24].

Japanese researcher H. Mori proposed a bifacial cell design with a collecting p-n junction on each surface of a silicon wafer, thus forming a p^+np^+ based double junction structure in 1960; the prototype cell was made by 1966 [25]. Bifacial PV modules were deployed first in the 1970s for satellite applications by the Russians. A bifacial single junction cell with dielectric passivation was proposed in 1997 by Chevalier and Chambouleyron. In 1980, Cuevas realized the effect of the albedo in increasing the power of bifacial solar cells. The bifacial solar cell was not adapted for industrial production due to its high cost of manufacturing.

Bifacial solar cells are slightly different to standard monofacial solar cells. They capture the light from both sides of the cell and convert it into energy. The rear side of the solar modules receive reflected or diffused light from the ground or surface on which the modules are mounted, and this contributes to increased power. The reflected or diffused radiation coming from the ground surface is called the albedo. The amount of the albedo depends on the surface where the modules are installed, the height of the mounting, and the inter row distance of the module mounting structures. The rear side of the solar panel can absorb between 5 and 30% of the light absorbed by the front surface of the solar panel.

Bifacial modules can be manufactured with and without frames. Some modules have dual-glass configurations and some use a transparent back sheet on the rear side. Most bifacial solar modules are made with monocrystalline solar cells, but there are designs with polycrystalline solar cells. The one thing that is constant is that power is produced from both sides. Bifacial solar cells have contacts/busbars and ARC on both the front and back sides and power is produced from both sides.

The commercial cell architectures—heterojunction (HJT), PERT, and PERC–are easily converted to the bifacial type. The bifaciality factor, which is the ratio of the rear side generated power to front side generation, depends on the type of structure and the amount of area the metallic grid covers on the rear side. HJT and PERT cells have a bifaciality of more than 90%, whereas PERC has a bifaciality of 70%. An IBC cell provides a bifaciality of about 80%.

A typical PERC structure that is still employing Al-BSF (local) in today's manufacturing is not bifacial. But in order to convert the cell into a bifacial device, only the aluminum BSF has to be replaced with a metallic grid. The contacting bus bars are screen printed with Ag paste and the grid lines are screen printed with Al paste. The structures of monofacial and bifacial solar cells with p-type PERC and PERT configurations are given in Fig. 4.28.

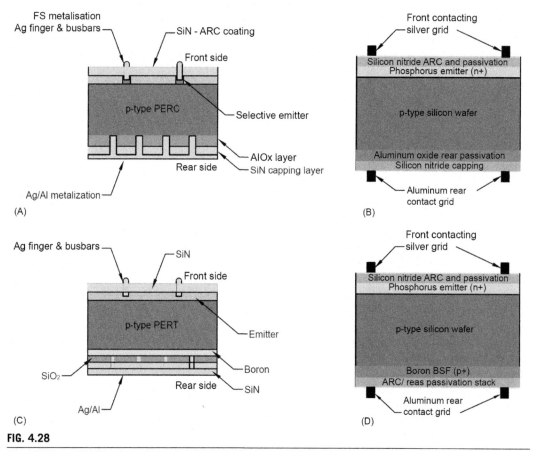

FIG. 4.28

The PERC and PERT solar cell structures and their corresponding bifacial cell structures: (A) p-type PERC cell, (B) p-type PERC cell in a bifacial structure, (C) p-type PERT cell, (D) p-type PERT cell in a bifacial structure.

Key drivers for bifacial solar cells/modules

- There is improved energy generation due to bifacial energy gains and reliable performance over the life of the system due to improved durability. The performance warranties for bifacial PV modules are extended to 30 years due to the reliable design.
- Unlike PV systems deployed with monofacial modules, bifacial PV systems can convert light that falls on the back side of the module into electricity. This additional back-side production increases energy generation over the life of the system. A bifacial PV system could generate 5–30% more energy than an equivalent monofacial system, depending on the mounting methodology of the modules and the characteristics of the ground where the systems are installed.
- The use of higher-efficiency bifacial solar cell-based modules not only reduces the area of the mounting system on a per kW basis, but also allows a developer to increase system capacity and energy harvest at a given site with fixed development costs. Due to this, the BoS cost will be reduced.
- The LCOE for a power generation asset is found by dividing the total lifecycle costs–both the up-front construction costs and the operational costs over time–by the total lifetime energy production. In the field, bifacial PV modules outperform their nominal power and efficiency ratings, which addresses the energy-generation side of the LCOE calculation. Due to increased generation, the levelized cost of energy (LCOE) of the system will become lower.
- Bifacial technology is primarily becoming popular due to the low levelized cost of electricity.
- The continued reduction of tariff rates for the systems pressures manufacturers to go for higher efficiency and low-cost solutions. At present, bifacial technology is the promising option.
- Mono PERC technology has become successful and every manufacturer is converting its standard cell manufacturing lines into mono-PERC. Every solar cell manufacturer producing PERC cells is also seriously evaluating bifacial technology. It is simple to upgrade mono-PERC cell manufacturing to bifacial cells.
- The recent technology developments in the PV industry such as HJT and PERT/PERL are intrinsically bifacial in nature. It is easy to adapt or upgrade the technology to bifacial.
- Glass-glass modules: With thin glass getting cheaper and the products being mature, glass-glass modules are becoming increasingly competitive, even more as they allow module manufacturers to offer longer warranties.
- Transparent back sheets: This shift actually complements bifacial technology, as replacing the opaque back sheet with transparent materials such as glass on the rear side is the prerequisite for bifacial technology at the module level. The situation became so favorable that bifacial is considered the natural technology progression in today's PV manufacturing.

p-Type bifacial cells

There are two important steps required to upgrade standard BSF solar cell processing to PERC. One step is adding a dielectric layer stack such as AlOx + SiN as a passivation layer on the rear side. The second step is opening the deposited dielectric stack by lasers to allow the formation of a rear contact. The rear polishing step can be done in the chemical wet-bench process along with the edge-isolation process. Thus, a passivation film deposition system and a film opening system—mainly accomplished with lasers—are additional tool sets typically hooked up to standard cell processing lines. This is the change in process from standard BSF technology to PERC technology. The present PERC structure uses Al-BSF locally, which is not bifacial. To upgrade the PERC to bifacial, the

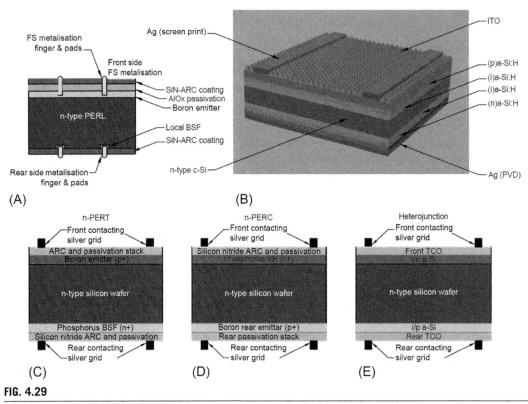

FIG. 4.29

PERL, PERC, PERT, and HJT solar cell structure and their corresponding bifacial cell structures. (A) n-type PERL cell, (B) n-type HJT cell in bifacial structure, (C) p-type PERT bifacial, (D) n-type PERC bifacial, and (E) n-type HJT bifacial structure.

　　back side BSF can be removed and metallization can be printed on the full area of the rear side. This alone makes it the most cost-effective for implementing the bifacial concept in PERC-based cells (Figs. 4.28 and 4.29).

- The PERC solar cell is based on the p-type substrate, which is the current mainstream wafer type. Many companies have moved toward PERC technology from standard BSF technology. The steps required to shift from PERC to a bifacial cell are negligible compared to the modifications needed to upgrade from standard BSF to PERC. So, more manufacturers are opting for PERC-based bifacial technology.

The bifacial solar cells upgraded with PERC technology show a lower bifaciality factor. This is due to the large area coverage of the rear surface with aluminum contact grid lines and bus bars. To get better conductivity compared to the silver contact, a wider grid lines with widths ranging from 200 to 250 μm are printed on the rear side. The rear metallization with wider fingers causes a shading of about 30–40%

on the rear side. To reduce the rear metal contact area, concepts such as more busbars and multibusbars, which lower the rear metallization resistivity, are becoming popular. This concept compensates for the low conductivity arising from the rear contact metallization of aluminum and facilitates bifacial PERC. The bifacial gain is much higher for the cells made with n-type wafers compared to p-type PERC-based bifacial cells. So, manufacturers are starting to look into this n-type wafer-based cell technology, although the n- type wafers are expensive.

From an electrical point of view, a bifacial solar cell is modeled, as shown in Fig. 4.30, with representation by two monofacial cells in parallel. A two-diode models is commonly considered for monofacial cells. The two-diode model is slightly more accurate than a one-diode model, especially in low irradiance conditions. Two two-diode models are mounted in parallel to represent the bifacial technology and the influence of each side on the other [26].

Several studies use two two-diode models in parallel to represent a bifacial cell. Subscript "f" represents the front side and subscript "b" the back side. R_{sh} refers to the shunt resistance: a low shunt resistance provides the light-generated current an alternate path and therefore induces power losses. It is generally due to manufacturing defects.

R_s refers to the series resistance: a high series resistance reduces the fill factor. The fill factor of a solar cell is the ratio of the solar cell's actual power output ($V_{pmax} \times I_{pmax}$) versus its ideal power output ($V_{oc} \times I_{sc}$).

The PV module is considered as two one-diode models in parallel, and the bifacial short-circuit current, the bifacial open-circuit voltage, the bifacial fill factor (FF), and the bifacial efficiency are defined as follows:

Bifacial short-circuit current:

$$I_{sc}.bi = I_{sc}.front + \chi I_{sc}.rear = RI_{sc} \times I_{sc}.front \tag{4.24}$$

where χ is the irradiance ratio and RI_{sc} is the gain in short-circuit current relative to monofacial front-side only illumination, respectively defined as the following:

$$\chi = \frac{Gr}{Gf}. \tag{4.25}$$

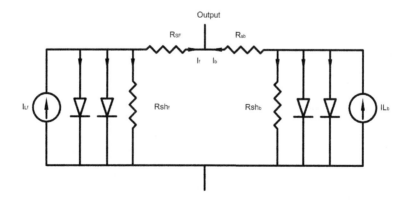

FIG. 4.30

Electronic model of bifacial solar cell.

Where Gf is the irradiance on the front side of the solar cell, and Gr is the irradiance on the rear side of the solar cell.

Bifacial open-circuit voltage

$$V_{oc}.bi = \frac{kT}{q} In\left(\frac{I_{sc}.bi}{I_o} + 1\right) \tag{4.26}$$

$$P_{mp}.bi = I_{sc}.bi x V_{oc}.bi \times FFbi \tag{4.27}$$

The resulting bifacial power and bifacial efficiency are:

$$\eta bi = \frac{I_{sc}.bi \times V_{oc}.bi \times FFbi}{A \times (Gf + Gr)} \tag{4.28}$$

$$Pbi = I_{sc}.bi \times V_{oc}.bi \times FFbi \tag{4.29}$$

To obtain the new "bifacial gain" and "bifacial efficiency," a single-side measurement for the front and the rear is needed. This model gives a predicted output power within 1% of the measured power for a set of five different irradiance conditions. To obtain these results, the nonilluminated side of the module must be covered with an extremely low reflectance black cover [27].

The bifaciality factor and the bifacial efficiency are some of the characteristics of a bifacial solar cell. The bifaciality factor is defined as the ratio of the rear side power/front side power.

The bifaciality factor depends on the design of the cell, the rear side texture, the type of rear side antireflection coating, and the percentage of metal coverage on the rear side of the cell. It also depends on the base resistivity, the lifetime of the cell, and the type of back surface field.

The bifaciality factor is usually <1 because the front and rear side configurations are not symmetric in nature and the metallization is optimized for front side efficiency only.

The current density to be generated from the rear side depends on the metallic contact grid covered area and the efficient charge carrier transport from the illuminated side to the other side. Charge carrier transport from front to rear during illumination of the front side is driven by the field, whereas diffusion drives the charge transport from the rear to the front side.

HJT solar technology can be adopted to the bifacial cell. It has indium tin oxide (ITO)-based contacts on both sides with the use of low-temperature process-based silver paste. It gives good bifaciality factor.

From PERC to the bifacial process, the localized BSF on the rear side has to be removed. SiN will be applied for antireflection. Using aluminum paste, the metallic grid will be applied on the rear surface.

The manufacturing processes of the p-type PERC cell and the p-type bifacial cell are shown in Table 4.5.

Full area metallic grid lines and busbar formation are the main changes in bifacial cell fabrication compared to the PERC cell.

For PERT solar cells, instead of localized metallization, the rear side will be covered with a full metallic grid using Ag/Al paste. The ARC will be optimized for antireflection purposes also. The process flow is shown in Table 4.6.

Fig. 4.29 shows the PERL, PERC, PERT, and HJT solar cell structures and their corresponding bifacial cell structures.

From PERL to the bifacial process, the localized BSF on the rear side will be replaced with a metallic grid using Ag/aluminum paste. The ARC will be optimized for antireflection purposes also. For

Table 4.5 Manufacturing processes of p-type PERC and p-type bifacial.

P-type PERC	P-type bifacial
Surface damage removal	Surface damage removal
Texturing on front side	Texturing on front side
POCl$_3$ diffusion	POCl$_3$ diffusion
Laser doping for selective emitter	Laser doping for selective emitter
PSG removal and rear polish	PSG removal and rear polish
Anti-PID SiO$_2$ process	Anti-PID SiO$_2$ process
Rear side AlOx	Rear side AlOx
Rear side SiN PECVD	Rear side SiN PECVD
Front side SiN PECVD	Front side SiN PECVD
Laser contact opening	Laser contact opening
Metallization rear	Full Metallization rear
Metallization front	Metallization front
Contact firing	Contact firing
LID regeneration	LID Regeneration on both sides
Testing and sorting	Testing and sorting

Table 4.6 n-type bifacial cell process in comparison with the n-type PERT cell.

N-type PERT	N-type bifacial (PERT)
Surface damage removal	Surface damage removal
Polishing	Polishing
BBr3 diffusion	BBr3 diffusion
Rear side SiN PECVD	Rear side SiN PECVD
Texturing on front side	Texturing on front side
POCl$_3$ diffusion	POCl$_3$ diffusion
Front side SiN PECVD	Front side SiN PECVD
Laser contact opening rear	Laser contact opening rear
Metallization rear	Full Metallization rear
Metallization front	Metallization front
Contact firing	Contact firing
Edge isolation	Edge isolation
Testing and sorting	Testing and sorting

HJT solar cells, instead of localized metallization, the rear side will be covered with a full metallic grid using Ag paste.

Currently, there is no standard characterization method or standard equipment for testing bifacial solar PV modules. The usual way of reporting bifacial PV is by covering the back side of the panel with a black sheet, measuring the electrical parameters when illuminating the front side, then doing the same for the other side. The ratio of the two efficiencies or the two maximum power values gives a factor of bifaciality. This bifaciality factor is then indicated along with the characteristics of the front side.

For measurement of the efficiency of bifacial cells and modules, the IEC standard is under development (IEC 600904-1-2).

The I_{sc} and P_{max} values are measured on the front side under standard test conditions by covering the rear side with a black sheet. In a similar way, the I_{sc} and P_{max} values of the rear side are measured under standard test conditions by covering the front side with a black sheet.

The bifaciality ratio is determined by

$$\frac{I_{sc} - rear}{I_{sc} - front} \tag{4.30}$$

whereas I_{sc}-rear and I_{sc}-front are I_{sc} values from and rear side of modules at STC, respectively. The compensated I_{sc}, I_{sc}. *comp*, is calculated using the formula.

$$I_{sc}.comp = (1 + 0.2(\ bifacialityratio) \times Isc.front \tag{4.31}$$

This I_{sc}. *comp* is equal to the product of the compensated irradiance and the front side I_{sc}.

$$I_{sc}.comp = G.comp \times Isc.front \tag{4.32}$$

The cell is measured under compensated irradiance (Gcomp) to determine the bifacial efficiency.

Factor 0.2 is the initial estimate for the rear side irradiance relative to the front side.

4.8 Advanced high efficiency cells and the manufacturing process

4.8.1 IBC solar cell

The standard conventional solar cell has an emitter on the front surface and contacts on both sides of the device. Different concepts have been developed to improve the efficiency of the solar cell to meet higher power ratings. One of the concepts is to keep both the contacts on the back side of the solar cell and shift the emitter to the rear side. This type of cell is called an interdigitated back contact (IBC) solar cell, as the contacts are alternately arranged on the rear side with the interdigitated format. The IBC cell was developed at Stanford University in the 1980s and has achieved an efficiency of 21.3% [28]. The IBC solar cell is one of the concepts to realize high efficiency, and it is considered a promising route for large-scale industrial production. Both the emitter and the BSF doping layer with their corresponding metallization grids are located in an interdigitated structure on the back side of the solar cell, as shown in Fig. 4.31.

The features and advantages of the cell structure are as follows.

The emitter of the cell is on its rear side. Both the top and bottom contacts are placed on the rear side of the solar cell. The absence of contact on the front side completely eliminates the optical shading losses on the front surface. So, the IBC solar cell has an increased absorption and provides increased short circuit current density compared to the cells with other configurations.

As there are no metal fingers on the front side, there is no need to consider the front side contact resistance. The absence of a metallic grid on the front side provides more area and potential for the optimization of the front surface passivation layer for better performance.

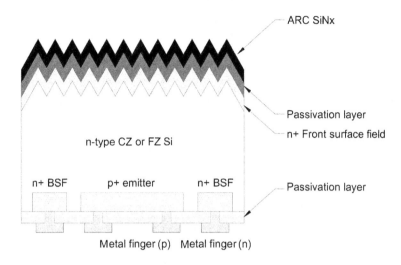

ARC SiNx

Passivation layer

n+ Front surface field

n-type CZ or FZ Si

n+ BSF   p+ emitter   n+ BSF

Passivation layer

Metal finger (p)   Metal finger (n)

FIG. 4.31

Structure of the rear emitter-based IBC solar cell.

On the rear side, to reduce the contact resistance, wider metallic grid fingers could be used. On the front side, to improve light absorption, a pyramid structure is implemented. An antireflection coating is also used to minimize the reflection losses and hence to improve the short circuit current density of the solar cell. The SiO_2 on the front side will act as a passivation layer that reduces the recombination velocity on the front surface. It also acts as a front surface field that repels the minority carriers reaching the front surface.

The p^+ emitter, which is on the rear side, forms a p-n junction with the bulk n-type silicon material. The rear side n^+ layer acts as a BSF to promote carrier collection and its high doping enables better ohmic contact. The rear side SiO_2 layer acts as a passivation layer to reduce the recombination velocity on the rear side. It also acts as a reflector for long wavelengths and improves the long wave response for the enhancement of current density. The rear metal electrode formed by screenprinting realizes the point contact with the silicon substrate by passing through the contact hole in the SiO_2 passivation layer, which can reduce the contact area between the metal electrode and the silicon wafer and thus further suppress the carrier recombination at the contact interface. The design of all rear contacts makes cell interconnection in modules more simple.

Compared with conventional crystalline silicon cells, the surface recombination of the front surface has a greater impact on the performance of the IBC solar cell because the front surface is far away from the p-n junction, which is located on the rear side. In order to suppress the front surface recombination, a better surface passivation scheme is required for the front surface.

The IBC cells are mainly manufactured using n-type silicon substrates and emitters are formed using boron diffusion. The Si wafer should be from the ingot grown by the CZ or FZ technique. It should be of very high quality with a lifetime of $100\,\mu s$ with a long diffusion length, which is often needed to ensure that the photon-generated minority carriers do not recombine before arriving at the back junction. The IBC solar with a rear side emitter manufacturing process is shown in Table 4.7. The process of the flow is from top to bottom.

Table 4.7 Process flow for an IBC cell with a rear emitter.

Saw damage removal by etching of n-type wafer

Cleaning of the wafer for prediffusion wet cleaning sequence

P$^+$ diffusion on both sides

Growth of SiO$_2$ layer

Application of photoresist on rear side

Mask printing on photoresist, UV exposure, curing and development to get pattern

Etching of thermal oxide and etching out of diffused P+ layer

Wafers diffused on both sides by n+layer

Stripping of photoresist

Oxide growth

Photoresist on rear side

Mask printing on photoresist, UV exposure, curing and development to get pattern

Front side texturization

FSF formation (POCl$_3$ +O$_2$)

Drive-in diffusion

Oxide growth on both sides

SiN ARC coating deposition on front side

Photoresist on rear side

Mask printing on photoresist, UV exposure, curing and development to get pattern

Etched out of oxide layer contact window opening

Contact window opening

Contact formation–all silver or Al, TiW, Cu

Photoresist mask removal

Contact firing

Laser isolation

Cell testing and sorting

The saw damage of the n-type silicon wafer cut by a diamond wire has to be removed by chemical etching. Alkaline etchants are used for saw damage removal for the monocrystalline wafers. Next, the wafer is cleaned and prepared for diffusion. The main issue in manufacturing the IBC solar cell is how to prepare the n-regions and p-regions with good quality and interdigitated distribution on the rear surface. There will be a nonuniform diffusion and high temperature damage in the traditional liquid boron diffusion process, and the process of operation is complex and the process can be simplified.

A p+layer with boron doping has to be created on the rear side to form a junction with the n-type silicon. As the emitter is not completely covering the full surface, a masking technique will be used to create locations for selective diffusion. Photolithography is used for the creation of pattern generation. The wafer is blanket diffused on both sides using the p+layer. The p+layer deposition can be done either by diffusion of BBr$_3$ and by deposition of boron-doped SiO$_2$ in the atmospheric pressure chemical vapor deposition (APCVD) technique. Immediate to p+layer formation, the SiO$_2$ layer is grown as a protecting layer for masking for the next processing step.

The photoresist is applied on the rear side of the solar cell and it is dried. Using screenprinting or an optical mask, the photoresist is exposed to UV radiation to get cured and the pattern is developed using a developer such as acetone. The openings are created in the required locations of the photoresist mask. The SiO_2 layer and the p + layer are etched out in the opening. Next, the n + layer is deposited and subsequently the SiO_2 layer is deposited as a protection layer for the next photolithography step. Next, the front surface is textured and n + layer is created for the front surface field. SiO_2 is deposited as passivation layer and SiN as antireflection coating is deposited by PECVD technique. In the next masking step, the emitter and BSF layer locations are opened and contact metallization is deposited. The contacts are fired, edge isolation is done by laser ablation, and the cell is tested for I-V and electrical parameters.

Several problems arise during the operation of multiple lithography. The costs of the process to avoid nonuniform distribution and high-temperature damage are becoming higher. But some problems still exist for the screenprinting process because of the accuracy of alignment and printing repeatability. Doping with ion implantation is widely used in the semiconductor industry. In recent years, ion implantation has been applied to the IBC solar cell because this technology can precisely control the doping concentration to achieve a uniform p-region and n-region with controllable junction depth. Furthermore, this doping technique allows for single side doping, which simplifies the fabrication process. However, it requires a high annealing temperature for drive in and activation of the dopants, which is thus a difficult problem in the PV industry.

As an alternative, laser techniques [29] are used to dope the silicon material. They give a uniform doping concentration and depth controllability as well as pattern ability in the doping area. The selective doping area protects the entire silicon from high-temperature damage.

Considerable efficiency gains can be achieved by using high-quality base material, improving light trapping via surface treatments, optimizing rear surface passivation, and alleviating edge losses. The IBC cell performance significantly depends on the pitch size (the period in which the alternating doping repeats) and the emitter fraction (namely, the ratio of the emitter size to the pitch size). The IBC cells with larger emitter fractions will perform better. However, larger emitter coverage often leads to more series resistance loss. Therefore, a balance between higher short-circuit current and more series resistance loss should be considered to obtain the maximum efficiency. In general, the emitter fraction is between 70% and 80% for the typical IBC cell designs. Furthermore, the smaller the size of the pitch, the shorter the mean transport distance to the base contact of the carriers, thereby reducing resistance and recombination loss. However, this would also result in a higher demand for manufacturing process accuracy.

The SunPower Corporation has been in a leading position in the research and development of IBC solar cells, and has achieved an average efficiency of mass production of 23%. In October 2016, SunPower announced a high conversion efficiency of 25.2% ($V_{oc} = 737\,mV$, $Jsc = 41.33\,mA\,cm^{-2}$, FF $= 82.7\%$) by further reducing the edge loss, series resistance, and emitter recombination based on its X-Series cell structure on a 130 μm thick, n-type CZ silicon substrate with a total cell area of 153.49 cm^2.

Apart from SunPower, different research organizations and solar cell manufacturers such as Sharp, Panasonic, IMEC, Fraunhofer ISE and ISFH, the Australian National University (ANU), Trina Solar, Bosch, Samsung, and Jollywood are developing commercial IBC solar cells. Some companies are combining the IBC concept with silicon heterojunction (SHJ) technology.

As a promising route to keep up with the ongoing trend of increasing commercial efficiency for large-scale industrial production, the IBC solar cell is also becoming more and more attractive for cell manufacturers.

Some companies are using low-cost industrial processing technologies such as tube diffusion and screenprinting while some are using ion implantation for doping silicon. Also, some are using the APCVD technique to deposit n- or p-doped SiO_2 and dopant driving to form emitter and BSF regions.

4.8.2 TOPCon solar cell

The highest solar cell efficiencies are possible with the use of charge carrier-selective contacts and implementing passivation techniques on the front as well as the rear side of the solar cell. Fraunhofer ISE achieved an efficiency of 25.3% for an n-type solar cell with full-area charge carrier-selective backside contact. It uses charge carrier-selective contacts on both sides. On the front side, the contact allows only hole transport, blocking the electrons from reaching the contact. On the rear side, the contact allows only electrons and blocks the flow of holes to reach the rear contact. To passivate the rear contact, the Fraunhofer ISE uses an ultrathin tunnel oxide in combination with a thin silicon layer; this is called tunnel oxide passivated contact (TOPCon). Due to the TOPCon contact, the dark current is reduced and the Fraunhofer ISE was demonstrated on an n-type cell with a V_{oc} of 718 mV, a current density J_s of 42.5 mA/cm^2, and an FF of 82.8% [30].

TOPcon is a newer solar manufacturing concept where a thin oxide layer is introduced to the silicon cell to reduce recombination losses and increase cell efficiency. The structure of a solar cell with TOPCon contact is shown in Fig. 4.32.

The manufacturing process flow of TOPCon in comparison with a p-type PERC cell is shown in Table 4.8.

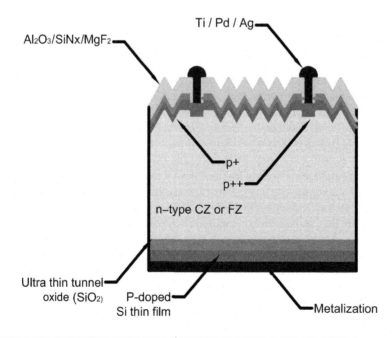

FIG. 4.32

Structure of a TOPCon solar cell.

Table 4.8 Comparison of processes for a p-type PERC and an n-type TOPCon cell.

P-type PERC	N-type TOPCon
Surface damage removal	Surface damage removal
Texturing on front side	Texturing on front side
POCl$_3$ diffusion	BBr$_3$ diffusion
Laser doping for selective emitter	Laser doping for selective emitter
PSG removal, edge isolation+rear polish	Edge isolation+polishing
Annealing	Tunnel oxide+a-Si deposition
Rear side AlOx	Annealing
Rear side SiN PECVD	Cleaning
Front side SiN PECVD	Rear side SiN PECVD
Laser contact opening	Front side SiN PECVD
Metallization rear	Metallization rear
Metallization front	Metallization front
Contact firing	Contact firing
LID regeneration	LID Regeneration on both sides
Testing and sorting	Testing and sorting

Different manufacturers worldwide are working on TOPcon technology. High-quality n-type wafers, selective doping technology, and advanced fine-line printing technology are required to achieve higher efficiency.

4.8.3 HJT solar cells

PERC solar cell technology is dominating the industry due to increased power and efficiency. Next to PERC solar cell technology, heterojunction technology (HJT) has been making big progress, as it has the potential to improve efficiency to satisfy the demand for higher module power ratings. HJT is an age-old technology. The cell was developed by Sanyo and later taken by Panasonic. The junction is formed between the crystalline silicon and amorphous Si materials, which have band gaps of 1.1 and 1.76 eV, respectively. The cell uses a (100) oriented high-quality n-type wafer as the base material with a higher minority carrier lifetime. It is becoming a promising candidate for low LCOE-based PV systems. HJT can be easily adaptable to a bifacial structure with minimum process and it shows the highest bifaciality factor. The manufacturing needs a much lower number of process steps than traditional solar cells. So, there is low-cost production potential for heterojunction technology.

The most important attribute of HJT is that it is a fusion of wafer-based solar cell technology and thin-film PV, taking the best features of each. It has the excellent absorption properties of standard silicon wafer-based cells and the passivation characteristics of amorphous silicon based thin film. Metal contacts formed in most of the approaches are highly recombination-active and cause losses. This can be avoided by electronically separating contacts from the absorber by the insertion of a wider band gap layer. Implementing this change is basically what HJT is all about. The cell architecture results in very high open-circuit voltages without the need for any patterning techniques.

With silicon HJT, the junction is formed between crystalline Si and amorphous silicon materials. The heterojunction is formed between a doped crystalline silicon substrate (n- or p-type, although the former is more widely used) and an amorphous silicon layer of opposite conductivity (p- or n-type, respectively). As the surface of the crystalline silicon wafer has abrupt discontinuity in the crystal structure, it has a high density of dangling bonds, creating a large density of defects. Amorphous silicon passivates the surface exceptionally well by reducing the dangling bond density on the surface. However, the junction of amorphous silicon and crystalline silicon creates interface losses. Thus, Sanyo proposed the insertion of an intrinsic amorphous layer between the doped amorphous silicon and the crystalline silicon surfaces. This has given birth to Sanyo's famous and proprietary heterojunction with intrinsic thin layer, commonly known as HIT. For practical reasons, the HIT structure consists of a crystalline silicon wafer, typically n-type, sandwiched between intrinsic and oppositely doped amorphous silicon layers deposited on both sides, plus a transparent conductive oxide (TCO) on top.

Advantages

It has a low temperature-based simple process for manufacturing. It has a low temperature coefficient of power due to the formation of the heterojunction with a:Si material. It has a lower annual degradation in power. There is no light-induced degradation in power due to the use of an n-type wafer.

The HIT solar cell has attracted a growing amount of attention year by year due to the following advantages:

(a) Compared to a conventional crystalline silicon solar cell, it has symmetrical structure, which reduces the mechanical stress. So, a solar cell of reduced thickness can be fabricated and the cell production cost can be reduced greatly.

(b) As the cell structure is symmetrical in nature, it can be easily converted to a bifacial solar cell structure with the fewest process steps.

(c) All the processes such as deposition of the amorphous Si and TCO layers and the annealing step for metallization are done at a temperature below 200°C, so a lot of energy can be saved.

(d) This low-temperature process prevents any degradation of the bulk quality of the wafer that happens with high temperature-processed low-quality silicon materials.

(e) The large band bending due to the heterojunction between the amorphous Si and the crystalline Si provides high open circuit voltage.

(f) The excellent surface passivation of the crystalline Si surface by intrinsic amorphous Si results in low dark current, enhancing the open-circuit voltage and conversion efficiency.

(g) As n-type silicon wafers are used as the substrate, the cell will not have light-induced degradation.

(h) As the junction is the heterojunction type, the HJT cell has a much better temperature coefficient compared with conventional diffused cells.

A typical scheme of the n-type Si HJT solar cell is shown in Fig. 4.33. Fig. 4.34 shows the band diagram of the HJT solar cell.

Process flow of HJT

The production of HJT cells requires a higher degree of cleaning than other cell technologies, such as standard BSF or even PERC. The most important step in manufacturing HJT cells is the deposition of amorphous silicon and TCO layers. Plasma-enhanced chemical vapor deposition (PECVD) equipment

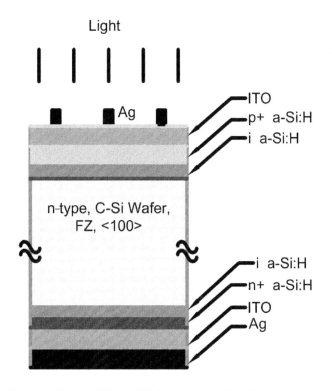

FIG. 4.33

Structure of the HJT solar cell.

is used to deposit doped and intrinsic amorphous silicon layers. Indium tin oxide (ITO) as the transparent conductive oxide (TCO) layer is deposited by sputtering tools.

The HJT cell is manufactured with simple process steps. Wafer texturing and cleaning with chemical treatment are done. RCA or ozone-based and hydrogen peroxide-based cleaning follows.

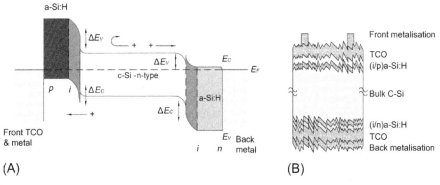

FIG. 4.34

HJT solar cell: (A) band diagram, and (B) solar cell structure.

Amorphous Si with an intrinsic layer and doped a:Si layers are deposited on both sides of the wafer. It requires a special type of silver paste and curing for metallization. A transparent conductive oxide such as ITO is applied on both sides of the cell. For the deposition of a:Si layers, PECVD technology is used by most of the manufacturers. But ULVAC has developed the CAT- CVD technique, whereas Lead-micro is offering PE ALD for core layer deposition. For TCO deposition, a PVD-based sputtering technique is being used. The solar cell manufacturing process is a mature technology with significant potential for performance improvements and cost reductions. Rapid thermal processing is used for annealing to achieve high efficiency. The screenprinting technique is used for the application of metallization, and some manufacturers may use additional plating for the contacts.

As with standard cell processing, the incoming silicon wafers are etched to remove saw damage, followed by texturing. The textured wafers are first deposited with intrinsic amorphous silicon layers on both sides, then deposited with doped amorphous silicon films of opposite polarities, also on the two sides. The deposition of amorphous silicon layers is typically accomplished using PECVD tools. Using stacks of intrinsic plus-doped amorphous layers to form both the emitter and the back surface field (BSF) not only results in a higher V_{oc} and efficiency, but it also opens the route to utilizing thin wafers with no risk of bowing associated with aluminum BSF.

The next step is to apply TCO films on both sides. TCO acts as a conductive electrode to the semiconductor to extract and laterally conduct the electrical current. TCO acts as an antireflective coating to minimize the reflection losses of light. TCO is applied mainly via sputtering using physical vapor deposition (PVD) tools. The majority of today's HJT cells are made bifacial by applying the TCO on both sides. Full-area metal film can also be applied on the rear side, which makes the cell bifacial. Finally, a metallic grid is deposited on top of the TCO to produce an HJT cell. These are the only steps needed to commercialize the technology successfully.

The n-type CZ Si substrate is first cleaned and textured for double sides. Afterward, a 5–10 nm intrinsic amorphous Si layer and a 5–10 nm p-type amorphous Si layer are deposited successively on the front side by the PECVD method to form the p-n junction. Silane, diborane, and phosphine gases are used for the deposition of intrinsic and doped amorphous layers. On the rear side, the BSF structure is composed of symmetrical stacking layers, namely a 5–10 nm intrinsic amorphous Si layer and a 5–10 nm n-type amorphous Si layer. Finally, TCO layers and metal electrodes are synthesized by sputtering and screenprinting methods, respectively, on both surfaces. All processes (including the metallization process) are performed at temperatures below 200°C. The HJT process flow is shown in Table 4.9.

4.9 Equipment used for advanced solar cell manufacturing

- Wafer inspection equipment–robotic based
- Alkaline texturing equipment with different chemical baths
- P-doper thermal diffusion chamber with quartz tube system
- B- doper boron diffusion equipment
- APCVD equipment to deposit P -or B-doped SiO_2 in the bulk Si to provide a source for diffusion
- Laser for selective doping of APCVD-grown doping source
- Ion implantation equipment for P or B doping
- Equipment with chemical baths for etching phosphosilicate glass and edge isolation

---

Table 4.9 HJT solar cell manufacturing process.

Wafer inspection

Saw damage removal of wafer

Texturing on front side of the wafer

Wafer cleaning

a-Si—intrinsic layer deposition front side by PECVD

a-Si—P-doped layer deposition front side by PECVD

a-Si—intrinsic layer deposition rear side by PECVD

a-Si—n- doped layer deposition rear side by PECVD

TCO front side deposition by sputtering

TCO rear side deposition by sputtering

Front contact printing

Rear contact printing

Firing/curing and edge isolation

Testing and sorting

- PECVD equipment to deposit the SiNx layer on the front
- Atomic layer deposition chamber to deposit the AlOx layer on the rear side
- Or PECVD equipment to deposit the AlOx layer on the rear side
- PECVD equipment to deposit the SiNx layer as the cap layer on the rear side
- Laser ablation equipment to create openings in the dielectric stack
- PECVD equipment to deposit intrinsic and doped amorphous Si layers for the HJT cell
- Sputtering equipment to deposit indium tin oxide coatings for the HJT cell
- RCA cleaning equipment for the wafers
- Screenprinting equipment
- Photolithography equipment for the IBC cell
- Firing furnace for contact sintering
- Rapid thermal annealing furnace for contact annealing
- Solar cell tester and sorter integrated with PL or EL equipment

---

4.10 Manufacturers of silicon solar cells and technology trends

Longi solar, JA Solar, Tongwei, Trina Solar, Jinko Solar, Hanwha Q-cell, Suntech, Canadian Solar, Aiko, and many Chinese companies are working in the area of solar cells. Kaneka, LG, Panasonic, SunPower, Sunpreme, and Hanergy are also working in the area of HJT solar cells.

There are tremondous efforts going on to increase the efficiency and power of solar cells. Different advancements have taken place in solar cell technology and in the manufacturing processes. Mono/multi based Al-BSF based technology ruled the industry for more than a decade. The solar manufacturers have switched to PERC-based technology. Very few companies are working with poly-PERC technology. The adoption of diamond wire cutting in wafer manufacturing paved the way for the dominance of the p-type monocrystalline technology. In 2019, mono-PERC dominated the industry. The n-type monowafer-based technologies such as IBC, HJT, and PERT are rising and contributed to a market share of 10%.

Some companies are working with IBC cells with emitters on the front side.

Some companies are working at incorporating contacts for the HJT cell on the back side.

As part of the efforts to develop passivation technology and charge carrier selective contacts, TOPCon-based cell technology is also coming up.

The efficiencies and electrical parameters for manufactured cells of different technologies are given in Table 4.10.

Bifacial solar cell technology is also rising. As the PERC, HJT, and PERT technologies can be upgraded easily to bifacial cells, the manufacturers are upgrading their lines for bifacial cells. Fig. 4.35 shows the structure of a bifacial solar cell.

The size of the solar is increased from 156.75×156.75 mm to 210×210 mm to increase the power output of the module.

To reduce the series resistance of the solar cell and reduce the consumption of silver in metallization, the number of bus bars has changed from 3 to 5, 9, and 12 numbers.

Figs. 4.36–4.38 show the solar cells with different numbers of bus bars.

Table 4.10 Efficiencies and electrical parameters of different types of cells.

Cell type	Efficiency	P_{mp} (W)	V_{oc} (mV)	I_{sc} (A)	FF (%)	Remarks
BSF- p-type Poly	19%	4.67	645	9.02	80.58%	156.75×156.75 mm, five busbars
BSF—p-type Mono	20.4%	4.98	645	9.54	80.93%	156.75×156.75 mm with psuedo square, five busbars
P-type PERC	21.5%	5.28	668	9.643	82.22%	156.75×156.75 mm, five busbars
P-type Mono PERC	22.5%	5.5	686	9.84	81.51%	156.75×156.75 mm, five busbars
N-type Mono PERT	22.5%	5.49	673	9.821	83.06%	156.75×156.75 mm, five busbars
N-type Mono HJT	23.5%	5.74	743	9.6	80.53%	
N-type IBC	23.5%	5.75	696	10.185	81.05%	
TOPCon	23.5%		718	42.5 mA/cm^2	82.8%	
N-type PERT	23%	5.75	700	10.1	81.28%	158.75×158.75 mm

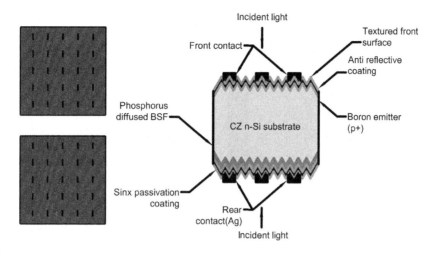

FIG. 4.35

Five Busbar-based N-type PERT-based bifacial solar cells and their structure.

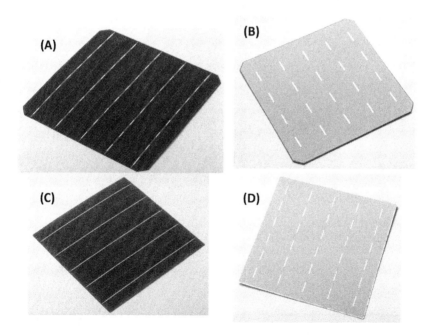

FIG. 4.36

Five Busbar-based p-type solar cells with front and rear sides. (A) Monocrystalline Si solar cell—front side; (B) monocrystalline Si solar cell—back side; (C) multicrystalline Si solar cell—front side; and (D) multicrystalline Si solar cell—back side.

Courtesy: Choudary, Risen Energy.

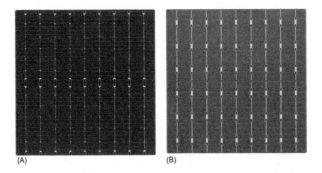

FIG. 4.37

Nine Busbar-based P-type PERC bifacial solar cells of size 158.75 × 158.75 mm full square. (A) Front side and (B) rear side.

Courtesy: Choudary, Risen Energy.

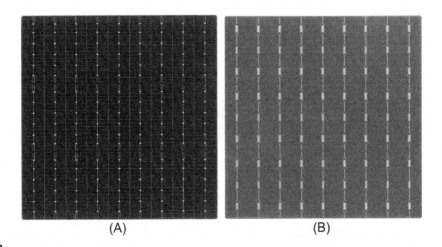

FIG. 4.38

Nine busbar-based p-type PERC bifacial solar cell of size 210 × 210 mm full square: (A) front side and (B) rear side.

Courtesy: Choudary, Risen Energy.

References

[1] E. Bellini, World now has 583.5 GW of operational PV. *PV Magazine*, April 6, 2020, https://www.pv-magazine.com/2020/04/06/world-now-has-583-5-gw-of-operational-pv/

[2] ASTM G159-98 Standard Tables for References Solar Spectral Irradiance at Air Mass 1.5: Direct Normal and Hemispherical for a 37° Tilted Surface.

[3] C. Battaglia, A. Cuevas, S. De Wolf, High-efficiency crystalline silicon solar cells: Status and perspectives, Energy Environ. Sci. 9 (2016) 1550–1576.

[4] F.A. Kröger, G. Diemer, H.A. Klasens, Nature of an ohmic metal-semiconductor contact, Phys. Rev. 103 (1956) 279.

[5] Air Products and Chemical, Inc, Process Guidelines for Using Phosphorous Oxychloride as an N-Type Silicon Dopant; 2016, (2016).

[6] J. Zhao, M.A. Green, Optimized antireflection coatings for high-efficiency silicon solar cells, IEEE Trans. Electron Devices 38 (1991) 1925.

[7] Y. Liu, O.J. Guy, J. Patel, H. Ashraf, N. Knight, Refractive index graded anti-reflection coating for solar cells based on low cost reclaimed silicon, Microelectron. Eng. 110 (2013) 418–421.

[8] M.A. Green, Solar Cells, Operating Principles Technology and System Applications, The University of New South Wales, Kensington, 1986.

[9] R. Sharma, A. Gupta, A. Virdi, Effect of Single and Double Layer Antireflection Coating to Enhance Photovoltaic Efficiency of Silicon Solar, J. Nano- Electron. Phys. 9 (2) (2017) 0200 (4 pp).

[10] O. Gabriel, S. Kirner, M. Klick, B. Stannowski, R. Schlatmann, Plasma monitoring and PECVD process control in thin film silicon-based solar cell manufacturing, EPJ Photovolt. 5 (2014) 55202.

[11] H. Yang, W. He, M. Wang, Investigation of the Relationship between Reverse Current of Crystalline Silicon Solar Cells and Conduction of Bypass Diode, Int. J. Photoenergy (2012) Article ID 357218, 5 pages.

[12] International Technology Roadmap for Photovoltaics (ITRPV), eleventh ed., April 2020.

[13] L.M. Frass, Chapter-1, History of Solar Cell Development, in: Low Cost Solar Electric Power, Springer Publications, 2014.

[14] H.Y. Tada, J.R. Carter Jr., B.E. Anuspaugh, R.S. Downing, Solar Cell Radiation Hand Book, NASA CR-169682, JPL Publication, 1982 82-69 N83-14687, November 1982.

[15] H.J. Hovel, Solar cells, in: R.K. Willardson, A. Beer (Eds.), Semiconductors and Semi Metals, vol. 11, Academic Press, 1975.

[16] A.u. Rehman, S.H. Lee, Advancements in n-Type Base Crystalline Silicon Solar Cells and Their Emergence in the Photovoltaic Industry, Sci. World J. (2013) Article ID 470347, 13 pages.

[17] D. Macdonald, L.J. Geerligs, Recombination activity of interstitial iron and other transition metal point defects in p and n-type crystalline silicon, Appl. Phys. Lett. 85 (18) (2004) 4061–4063.

[18] J. Benick, TOPCon – Overcoming Fundamental Bottlenecks to a New World-Record Silicon Solar Cell, https://www.ise.fraunhofer.de/en/research-projects/topcon.html.

[19] A.W. Blakers, A. Wang, A.M. Milne, J. Zhao, M.A. Green, 22.8% efficient silicon solar cell, Appl. Phys. Lett. 55 (13) (1989) 1363–1365.

[20] T. Dullweber, R. Hesse, V. Bhosle, C. Dubé, Ion-implanted PERC solar cells with $Al_2O_3/SiNx$ rear passivation, Energy Procedia 38 (2013) 430–435.

[21] J. Schmidt, A. Merkle, R. Brendel, B. Hoex, M.V. de Sanden, W.M.M. Kessels, Surface passivation of high-efficiency silicon solar cells by atomic-layerdeposited Al_2O_3, Prog. Photovolt: Res. Appl. 16 (6) (2008) 461–466.

[22] S.K. Chunduri, M. Schemela, PERC+: How to improve high efficiency crystalline solar cells, PERC Solar cell Technology Taiyang News Edition (2018).

[23] A White paper by Solar world "How to maximize energy yield with bifacial technology" https://solarkingmi.com/assets/How-to-Maximize-Energy-Yield-with-Bifacial-Solar-Technology-SW9001US.pdf

[24] A. Cuevas, The early history of bifacial solar cells, in: 20th European Photovoltaic Solar Energy Conference (EU PVSEC) Proceedings, 2005.

[25] C. Duran, Bifacial Solar Cells: High Efficiency Design, Characterization, Modules and Applications, Ph.D thesis, University of Konstanz, 2012.

[26] M. Chiodetti, Bifacial PV Plants: Performance Model Development and Optimization of Their Configuration (Thesis), KTH Royal Institute of Technology, 2015, https://doi.org/10.13140/RG.2.2.33824.33280.

[27] J. Lopez-Garcia, A. Casado, et al., Sol. Energy 177 (2019) 471–482.

[28] Y. Lee, C. Park, N. Balaji, Y.J. Lee, V.A. Dao, High-efficiency silicon solar cells: A review, Isr. J. Chem. 55 (10) (2015) 1050–1063.

[29] G. Masmitja, P. Ortega, I. Martín, G. López, C. Voz, R. Alcubilla, IBC c-Si (n) solar cells based on laser doping processing for selective emitter and base contact formation, Energy Procedia 92 (2016) 956–961.

[30] S. Chowdhury, M. Kumar, S. Dutta, J. Park, J. Kim, S. Kim, J. Minkyu, Y. Kim, Y. Cho, E.-C. Cho, J. Yi, High-efficiency crystalline silicon solar cells: A review, New Renew. Energy 15 (3) (2019).

Manufacturing of crystalline silicon solar PV modules

5

5.1 Introduction

The demand for solar photovoltaic (PV) modules is continuously increasing globally. A record of 104 GWp was added during 2018, and as per forecast, the total global installed capacity would exceed 1270 GWp by the end of 2022 [1].

As per the world annual solar PV market scenarios, the total installed capacities reached GW in 2019, putting the industry on track to reach around 125 GW per annum by 2020 [2]. Solar photovoltaic electricity has already reached grid parity in many countries. Crystalline Si solar cell has been the workhorse since the inception of the solar PV industry. Silicon has the advantage in use as a photovoltaic material due to its stability in high-temperature processing, nontoxicity, high abundance, and matured technology compared to other solar photovoltaic technologies.

A single solar cell as a component cannot generate the required output power. So, to provide the required output power for a solar PV system, a number of solar cells must be connected in series and parallel combinations to get the required voltage and current.

The solar module can be designed to provide output power from a level of a few mW to 600 W for different solar applications. The solar PV module constitutes a main component of a solar photovoltaic power system. Fig. 5.1 is the assembled view of a glass—back sheet-based crystalline silicon solar module, and the major components are shown.

The solar cells are connected in series and a circuit is formed. The solar cell circuit is sandwiched between a glass and back sheet, which is a polymer using an encapsulant such as a thin sheet of ethyl vinyl acetate (EVA), and the whole assembly is laminated. An aluminum frame is fixed on all four sides of the laminate and a junction box with cables and connectors is fixed on the rear side of the module.

Design of a solar PV module

The solar module has to work for 25–30 years and be able to withstand harsh weather conditions such as high and low temperatures, high wind storms, rains, dusty environments, etc. The design of a solar module depends on the selection of the materials. It should provide maximum energy conversion efficiency and operate reliably. Cell to module conversion losses are to be minimized. All materials should withstand −40°C to 85°C. The solar module should withstand a high voltage bias test and a salt spray test. The solar module should pass design and type approval tests as per IEC 61215:2016

Solar PV Power. https://doi.org/10.1016/B978-0-12-817626-9.00005-8

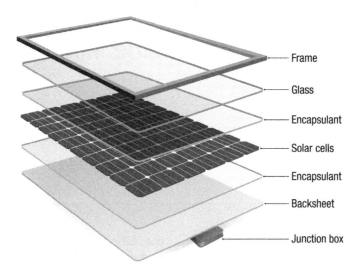

FIG. 5.1

Assembly view of a solar PV module and its components.

and construction and safety tests as per IEC 61730:2016. These standards specify the environmental, mechanical, electrical, and safety tests to ascertain the performance and withstandability of a solar module for hostile weather conditions.

The use of high-efficiency solar cells gives a module of higher wattage. The solar cell should have good mechanical hardness, a suitable refractive index, and a denser antireflection coating to mitigate the potential induced degradation (PID) problem in the cell level. The solar cell should have lower series resistance and higher shunt resistance. The cell should have better optical coupling with EVA and glass to maximize the harvest of incident energy. The cell should have low light-induced degradation, better spectral response, and low temperature coefficients of power and voltage for better performance in the field.

The glass should be tempered to provide hardness, have a low iron content with a reliable antireflection coating, and suitable thickness with high transmittance. It should have suitable mechanical properties to protect the module from the harsh environment and withstand static and dynamic loads.

An encapsulant such as EVA should have high transparency, minimum shrinkage, a higher water moisture barrier, and good adhesion while providing a cushioning effect to the embedded materials with low temperature processing properties. High volume resistivity EVA is preferred on the glass side to protect the module from PID problems. It should have weather and UV resistance and work in high temperature and high humidity environments. It should have better optical coupling properties with glass and solar cells to transmit all the light energy to the solar cell. The EVA formulation should be such that it should not cause delamination, discoloration, corrosion, or snail trails in the module during its life of operation.

A UV resistant single/dual or trilayer back sheet should provide high electrical insulating properties, a higher water moisture barrier, and light reflection from the cell side. It should protect all the solar module components and should be waterproof, airproof and electricproof.

To get the proper voltage and current and establish the flow of electric current externally, an interconnector is used to join the solar cells. The interconnector or solar PV ribbon should have a better thermal expansion coefficient matched to the Si substrate and offer lower ohmic loss. Also, it should not cause cracks to solar cells during the process of solar cell interconnection. Electrically conductive, solder-coated copper strips that are ultrasoft and with a lower thickness and width as per the requirements of the solar cell busbar are to be used as solar PV ribbons. The PV ribbon should have thermal fatigue withstanding characteristics. To join the substrings at the U-turns of the solar cell string, a thicker and wider solder-coated copper strip called a bussing interconnector is used. The bussing interconnector should be suitably thick and wide enough to carry the current of the solar cell.

The frame should be lightweight with good mechanical and thermal properties. It should withstand static and dynamic loads of the installed environment and should work in a corrosive atmosphere also. Aluminum oxide anodized Al 6063 alloy extrusions with different profiles are being used as frames for the solar module. The frame will have mounting holes to fix the module firmly to a structure. It will have grounding holes to provide grounding to the solar module.

The junction box houses the electrical connectors coming out of the solar cell strings and bypass diodes while having cables and connectors with IP65/IP67 ingress. It should have good mechanical, insulating, and UV resistance as well as better thermal energy handling capabilities. The Schottky diode is being used as the bypass diode due to its low forward bias voltage and fast switching characteristics. The junction box will have suitably rated cables connected with MC4 connectors.

An adhesive or adhesive tape is required to fix the frame to the laminate and mount the junction box on the rear side of the module. The adhesive is also required to fill the junction box to seal against moisture ingress. The adhesion strength, the moisture barrier properties, and the curing process decide the selection of material.

The front and rear views of the solar cell module layout for 72 cells are shown in Figs. 5.2 and 5.3.

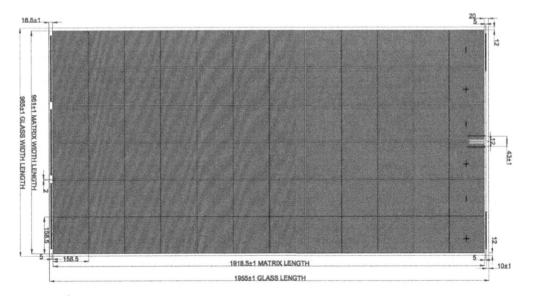

FIG. 5.2

Circuit drawing of 72 cells of a solar cell-based module.

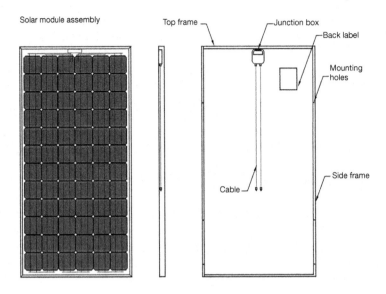

FIG. 5.3

Front and rear side drawing of the 72-cell solar module.

The cell-to-cell gap has to be 1.5–2 mm to provide a gap for the contraction and expansion of the solar cell interconnector due to extreme temperature variations.

The series of cells forming a column is called a string. The gap from string to string can be maintained as 2–3 mm. This is to provide a sufficient gap for the alignment of strings in the layup. The distances from the cell edge to the interconnector busbar and the busbar to the next busbar are normally maintained at 5 mm for ease of production.

The gap between the current carrying cell and the bussing interconnector edge to the glass edge is to be 8 mm for a 1000 V system voltage and 11 mm for 1500 V as per the creepage and clearance requirements of the latest IEC 61730-1 standard. The creepage and clearance distances are decided based on the used polymer material group and its pollution degree [3]. If Comparative Thermal Index which is the measure of electrical breakdown property of encapsulant and back sheet is between 400 and 600, it is classified as Material group-II. For Material group-II and pollution degree of 2, the creepage distances are 7.1 and 10.4 mm, respectively, for system voltages of 1000 and 1500 V. These gaps will decide whether the module is suitable for a system voltage of 1000 or 1500 V. A sufficient gap has to be provided from the solar cell circuit to the frame to facilitate the easy accumulation of dust at the edge.

The interconnectors coming out of the solar cell strings are shown in Fig. 5.2. Four ribbons come out on the rear side through the cutting hole of the EVA and back sheet and are connected to three bypass diodes housed in the junction box.

The bypass diodes are connected across 24 cells, and hence a 72-cell module has three bypass diodes. For redundancy, some junction boxes have three more diodes in parallel to the existing three, making a total of six diodes.

There are three strings each of 24 cells in a 72-cell solar module. For solar modules of 96, 60, 54, 48, and 36 cells, the bypass diodes are connected across 24, 20, 18, 16, and 18 cells, respectively. Nowadays, with the usage of half-cut cells, a bypass diode is connected across 26 cells.

Bypass diodes

When a solar cell in a module is generating lower current due to partial shadow or a local soiling effect, the other cells that are in series and generating higher current force the shadowed cell to operate at a reverse bias. The shadowed cell starts dissipating the power instead of generating, a phenomenon called the hot spot problem.

Due to the inherent structure, each solar cell can dissipate a certain amount of power and the power that should not be exceeded is called P_c. The power dissipation depends on the shunt resistance and the reverse current voltage characteristics of the solar cell as well as its material structure and defects, area, and maximum operating temperature. When its reverse dissipation exceeds the critical dissipation power, a shaded cell shall be damaged due to the hot spot effect. The bypass diode is blocked when all cells are illuminated, and conducts when one or several cells are shadowed.

The maximum number of solar cells across which a bypass diode is connected is decided by the breakdown voltage of the solar cell (V_c). The breakdown voltage of a solar cells depends on its resistivity, the doping concentration of the base, and the defect concentration [4]. The literature gives a range for the polycrystalline silicon cells from 12 to 20 V. For monocrystalline silicon cells, the breakdown voltage extends up to 30 V [5].

For the efficient operation of a solar module, there are two conditions to fulfill, as shown in Fig. 5.4.

A bypass diode has to conduct when one or more cells is shadowed. The solar cell voltage, V_s, when a cell is shadowed shall have to be within the breakdown voltage (V_c). It is specified by the respective solar cell manufacturer. The maximum number of solar cells (n) to bridge is calculated using both these conditions and with the following formulae [5]

$$V_{bypass} = V_s - (n-1)*V_{oc};$$

where V_{bypass} is the forward voltage of the cell during the bypass condition and V_{oc} is the open circuit of the solar cell.

With $V_s < V_c$; the V_{oc} of the cell under illumination is 0.64 V for a normal polycrystalline solar cell; $V_{bypass} = V_F$

$$V_F < [V_c - (n_{max} - 1)*0.64]$$

$$n_{max} < \frac{V_c - V_F}{0.64} + 1$$

For a bypass diode of 0.5 V forward voltage and a cell breakdown voltage of 16 V, the maximum number of solar cells across which the bypass diode is to be connected is 24.

If a 60 solar cell module is considered, the diode will be connected across 18 solar cells and the cell will not go to reverse breakdown.

Cell to module loss (CTM)

The P_{max} of a 156.75 mm × 156.75 mm multicrystalline solar cell of efficiency 18.6% is 4.57 W. The total wattage of 72 cells in air is 329 W. But, when 72 solar cells of the above type are used for making a 72-cell solar module, the solar module output will be less than 329 W and it will be around 325 W. The

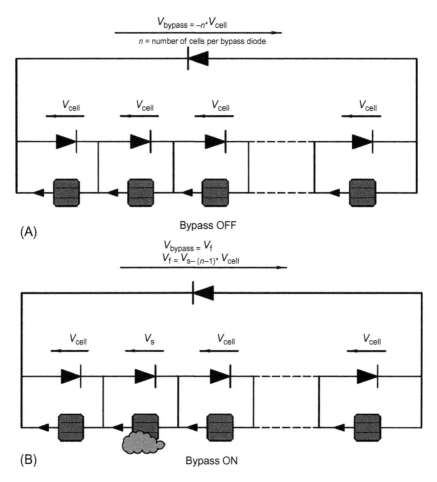

FIG. 5.4

Hot spot phenomenon due to shadow effect.

difference in the module power and the 72 cells' power in air is represented in percentage and is called the cell to module (CTM) loss. The module design should be aimed at achieving a minimum CTM loss. In the present example, the CTM loss is 1%.

The Fraunhofer institute has developed SmartCalc CTM software [6] to calculate the cell-to-module efficiency ratio for different types of solar cells used for modules.

The solar cells are tested in the air, connected with a PV ribbon, and sandwiched between a glass and polymer sheet using an encapsulant to form a solar module. In the solar module, the solar cells operate in the environment of EVA and glass.

When light is incident on the solar module, there are various loss mechanisms that reduce the amount of light reaching the cell. Some portions of light are reflected from the glass, the glass/EVA interface, and the interconnector. Some portions of light are absorbed in the glass, in the

EVA, and in the solar cell. Some portions of light are reflected between the gaps of the cells, some portions reach the back sheet, and some portions are transmitted out of the module. The cell-to-module power factor (k) represents the ratio of the module power to the initial power of the solar cells.

The light energy losses are influenced by geometrical effects, such as the inactive areas at the module edge and the gaps between the solar cells; optical effects, such as reflection and absorption within the front cover; and electrical effects, such as ohmic losses within the cell and string interconnections.

However, not only losses occur, the additional reflection of light onto the solar cell or reduced reflection from the encapsulated cell also results in gains, as depicted in Figs. 5.5 and 5.6. There are several optical effects interacting with each other between the layers of the module.

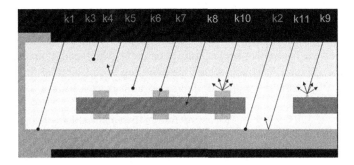

FIG. 5.5

Geometrical and optical phenomenon of incident energy in a solar cell module.

Courtesy: Fraunhofer Institute.

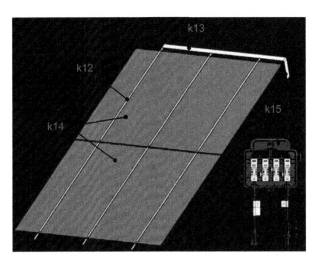

FIG. 5.6

Electrical effects/ohmic losses of incident energy in a solar cell module.

Courtesy: Fraunhofer Institute.

The effective reflectivity of the air-glass interface results in optical losses. It relies upon the refractive index of the material, the surface structure, and the antireflective coating. The bulk absorptions of the ethylene vinyl acetate and glass can generate optical losses. The optical coupling gains may be separated into direct and indirect gains.

A direct coupling gain may arise because the cell was previously characterized in an air environment with a refractive index of 1 while it is optically coupled to an encapsulant with a refractive index around 1.5 in a common module built-up.

The coupling helps to reduce the index gap at the top interface of an AR-coating with a 2.1 refractive index, hence reducing the reflectance. The refractive index of the encapsulant influences the reflection losses at the glass-encapsulant interface and at the silicon-antireflective coating (ARC)-encapsulant interface. The coupling also reduces the entire stack (encapsulant/AR-coating/wafer) reflectance. The direct coupling gain is significantly dependent on the cell's AR coating design and the cell surface's light-trapping texture.

The indirect coupling demonstrates the effect that light gets reflected on the active solar cell, the finger, the ribbons, and the back sheet between the solar cells, which could be redirected back to the active solar cell at the glass/air interface, which happens due to full internal reflection. For a nonembedded solar cell, the reflected light at the solar cell front can not be recoupled and is subsequently lost.

Fig. 5.7 shows a waterfall chart with the loss or gain in the solar module efficiency with the effects of incident energy, and the contribution of loss/gain for each phenomenon is given.

The optical gain due to coupling the EVA, solar cell, and glass refractive indices is more for acidic-etched multicrystalline solar cells. It has been reported by Haneef et al. [7] that the monocrystalline solar cells that undergo alkaline etching show higher CTM ratios due to the optical coupling gain. Even solar modules made with half-cut solar cells also benefit in the efficiency due to optical coupling of the energy. If a suitable EVA and reflective-type back sheet are used, the CTM loss can become zero, or sometimes it can become positive for multicrystalline Si solar modules.

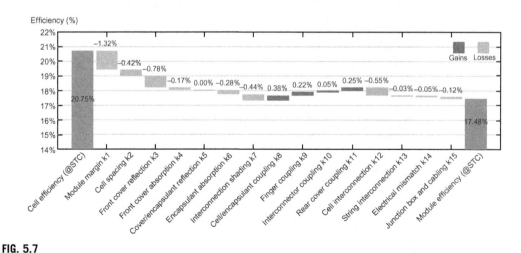

FIG. 5.7

Waterfall chart showing the loss or gain in the solar module efficiency with the effects of incident energy.

Courtesy: Fraunhofer Institute.

5.2 Specifications/requirements of solar PV modules

– The solar module should withstand harsh environments of hot, cold, rain, wind, and dust for 25–30 years.
– Solar modules have to be tested for design and type approval tests as per the latest IEC 61215 standard or local standards of countries.
– Solar modules should meet construction of materials requirements as per IEC 61730-1 and should pass safety tests as per the latest IEC 61730-2 standard.
– Solar modules should pass the salt mist test as per IEC 61701 for use in a marine atmosphere.
– Solar modules should pass the potential induced degradation test as per IEC 62804, for system voltage of 1000 and 1500 V.
– Solar modules should pass the sand, dust, and ammonia tests as per the standard.
– Solar modules must withstand wind loads of 2400 Pa and snow loads of 5400 Pa.
– Solar energy conversion efficiency at standard test conditions is one of the key metrics of the solar module. It should have higher efficiency. A higher power rating will depend on the size, number, and efficiency of the solar cells used in the module.
– The solar module should have zero cell to module (CTM) loss and its CTM should be positive.
– The solar module should have low light-induced degradation.
– The solar module should have minimum annual degradation in power.
– Solar modules should have low NOCT/NOMT and low temperature coefficients for power and voltage.

The data sheets contain the electrical and mechanical data of the module.

Dimensions

The module length, width, and height with mounting hole size and their location with a drawing are provided in the data sheet. This size can be used for layout making, wind pressure calculations, and weight loading on the structure. The module area is also useful to calculate the energy conversion efficiency of the solar module. The frame thickness determines what rack components are to be used for mounting. The vertical and horizontal mounting holes help in designing the module mounting structure.

Weight

The weight is required for structural design engineers to design the module mounting structure and for wind load studies.

Cells

Solar cells will be either monocrystalline, polycrystalline, SunPower's interdigited back contact (IBC), or Sanyo's HIT type. The cells are available with different technologies such as BSF, PERC, PERT, HJT, and Bifacial. The cell decides the electrical characteristics, efficiencies, and appearance of the module. Solar modules can have variable numbers of cells such as 36, 48, 54, 60, 72, and 96 or any number of cells in series, depending on the wattage and voltage requirements.

Cell dimensions

The solar cell dimensions are available in full square and pseudosquare sizes of 125×125, 156×156, 156.75×156.75, 158.75×158.75, 161×161, and 210×210 mm. The size of the cell will determine the current output of the cell, with larger cells producing higher current and higher wattage.

Glass

The glass is used as a superstrate and the type of glass and its thickness shall be specified in the data sheet. In general, crystalline solar PV modules use low-iron, high-transparency tempered glass with an antireflection coating.

At present, the commonly used thickness of the glass is 3.2 mm. For glass-glass or bifacial modules, the front glass thickness will be 2–2.5 mm. Some customers may ask for 4 mm thick glass for a 72-cell module.

Back sheet

Glass/back sheet-based solar modules have a plastic backing material that seals the cells against direct exposure to environmental conditions. The type of back sheet tells the quality and long-term performance of the module. For glass-glass and bifacial modules, glass is used as a back cover. Transparent back sheet is also used as back cover for bifacial modules.

Encapsulation

A glue laminate, such as ethylene vinyl acetate, is used to seal and protect the solar cells and interconnectors within the module and the back sheet. In the long-term reliability of the solar PV module, the composition and quality of the encapsulation material shall play a very crucial role.

Frame

Crystalline Si modules have anodized aluminum frames, with aluminum oxide anodization with silver or black colors. Some crystalline modules such as glass/glass and bifacial are frameless. The frame has a profile to accommodate the laminate. The load-bearing capacity of the frame depends on its wall thickness, height, width, and mechanical properties.

Cable and connectors

The type of connector used in the module is important. Factory-attached solar module cables with cable size, type of insulation, and cable length are generally listed in the data sheet. The diameter of the cable generally is 4 mm^2 and the length is 1200 mm and the material is copper with XLPE insulation. Pin and socket type lockable type MC4 connectors with IP67/68 ingress are connected to the cables.

Junction box

A junction box is factory-installed housing on the back of the module for the connections. Many are sealed and inaccessible to the end user. If it is specified as field-serviceable, the junction box can be opened, and leads and bypass diodes can be installed or replaced. Also mentioned is the degree of ingress protection such as IP 67, etc.

Bypass diodes

Partial shading of the solar PV module can have a high impact on its performance. In the event of a shaded solar PV module, the current that is maintained through the solar module is likely to change direction, which causes hot spots. This eventually will lead to damage of the solar cell, the internal connections, and the back sheet of the solar PV module. The bypass diode prevents the reverse flow and also directs current around the solar module's shaded area. Almost all solar PV modules come with factory-fitted bypass diodes. A standard 72-cell solar PV module will have three bypass diodes with all the solar cells in sequence, each protecting a series of 24 cells that can be bypassed in the event of partial or complete shading of any of the 24 solar cells.

Depending on where they are located on the module and the type of junction box, diodes may be field-accessible. Regardless of the benefit of diodes, shading should be avoided whenever possible.

Electrical data

I-V curve

The solar PV module manufacturer conducts tests of the solar PV module under standard testing conditions (STC), which are $1000 \, \text{W/m}^2$ of solar irradiance, 25°C cell operating temperature, and 1.5 air mass spectrum. The current versus voltage or *I-V* curve has five main parameters (V_{mp}, V_{oc}, I_{mp}, I_{sc}, and P_{max}) used for solar PV system design, quality checks, and other purposes. The *I-V* curves for any operating temperature and the solar irradiance can be determined using standard procedures and following relevant standards.

Open circuit voltage (V_{oc})

It is the maximum voltage available under open circuit conditions of the solar PV module (+ve and −ve in open condition with no connections) while it is generating zero current, called V_{oc}. It is an important parameter that signifies the maximum voltage that a solar PV module can generate under STC and is generally used to calculate the maximum number of modules in a series string. If the solar PV module temperature increases, the V_{oc} decreases.

Short circuit current (I_{sc})

It is the maximum current that the solar PV module can generate and is designated as I_{sc}. In the event of a solar PV module getting short-circuited, there is zero voltage. I_{sc} helps to size the cable and also design the overcurrent protection while performing system design. I_{sc} is dependent upon solar irradiance, and the solar cell area.

Maximum power point (P_{max})

The maximum power point (P_{max}) is the highest possible power that can be generated by a solar PV module and is a multiplier function of V_{mp} times I_{mp}. In a typical *I-V* curve of a solar PV module, it is located at the point where it corresponds to the maximum voltage and current.

Voltage at maximum power point (V_{mp})

The V_{mp} is the voltage point on the *I-V* curve when the power output of the solar PV module is the maximum. V_{mp} is used to calculate the number of modules in series and is dependent on module temperature and the type of PV ribbon and the solar cell interconnections used in the module.

Current at maximum power point (I_{mp})

The I_{mp} is the maximum current a module can generate when the power output from the solar PV module is the highest. I_{mp} is dependent on solar irradiance and is used in voltage drop calculations to determine the cable sizes for solar PV systems.

NOCT/NMOT

The nominal operating cell temperature (NOCT) value is generally provided by the solar PV module manufacturer. NOCT is the stabilized temperature of the module when it is mounted at an angle of 37 degrees with the following weather conditions. The ambient temperature is 20°C and sunlight intensity is 800 W/m^2 and the wind speed is of 1 m/s and the module is not connected to any load. When the module is connected to the load, the stabilized temperature is called the nominal module operating temperature (NMOT). The NOCT/NMOT value is used to calculate the temperature of the module for different intensities and different ambient temperature conditions. The NMOT will be 3–4°C lower than the NOCT. If the ambient temperature is 40°C, the intensity of radiation is 1000 W/m^2, then the module temperature can be calculated as follows:

$$\text{Module Temperature} = \text{Ambient Temperature} + \left(\frac{\text{NOCT} - 20}{800} \times 1000 \right) \times 0.9$$

If the NOCT is 45°C, the calculated module temperature is 68.125°C. The 0.9 is a correction applied for nonconnection of the module to the load. There are different methods to calculate the module temperature using NOCT/NMOT value.

PTC rating

PTC is known as PVUSA test conditions. PVUSA means photovoltaics for utility-scale applications and the rating was introduced by the California Energy Commission to know the realistic power of the solar PV module. This is the power output of the solar PV module when the module is exposed to 1000 W/m^2 solar irradiance under an ambient temperature of 20°C, and a wind speed of 1 m/s. The ratio of PTC to STC rating of the solar PV module should be at the higher end of the scale. If the NOCT is 45°C, the temperature at the PTC is 48.125°C. By applying the temperature coefficient, the power at the PTC can be calculated.

Power tolerance

Power tolerance is the range specified by the solar manufacturer that the solar PV module can change from its STC-rated P_{max}, and it is specified by the solar PV manufacturer for providing a power warranty. Solar module manufacturers specify values of $\pm 3\%$, $-0\%/+5$ W. The loss factor for module quality required for PVSYST like energy simulation programs is derived from the power tolerance value. The positive tolerance is preferred.

Module efficiency and cell efficiency

Solar PV module efficiency is defined as the solar PV module power output upon solar power input at STC, solar irradiance is 1000 W/m^2, cell operating temperature is 25°C, and air mass is 1.5. For example, a solar PV module is 19% efficient, which means that the solar module can generate 190 W for a square meter and this is used to estimate the required area to set up the solar power plant at a given location.

Temperature coefficients

The solar PV module power output is directly affected by both solar irradiance and temperature. The voltage of the solar PV modules reduces with temperature increase, and the current increases in the order of 15–25 µA/cm^2 area per degree rise of temperature. The V_{oc} will reduce by 1.7–2.3 millivolts (mV)/degree rise of temperature depending on the used technology of the solar cell. Temperature coefficients are used to determine the power, current, or voltage of a solar PV module that will generate at various temperatures.

Maximum system voltage

Residential PV systems are allowed to operate at 600 and 1000 V in different countries. The rooftop- and ground-mounted systems are designed to operate at 1000 and 1500 V. So, the module has to be designed to meet the requirements of system voltage.

Maximum series fuse rating

This is the maximum current a solar PV module is designed to carry through the cells and conductors without damage. While modules themselves are current-limited, excess current can come from other sources (series strings) in parallel, or from other equipment in the system such as some inverters or charge controllers. A fuse or breaker for a series string must be no larger than the maximum series fuse specification.

Design load

This is the weight that a solar PV module has been tested to hold without damage. In areas with heavy snow loads, modules with a higher design load such as 5400 Pa should be used and may be required by the permitting authority.

Maximum wind speed

This is the maximum wind speed a module can handle without damage, and 2400 Pa is a common rating. 2400 Pa means that the module can withstand an uplift force of wind with a velocity around 200 kmph. A wind speed of 200 kmph which is 55.5 m/s creates a wind pressure of $0.6 \times 55.5 \times 55.5$ is 1858 Pa on a module mounted in horizontal condition.

Certifications and qualifications

To use solar modules for government-approved or private projects, it is mandatory for the modules to be tested as per the UL standard 1703 in the United States and Canada and the IEC 61215 and IEC 61730 certifications for Europe and Asian countries. Modules often list other compliances and qualifications, including the International Standard for Organization (ISO) 9001:2008, which is an international standard for a quality management system. The module should pass PID, salt spray with different severity factors, ammonia tolerance, and sand and dust tests.

Fire safety class

Glass/back sheet-based modules are listed under Fire Safety Class C, which means they are potentially energized electrical equipment and no conductive agents such as water should be used to fight the fire.

Warranty

Solar PV module manufacturers specify the power and manufacturing warranties. The manufacturing warranty is a limited warranty on the solar PV module materials and quality under normal application, installation, use, and service conditions. Certain parts of solar PV modules, including connectors and some junction boxes, have only short warranties from their manufacturer, and this is reflected in the overall workmanship warranties of 1–12 years. Manufacturers may provide replacement or servicing of a defective solar PV module under the manufacturing warranty. The performance warranty is 90% of power for 10 years and 80% of power at the end of 25 years or as may be specified by individual solar PV module manufacturers. Warranty conditions do not protect module buyers, if module manufacturers become insolvent or becoming nonresponsive to claims. It is very difficult to measure the degradation of module power in the field with precision. So, most successful warranty claims are for excessive underperformance or total failure of the modules. Warranties typically consider −3% for uncertainty in measurement.

5.2.1 **Bill of materials of solar PV modules and their requirements**

Solar cells—Multicrystalline Si-Al BSF-based, multicrystalline Si- PERC-based, p-type monocrystalline Al-BSF-based, p-type monocrystalline PERC/PERL-based, n-type monocrystalline IBC, n-type monocrystalline HIT, n-type monocrystalline HJT, and bifacial cells.
Encapsulant—Ethyl vinyl acetate (EVA) or polyolefin elastomer (POE) or polyvinyl butyryl (PVB)
Glass—Antireflection-coated (ARC) textured tempered glass, ARC tempered glass
Back sheet—Fluoropolymer-based or nonfluorine-based, trilayer type, dual-layer type
Frame—Anodized Al 6063 alloy with T5 or T6 tamper
Corner keys or screws—To fix the corners of the frame
PV ribbon—Solder-coated copper strips with and without lead-free solder with different widths and thicknesses
Interconnector/bussing ribbon—Solder-coated copper with different widths and thicknesses
Adhesive/sealant—Silicone adhesive, double-sided adhesive tapes, two-component adhesive for sealing
Junction box—3 Terminal/4 terminal with three or six bypass diodes with and without sealing type. With cables and connectors
Solder wire—For soldering the interconnector of substrings for bussing the solar cell circuit
Flux—To remove oxides on the contacting area of either solar cell or PV ribbon
Adhesive tape—To fix on some of the cells in the circuit for proper aligning

The materials are shown in Fig. 5.1 and each material is explained below.

5.2.2 **Superstrate**

The top layer of the solar module that faces the sun is called the superstrate. Normally, glass has been used as the superstrate since the inception of the solar PV industry for terrestrial applications. Ethylene tetra flouro ethylene (ETFE) foils have been used to make flexible solar PV modules. The following are the types of glasses used in the industry:

– Tempered or toughened glass.
– Tempered glass with texturization.
– Tempered, texturized glass with antireflection coating.

The following are the requirements of glass:

The glass should have very high transmittance in the wavelength range of 0.3–2 μm to harvest the maximum incident energy into the solar cell and thereby increase the generation of power.

It should have very low iron content, as the iron content in the glass absorbs the light and reduces the transmission of the glass. Iron commonly appears as a mixture of ferrous ions Fe^{2+} and ferric ions Fe^{3+} in glass. Fe^{3+} is a strong absorber of ultraviolet light, whereas ferrous Fe^{2+} ions absorb light in the near-infrared, causing a reduction in the performance of the PV modules [8].

The glass should be mechanically strong to withstand hail stones, a snow load of 5400 Pa, and a wind load of 2400 Pa. So, the glass is heat-treated to make it tougher. The annealed glass that is used for architectural purpose shows low levels of surface compression and low characteristic bending tensile strength, due to its slow cooling process in manufacturing. So, the glass is thermally strengthened by controlling the sudden cooling from 620°C. With this heat-treatment process and delayed cooling, there will be an improvement in the bending tensile strength of the glass due its surface compression and associated contraction of inner layers. So, the mechanical strength will be increased above 80 MPa, and this is responsible for the reduction in the breakage rate under hail impacts or heavy mechanical and snow loads. So, thermally toughened glass is used as the solar glass for solar modules [9].

Tempered glass should withstand critical temperature gradients that occur over the area of the solar module during field operations due to partial shadow and hotspot conditions. Hence, the glass has to be thermally strengthened to get a typical bending strength of about 125 N/mm^2.

The thickness of the glass influences the light absorption. If thinner glass is used, the absorption will be reduced, thereby transmittance will increase and the efficiency of the module increases. In glass-back sheet-based module configurations, the preferred thicknesses of the glass are 3.2 and 4 mm. For glass-to-glass module configurations, the front and rear side glasses are 2.5 or 2 mm thick based on the mechanical load requirements.

The mechanical strength is a very important feature of glass. A glass is a brittle material, its strength is limited by surface flaws, and tensile stress is more critical than compressive stress [10]. Due to inherent mechanical stress, the strengthened glass cannot be drilled or cut once it has been strengthened, or the entire glass will be shattered into small pieces.

It should have good adherence with solar cells as well as back sheets with the usage of encapsulants during lamination, so the cell side surface is textured to reduce the diffused and specular radiation.

It should have smooth edges so as not to cause cuts to the personnel during handling.

It should be able to withstand harsh environments of rain, wind, hot and cold temperatures, and high humidity while acting as a moisture barrier and protecting all the components of the laminate.

Patterned glass scatters the reflected light and thereby reduces its luminance when compared to specularly reflecting float glass. This effect may reduce the glare of PV modules in the field of vision. Only special patterns are able to reduce the reflectance of the glass surface. Fig. 5.8 shows the image of the solar glass.

There are two types of glass used in the photovoltaic industry. First, there is rolled glass, which is also referred to as cast glass. This form of production causes the glass to have a specific surface structure, and it is often used in crystalline modules. As the name suggests, it is rolled into the required form. The alternative to rolled glass is float glass, which has a coplanar surface. This type of solar glass is floated in a tin bath, which gives its characteristic smooth surface.

Antireflection-coated (ARC) glass

To harness maximum energy, efforts were made to increase the transmittance characteristics by reducing the reflection at the air-glass interface. Hence, antireflection-coated (ARC) glass has been developed.

The basic principles of ARC technology are shown in Fig. 5.9. When a light ray is incident on the glass surface coming from air, some part is reflected from the air-glass interface, some part is absorbed in the glass, and some part is transmitted into the glass. If two reflected rays from the air-glass interface interferes destructively the reflection will be zero. The reflection at the air-glass interface can be reduced by proper selection of the refractive index of the top glass layer. To get the destructive interference to reduce the reflectance of the glass, a thin layer coating of suitable refractive index is deposited on the glass; this is called the antireflectance coating. The transition layer thickness is about 100–200 nm. The reflectance of the interface between the two materials is determined by the square of the difference of their refractive indices [11]. The total transmittance of the glass is determined based

FIG. 5.8

Picture of the glass.

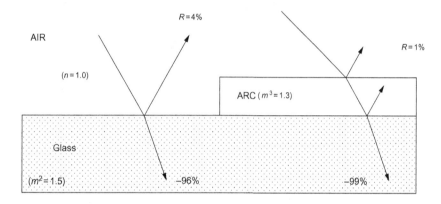

FIG. 5.9

Basic principle of antireflection coating.

on the thickness and refractive index of the coated material and the absorption/scattering effects of the glass.

There are different approaches to achieve ARC on the glass. They are thick layer deposition, refractive index gradient, and deposition of a quarter wavelength layer. A different approach for reducing reflectance is given by inverted pyramid structures on the glass surface of patterned glass. The inverted pyramid structures increase the number of ray intercepts on the glass surface by generating multiple reflections.

Antireflection coating is done optimally to increase light over the complete solar spectrum. In view of the broad wavelength, multilayer coatings shall have to be done. The angle of light incidence on the surface determines the light reflected from the surface. With a change of the sun's position during the day, the incident angle changes. Thus, a reduction of reflection shall have to be ensured by proper antireflective coating while increasing transmission during the day.

There are three types of technologies for producing antireflection-coated glass: (i) nano porous silica-based sol-gel technique, (ii) nanoetching, and (iii) sputtered AR coatings.

The nanoporous silica-based sol-gel technique is the dominant technology with a low investment cost to coat the glass and it is easy to integrate with glass production lines. The main coating technologies are the open porosity type with solid silica particles and the internal porosity type with core-shell technology. Conventional single-layer ARCs consist of solid silica nanoparticles glued together with a binder so that the spaces create nanopores. Extra binders can reduce nanoporosity, which results in antireflective performance reduction. Poor quality can lead to poor mechanical strength and short-term durability. The surface open structure makes the silica layer coating vulnerable to hydrolysis and that deteriorates the mechanical and optical properties. Some manufacturers use a polymer core and a silica shell as core particles. A nanometer layer of approximately 100–150 is deposited on the glass surface. Subsequently, a silica binder is deposited in the core shell particles in between the void spaces. In the process of thermal hardening, isolation of the polymer core happens, leaving a silica layer with a greater proportion of binder and internal nanoporosity. This coating has been the contributing factors for the increased efficiency of multicrystalline Si solar modules.

Table 5.1 gives the data sheet of the 3.2 mm thick ARC glass from Borosil [9].

Market players

Some of the manufacturers in the market, which make an impact on the photovoltaic cover glass market globally, are given below:

Saint-Gobain, NSG, AGC, Guardian, PPG, Interfloat, Trakya, Taiwan Glass, FLAT, Xinyi Solar, AVIC Sanxin, Almaden, CSG, Anci Hi-Tech, Irico Group, Huamei Solar Glass, Xiuqiang, Topray Solar, DSM, and Borosil.

5.2.3 Encapsulant

An encapsulant is required to sandwich the solar cell circuit between the glass and the back sheet. The encapsulation of solar cells into a solar module ensures good mechanical properties and protection against moisture that would rapidly lead to contact corrosion and electrical characteristic degradation. The encapsulant in the solar modules is in contact with different materials, such as the solar cell, tinned copper conductors, glass, and fluoride materials like the layers of the back sheet. Therefore, it should have good compatibility with these materials, and at the same time, maintain its excellent age-resistance performance, good transmittance, and thermal stability. The properties of the encapsulant have a bearing on the long-term performance of solar PV modules.

The majority of manufacturers use glass/back sheet-based crystalline silicon solar modules using glass at the front and a polymer-based insulating sheet at the back. The mono or polycrystalline Si solar cells with five or more busbars are electrically connected with tinned copper ribbons using a high-temperature soldering process. EVA is a commonly used encapsulant material for solar PV modules. EVA as an encapsulation material has been deployed in the field for more than 35 years. Several failures have been documented with detailed investigations [12]. Moisture ingression through the back sheet and encapsulant is the principal reason for solar PV module failure. When subjected to high humidity, high temperature, and exposure to ultraviolet radiation, EVA decomposes to produce acetic acid, which accelerates the corrosion of solar cell metallization and the metallic conductors used for solar cell interconnection [13]. The solar PV module experiences delamination as well as color changes due to moisture ingress. Thus, the solar PV module power degrades. EVA is associated with potential induced degradation (PID). PID has an impact on the solar PV module performance and reliability [14].

Table 5.1 Data sheet for 3.2 mm thick ARC glass from Borosil.

Sl. no.	Characteristic	Specification	
1	Description	Antireflective coated textured tempered solar glass	
2	Process	Single side roller coating	
3	Coating material	**Material**: Nano-SiO_2 particles sol-gel form	
Glass properties			
4	Glass quality	**Ref**—SG/QP/8/02 ANNX-3, ANNX-3B, ANNX-3C, ANNX-3D	
5	Size and thickness tolerance		
6	Tempered glass		
7	Light transmittance T_v % (380–780 nm) Measurement by UV-Vis—spectrophotometer by acc. ISO 9050:2003	2.1 mm, 2.5 mm 2.8 mm, 3.2 mm	$\geq$94.0
		4.0 mm	$\geq$93.8
Coating properties			
8	Test criteria: (EN 572-5:2012)/5.2.1 The glass to be examined is illuminated in conditions approximating diffuse daylight and observed in front of a matte gray screen. Place the pane of glass to be examined vertically 3 m in front and parallel to a matte gray screen. (Conditions approximating diffuse day light). Arrange the point of observation 1.5 m from glass, keeping the direction of observation normal to the glass surface	**Quality of coating on the edges of the glass surface**	
		Distance from edges $\leq$12 mm	Allowed
		Distance from edge >12 mm	Not allowed
		Minor aberrations in the coating	
		Spot diameter up to 10 mm	Allowed
		Spot diameter >10 mm	Not allowed
		ARC border area appearance (residues and color gradient)	
		Distance from edge $\leq$7 mm	Allowed
		Distance from edge >7 mm	Not allowed
		Coating scratch:	
		$W\leq$0.3 mm, $L\leq$60 mm	4 nos/m^2, with an interval of not less than 100 mm
		$W>$0.3 mm, $L>$60 mm	Not allowed
		Surface contamination with coating liquid polluted by foreign substance	
		Diameters $\leq$1.2 mm	Number of clusters (less than 20 within an area of diameter 100 mm)
		Diameter >1.2 mm	Not allowed

Requirements for encapsulation materials

- The encapsulant has to maximize the light transmission into the solar cell with low light absorption.
- High transmission value in the wavelength range of 0.3–2 μm.
- Suitable refractive index to minimize interface reflectance by optical coupling with the solar cell and glass materials.
- Good thermal conductivity to reduce the operating temperature of the solar module to get better energy generation yield.
- High volume resistivity to have very low leakage currents with the application of voltage and should protect the module from PID.
- Properties of the encapsulant should be such that when it is used in the solar module, it should withstand all standard type-approval testing in accordance with IEC 61215.
- Protecting the solar cell from the effects of electric potential relative to the ground.
- Should transmit UV but resistance to UV irradiation, to avoid yellow browning.
- To withstand high humidity levels, extremely low or high ambient temperature cycles.
- To withstand all mechanical loads and provide protection and cushioning to the embedded materials such as solar cells, metallic interconnects during static and dynamic load conditions.
- Providing structural support to the embedded materials, physically isolating the solar cell, and providing high damping capacity.
- Encapsulant has to maintain strong adhesion to glass, solar cells, interconnectors, and to the plastic back sheet while protecting the cell and metallization from external impacts.
- To meet the requirements of the module manufacturer with a lower material cost as well as lower processing cost and processing time along with a higher shelf life and minimum quality assurance issues.
- Suitable glass transition temperature to operate reliably at low temperature sites.
- Low melting temperature for lower electrical energy consumption during the encapsulation process of solar modules.
- Low water vapor transmission rate to avoid moisture penetration to the solar cell and other materials.
- High oxygen transmission rate.

On the basis of the requirements stated above, there are several crucial parameters that have to be taken into account while selecting a suitable PV encapsulant.

Different types of encapsulants

The commonly used encapsulants are EVA, silicone adhesive, polyvinyl butyral (PVB), thermoplastic silicone elastomers (TPSE), ionomers, and polyolefins. The EVA and two component systems such as silicone and urethane (TPU) materials have to be subjected to a cross-linking process, which can be induced by high temperature levels, UV irradiation, or a chemical reaction. The thermoplastic or TPE materials, polyvinyl butyral, TPSE, ionomers, and modified polyolefins (PO) melt during the module manufacturing process without forming chemical bonds between the polymer chains. There is no cross-linking between the polymer chains involved. Fig. 5.10 [15] shows the structure of different encapsulants.

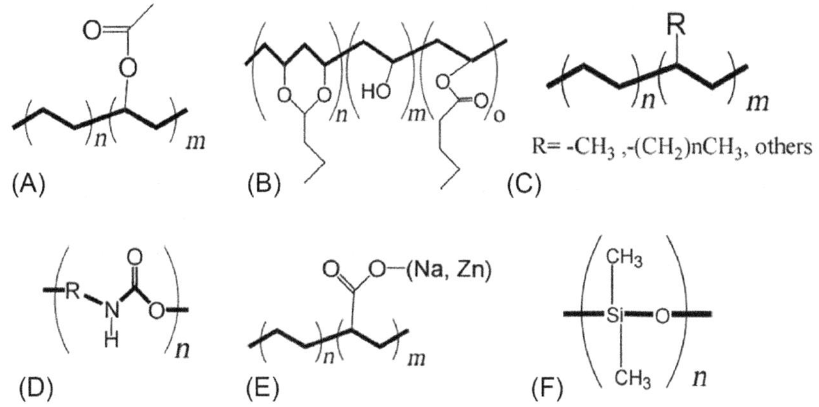

FIG. 5.10

Structure of different encapsulation materials. (A) Ethylene vinyl acetate copolymer (EVA), (B) polyvinyl butyral (PVB), (C) thermoplastic polyolefin (TPO), (D) thermoplastic polyurethane (TPU), (E) methacrylic polyethyleneco-acid (Ionomer), and (F) polydimethyl siloxane [15].

PVB

Polyvinyl butyral (PVB) is a well-known thermoplastic (noncross-linked) encapsulant. It has been used for a long time in architecture for safety-glass laminates as well as in the PV industry for building-integrated photovoltaics (BIPV) and for thin-film technology with a glass-to-glass configuration. One disadvantage of PVB is that it is very sensitive to hydrolysis because of high water uptake. So, it is not used for glass/back sheet-based crystalline Si solar modules. It was used in glass-glass modules and thin film-based solar modules.

Silicone adhesive

Liquid silicone demonstrates excellent resistance to oxygen, ozone, and UV light as well as a wide range of temperature stability, excellent transparency in the UV-visible range, and low moisture uptake. Although very promising as an encapsulant material, silicone is only rarely used owing to its high price and the need for special processing machines. The Dow Corning DC 93500 adhesive is extensively used for thin cover glass bonding of space solar cells. Wacker-Chemie's RTV-S-691 is used for solar cell circuit bonding with different substrates for space solar panel applications [16].

Thermoplastic silicone elastomers (TPSEs)

Thermoplastic silicone elastomers (TPSEs) combine silicone performance and thermoplastic processability. The fast curing and the additive-free physical cross-linking make TPSEs suitable for continuous lamination processing. In view of the high price, they are only used in special applications.

Ionomers

Belonging to the category of thermoplastic materials, ionomers represent a different class of photovoltaic encapsulant, demonstrating good UV stability. No formation of acetic acid has been observed during weathering and a much longer shelf life is achieved, but the production cost is very high.

Even more than the technical requirements, the main driving force that governs the selection of the encapsulant material suitable for PV module design is the intense and ever-increasing pressure to reduce module costs. From cost and performance points of view, EVA is considered for glass-back sheet-based modules. EVA, PVB, and TPO demonstrate a cost that is affordable when considering promising encapsulants for the PV module design based on glass-to-glass modules with high-efficiency solar cells.

The EVA and POE materials are discussed below.

EVA

EVA comprising ethylene and vinyl acetate (VA) is one of the most widely used encapsulating materials. The VA groups are randomly distributed along the backbone. The EVA film has a series of advantages as a solar cell encapsulation material, namely a high transmission value (>91%), low-temperature toughness, resistance to UV radiation, high volume resistivity ($0.2–1.4 \times 10^{15}$ V cm^{-1}), and good adhesion strength to glass (peel strength of $9–12$ Nm^{-1}). It also offers weather resistance and long-term reliability under long periods of exposure to different climates. However, there are still some problems to be solved with the solar cell encapsulation application of EVA film, such as high energy consumption, short service life period, etc. Researchers have done lot of work to cut the cost of the EVA encapsulant while improving the performance and meeting the requirements of service life.

The virgin type of EVA is not stable under different temperature conditions. Different kinds of additives need to be added to ensure stability. The UV radiation causes chain scission and photodegradation. EVA absorbs UV light in the range of 260–360 nm with a peak absorption intensity at 280 nm. Therefore, the addition of UV absorbers and stabilizers becomes essential.

The UV absorber's function is to absorb UV light and quench the reactant's excited states. Scavenging free radicals is the function of the UV stabilizer. UV absorbers and stabilizers begin to diminish over time.

Thermal oxidation can also cause EVA to deteriorate. Alkoxy or alkyl peroxide radicals can be formed in an oxidative environment. The radicals formed remove hydrogen from other products, resulting in freer radicals that are more active. Such free radical reactions to the base material can lead to different degradation of polymers such as chain scission, chain branching, and cross-linking. Antioxidants are needed to prevent these oxidation reactions by generating inactive radicals by decomposing peroxides or reacting with the active free radicals [17].

The basic EVA material melts to a viscous state at temperatures above 75°C while the actual operating temperatures of solar PV modules may be higher. If used directly for encapsulation of the PV module, it will soften at high temperatures to a viscous melt and will shrink and stiffen under cold weather. This form of thermal expansion and contraction can lead to cracking and delamination of cells. This can be stopped from creating a temperature stable elastomer by cross-linking EVA. Cross-linking is a method of transforming a thermoplastic material into a thermosetting material in the form of a network so that the substance does not move at high temperatures.

To cause a cross-linking reaction in EVA, peroxide is added within EVA. The peroxide is inert below 90°C or with minimal activity, so when EVA is extruded to a film format at temperatures below 90°C, no curing reactions occur. The peroxide can decompose at high temperatures above 100°C to create radicals that react with the polymer to form the cross-linking reaction. The degree of cross-linking can be represented as a gel substance that is the percentage by mass of EVA's three-dimensional components.

The solar module's performance depends on the strength of bonding between the EVA film and its other components. The adhesion property is significant for EVA film, and researchers found that the influencing factors included the form and structure of EVA, the type of back sheet and its pretreatment surface, and the type of glass and its pretreatment surface.

EVA is a low polarity polymer material while glass is a smooth, inorganic material. It's hard for a long time to glue them together. Because it is also difficult to bind to the fluorine-containing back sheet, it is necessary to add a coupling agent to enhance the polarity of EVA.

It was found that film containing a silane coupling agent gave a higher peel strength. The main reason for this is that the coupling agent can enhance EVA polarity and react with inorganic chemical groups on the glass surface, forming strong bonding. At the same time, the unsaturated bonds of the silane coupling agent react with EVA molecules, forming polarity branched chains on EVA molecules. In order to achieve strong bonding between the EVA film and the other components of the PV modules, an adhesion promoter, normally in the form of trialkoxysilane, is added to improve the adhesion between EVA and glass. Due to the addition of adhesion promoters, the EVA encapsulant gets an excellent adhesion property.

EVA is a copolymer consisting of ethylene and vinyl acetate. EVA chemical properties are influenced by vinyl acetate content. Crystalline regions formed by the polyethylene segments get disrupted in the initial effect while the next effect results from the polar nature of the acetoxy side chain. The EVA shows different properties by the VA content variations. The crystallinity of the EVA determines the EVA property.

The EVA cross-linking curing reaction also increases the bonding strength. VA content also affects the EVA peel strength. When the VA content is low, it shows good heat resistance, but a poor adhesion property and low temperature flexibility. When the VA content is high, it shows better low temperature flexibility and adhesion property. The appropriate VA content is generally fixed between 28% and 33%.

From the above discussion, it can be seen that EVA contains a complex formulation of additives with different functionalities. A stabilized EVA used for PV module encapsulation material usually contains a mixed composition including the bulk EVA copolymer, a UV absorber, a UV stabilizer, an antioxidant, a curing agent, and an adhesion promoter.

The durability of EVA is mainly influenced by the additive elements, and this has been improving in recent years. Multiple solutions have been proposed with respect to the degradation and yellowing problem, but other degradations (acetic acid production) are still prevalent for this type of encapsulant. The cross-linking additives in encapsulants create problems in connection with the processing time of the solar PV module as well as storage.

The solar modules that are mounted on open spaces are subject to harsh environments such as rain, hot and cold temperatures, UV light, dust, wind, and high humidity, and they have to work for 25–30 years. The harsh environment induces stresses that may decrease the solar PV module stability and performance. Additional losses in performance may be caused by rain, dust, wind, hail, condensation and evaporation of water, and thermal expansion mismatches. So the use of an encapsulating material made from EVA still has some limitations; that is, the polymer tends to degrade after being exposed to a high temperature and harsh environmental conditions. Degradation of the EVA solar encapsulating material is a serious issue and deserves consideration because the degradation is usually accompanied by some changes in the color of the polymer film, from colorless to yellow and/or brown. As a result of the aforementioned EVA browning effect, the performance and/or power conversion efficiency of the solar cell declines. Fig. 5.11 shows where EVA has shown failure modes.

The cross-linking degree is an important indicator for the EVA encapsulant. If the cross-linking degree is too high, the EVA will turn out to be crisp and will not be able to afford the outside stress and protect the silicon solar cell. If the cross-linking degree is too low, the aging resistance will be reduced, meaning it cannot meet the requirements for creep resistance. A suitable cross-linking degree is commonly around 75%–90%.

Different EVA manufacturers have different types of EVAs in the fast curing type, including normal EVA, anti-PID EVA, UV-transparent and UV-resistant EVA, and white EVA are available with different formulations. The disadvantages are different additives are added to virgin EVA for cross-linking, UV absorption, stabilization, and to avoid oxidation. When exposed to light, high temperature, and high humidity, the interactions among the additives take place and due to the photooxidation of EVA, acetic acid is produced. The lower water moisture barrier property enhances the browning and corrosion and is responsible for snail trails, etc.

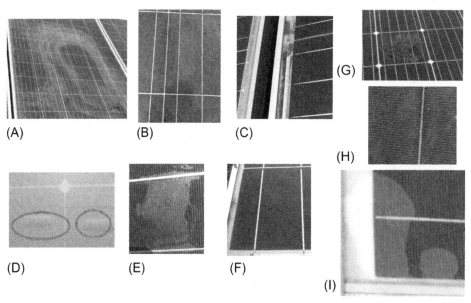

FIG. 5.11

Defects observed with EVA from the solar PV field. (A) Discoloration, (B) browning of EVA, (C) corrosion of interconnector, (D) bubbles in the back sheet, (E) EVA shrinkage, (F) snail trails, (G) browning coloration, (H) PV ribbon yellowing/corrosion, and (I) delamination.

As part of the quality checks, the gel content of the cured EVA and the peel strength with the glass and back sheet are checked. The gel content will provide a degree of polymerization of EVA during the lamination process and the value should be in the range of 75%–95%.

The EVA has to be stored in a controlled temperature and humidity environment. The temperature should 20–30°C and the relative humidity should be about 50%.

Table 5.2 gives the typical properties of a normal EVA being used for glass/back sheet modules.

Table 5.2 Typical properties of normal EVA being used for glass/back sheet modules.

Properties	Items		Test methods		Units	Value
Physical	Thickness (uncured)		ASTM F2251		μm	450–500
	Density (uncured)		ASTM D792		g/cm^3	0.9
	Tensile strength (cured)	MD	ASTM D882		MPa	>18
		TD			MPa	15
	Elongation (cured)	MD			%	>800
		TD			%	600
	Adhesion to glass		ASTM D903		N/cm	>75
	Water Absorption (cured)		ASTM D570		wt.%	0.1%
	Hardness (cured)		ASTM D2240		Shore A	80
	VA content		–		%	28
	%Gel content		–		%	80
Electrical	Dielectrical strength (cured)		ASTM D149		kV/mm	36
	Volume resistivity (cured)		ASTM D257		Ωcm	>1 × 10^{14}
	Surface resistivity (cured)		IEC61340-2		Ohms	–
Optical	Refractive index (cured)		ASTM D542		–	1.48
	Transmittance (cured)		ASTM D1003		%	>92
	UV-cut off (cured)				μm	360
Thermal	Dimensional stability (uncured)		ASTM D882	ASTM D1003	%	<2
			TD		%	<1
	CTE (cured)		MD	ASTM E831	%	–
			TD		%	–
Durability	UV resistance (32 kWh/m^2)		$\Delta b*$	ASTM G154	–	–
	UV resistance (32 kWh/m^2)		$\Delta T\%$		–	<3
	Damp heat resistance (85% RH, 85°C, 1000 h)		$\Delta b*$	IEC 61215	–	–
			$\Delta T\%$		–	<5% power loss (TUV)

Solar encapsulant market competitive landscape

Some of the players operating in the global solar encapsulant market are Hangzhou First Applied Material, Changzhou Sveck Technology (Sveck), RenewSys India (RIPL), STR Holdings, Encapsulantes de Valor Añadido S.A. (EVASA), E.I. du Pont de Nemours and Company, Jiangsu Akcome Science and Technology, Saint-Gobain S.A., ISOVOLTAIC AG, Eastman Chemical Company, Bridgestone Corporation, Mitsui Chemicals, 3M, Solutia, Bridgestone, Dow Corning, MITSUI, JGP Energy, Etimex, and Hanwah.

Polyolefin elastomers (POE)

Encapsulants based on a thermoplastic polyolefin (TPO) or thermoplastic elastomer (TPE) are entering the solar market because of their high electrical resistivity and hydrolysis resistance. These properties make TPO a good encapsulant material for solar PV modules.

Polyolefin-based encapsulants are of two types—cross-linked, that is, polyolefin elastomer (POE) and noncross-linked, that is, thermoplastic polyolefin (TPO).

POE has enhanced electrical properties. There is no chemical cross-linking, hence a lesser processing time in lamination. Higher volume resistivity allows the module to have low leakage current levels to help enhance electrical insulation, reduce the effects of PID, and further improve module reliability and service life. POE offers higher dielectric strength, facilitating operation at higher voltages.

The low WVTR of POE-based films also contributes to increased PID resistance, reduced corrosion damage or delamination, and opportunities to maintain high levels of power output and extended service life. The higher thermal conductivity of encapsulant films made with POE also reduces module operating temperatures, allowing for increased module efficiencies in high-temperature environments. POEs are not chemically cross-linked and contain no liquids that can cause hydrolysis or lead to bubble formation, thus eliminating the formation of acetic acid. Due to this, corrosion, decolorization, PID, and snail trails go to zero. POEs are being used for glass/glass-based solar modules that are monofacial cell-based as well as bifacial solar cell-based. Presently, POE has lower penetration into the market due to a higher price. But, glass-glass solar modules and bifacial modules are effectively using this material.

5.2.4 Solar cell

There are different type of crystalline Si solar cells:

P-type multicrystalline, p-type monocrystalline, and n-type monocrystalline with Al-BSF, PERC/PERL, HIT, or HJT configurations.
Bifacial solar cells and half-cut solar cells of different configurations are also being used.

The solar cells are available in different sizes including $125 \times 125\,mm^2$—pseudosquare; $158.75 \times 158.75\,mm^2$—pseudosquare and full square; $156.75 \times 156.75\,mm^2$—pseudosquare and full square; $161 \times 161\,mm^2$ and $210 \times 210\,mm^2$—full square. The solar cell can be cut by a laser for the required size to make lower-wattage solar PV modules.

- The solar cell should have higher energy conversion efficiency.
- Resistivity and base doping concentration will decide the reverse breakdown voltage.

- The solar cell should have lower series resistance and higher shunt resistance.
- The solar cell should have lower temperature coefficients of power and voltage.
- The solar cell should have higher current density and higher open circuit voltage.
- Lower dark saturation currents I_{o1} and I_{o2} and unity ideality factor.
- Thickness should be around 200 μm and should have good mechanical hardness.
- Solderable ohmic contacts on both sides.
- Anti-PID property of antireflection coating.
- Better spectral response.
- Better optical coupling with EVA and glass to maximize incident energy.
- No electrical degradation due to temperature cycling and light.
- Mechanically strong to withstand handling and module processing stresses.
- No spikes of the metallization on the surface of the solar cell.

Table 5.3 is the comparison of solar cell parameters.

Table 5.3 Comparison of solar cell parameters.

Parameter	Multicrystalline	Monocrystalline	Monocrystalline
Base material	P-type	N-type	P-type
Type of structure	Al-BSF	Interdigited back contact	PERC
Short circuit current (I_{sc})	9.02	6.18	9.847
Current at maximum power point (I_{mp})	8.513	5.89	9.342
Voltage at maximum power point (V_{mp})	548	632	578
Open circuit voltage (V_{oc})	645	730	679
P_{max}	4.67	3.72	5.38
Fill factor			
Efficiency	19	24.3	22
Contacts			
Front contact material	0.7 mm wide, five busbars with silver	No contact	0.7 mm wide five busbars with silver
Rear contact material	5 * 1.8 wide discrete bads with silver	Tin-coated metal grid	5 * 1.8 wide discrete bads with silver
Temperature coefficient of V_{oc}	−0.36%	−0.2380%	−0.36%
Temperature coefficient of P_{max}	−0.36%	−0.29%	−0.38%
Temperature coefficient of I_{sc}	0.06%	0.0469%	0.07%
Cell dimensions	156.75 × 156.75 mm	125 × 125 mm psuedosquare (153 mm^2)	156.75 × 156.75 mm psuedosquare
Cell thickness	200 ± 30 μm	150 ± 30 μm	

5.2.5 Substrate

The performance and reliability of a PV module is highly affected by the degradation behavior of the encapsulant and the back-sheet material. The back layer of the module is called the substrate. Usually for the back sheet, a polymeric material is used for glass-back sheet-based solar modules.

The main function of the back sheet is to provide protection from the rear side to the PV cells, metallic interconnects, busbars, and solder joints that are embedded in the laminate.

As the back sheet is exposed to the outdoor environment, it must be resistant to moisture, humidity, rain, and condensation.

The back sheet should have adequate UV stability, as it will be exposed to UV radiation through the module package and from diffused light on the back surface.

It should have sufficient mechanical properties in order to be scratch-resistant and withstand punctures during handling of the laminate and module in different stages of manufacturing, testing, and mounting on the structure.

The back sheet should withstand the lowest and highest temperatures seen by the module environment.

It should be able to withstand corrosive environments such as atmospheric chemicals, ammonia, and marine/coastal atmosphere

Additionally, the internal surface of the back sheet must allow for adequate adhesion to the encapsulant and the external side must provide a surface in which a junction box can be attached and sealed to prevent water penetration.

The back sheet should be airproof, waterproof, and electricproof.

A standard back sheet is a trilayered structure in which the PET core film is sandwiched between two protective layers. The back sheets can be divided into two categories: fluoropolymer and nonfluoropolymer-based back sheets, depending on the chemistry of the protective layers. Fluoropolymer-based back sheets contain atleast one fluoropolymer film to protect the PET core while nonfluoropolymer-based back sheets are without halogen components. The configuration of the trilayer back sheet is inner side layer/middle layer/air side layer. The air side layer should face the environment and should meet moisture ingress, electrical insulation, and stability against UV irradiation. High-energy UV photons can break polymer chains, causing decrease in optical transmission and degradation in mechanical properties. A reduction in optical transmission leads to reduced output current of the PV module. Degraded mechanical properties can result in delamination or cracks. The middle layer is the core material that acts as the main permeation barrier and provides the required electrical isolation. The innermost layer is chosen to have good adhesion to the encapsulant. These layers are typically bonded together with adhesives such as epoxy or polyurethane.

The core layer has traditionally been a polyester such as polyethylene terephthalate (PET) because it provides adequate performance at reasonable costs. Other possible core layers are thermoplastics such as polyamide (PA) or polyolefin (PO) or polystyrene materials. Hence, the material is selected such that it should provide reliable performance without causing any detrimental effects on the solar module.

The drive to reduce the cost of the solar module had led some manufacturers to go for alternative back sheet materials. The following materials are used for different layers of back sheets.

Outer (air) side layer

The key functions and requirements for the air side film of the back sheet are:

- Ability to resist UV radiation and not to allow moisture ingress inside the module.
- Ability to resist abrasion of sand and airborne particulates.
- Resistance to tear, cracks, or physical deformations.
- It should provide electrical insulation to the inside components such as solar cells, solder joints, and metallic interconnects.
    It should have good adhesion property with the junction box to enable good bonding.
    It should be thermally stable at laminating and operating temperatures.
- It should protect the inner core PET film.

The following materials are used as outer layers:

A polyvinyl fluoride (PVF) film called Tedlar, a trademark of Dupont, is used extensively as an outer layer of the back sheet material in solar PV module construction. It has proven to be reliable and durable in keeping the integrity of solar PV modules for more than 30 years.

A polyvinylidene fluoride (PVDF) called Kynar, a brand name of Arkemia, is a trilayer PVDF that is 5 μm thick on both sides while the middle is a layer with polyester mixing.

The PVDF film is promoted as a lower-cost alternative back sheet material. Fluoroethylene and vinyl ether copolymer (FEVE) coatings are new and yet to be proven. Long-term studies are yet to be completed on the field performance.

The fluoropolymer-based materials contain elemental fluorine. The fluorine resin is weather resistant, heat resistant, high-temperature resistant, and chemical resistant. This is because it has large electronegativity, a small van der Waals radius, and a very strong carbon-fluorine bond. Also, its unique overall the structure of the fluorinated chain with spiral rod-like close molecules, rigid and smooth surface, responsible for fluorine resin to be weather resistant. The excellent characteristics of fluorine-containing materials are responsible for the excellent weathering properties to safeguard the reliability of the material for long-term outdoor use.

Hydrolysis-resistant and UV-durable modified PET material is also one of the materials used for the outer layer of the back sheet. This has been used by many companies such as Covame.

Middle layer

The key functions of the middle layer are to provide electrical insulation and mechanical stability while acting as a vapor barrier. It must also be hydrolytically stabilized and resistant to sand abrasion. Polyethylene terephthalate (PET) film is widely used as a core inner layer of the trilayered back sheet.

The PET film has unrestricted supply. High-grade PET is already present as a core material and it has proven to have high mechanical strength, excellent electrical properties, and low moisture permeability. Because of its poor stability for UV irradiation, it has become a passive layer as a core material in the back sheet. Advances in polyester chemistry engineering have made the PET film highly UV-durable and hydrolysis-resistant.

Fluoropolymer films are superior in UV resistance and poor in moisture resistance, whereas PET films are excellent for moisture regression as well as sand and abrasion resistance. PET is made stable by adding titanium oxide and other UV-resistant additives. When PET is exposed to UV light, the titanium oxide is responsible for the higher yellowing index. Fluoropolymer-based back sheet manufacturers use a standard PET as the core layer to reduce cost, as hydrolysis-resistant PET costs more.

Apart from coextruded films as the inner core film, all other back sheet designs use hydrolytically stabilized PET film of various thicknesses such as 1000 V or more recently 1500 V, depending upon the end use application. There is no standardization between PET films, leading to inconsistent field performance and a high rate of field failures that leads to yellowing and cracking.

Inner side (cell side) layer

The cell or EVA side back sheet surface is commonly referred to as the "E-layer." It also plays a critical role in the long-term reliability of the PV module. The primary functions of the E-layer include:

- Protection of the inner core PET film.
- Should have thermal stability against deformation and softening.
- UV-stable, nonyellowing.
- Should have good electrical insulation.
- Adhesion to core film and encapsulant such as EVA.
- Should have high reflectance to harvest the maximum incident energy to enable enhanced energy conversion.
- Better heat dissipation properties to conduct the heat.

Materials used for inner layer

The polyvinyl fluoride (PVF) film called Tedlar as well as Kynar (PVDF), polyvinylidene fluoride (PVDF), fluoroethylene and vinyl ether copolymer (FEVE) coatings, PET, fluorine skin, polyolefin, ethylene, and the E-layer are the materials used as the inner layer for back sheet design.

Typically, the E-layer is made of low melting adhesives such as polyethylene or ethylene vinyl acetate. Both are thermoplastic materials, thus deforming and softening during heat exposure. This is problematic on two fronts: during vacuum lamination and during the lifetime in high-temperature installations. During the vacuum lamination process, typically conducted at 140–160°C under vacuum, the E-layer melts, pushing ribbon wires toward and in direct contact with the inner core PET layer. This reduces the electrical insulation characteristics of the back sheet. In addition, these olefin-based layers may yellow under UV exposure and have been reported to crack, pulverize, and disintegrate. This exposes the inner core films to a damaging UV light, thus marginalizing the safety and reliability of the PV module itself.

The major back sheet configurations in today's market are manifold (TPT, TPE, TPC, KPK, KPE, KPf, KPx, PPE, CPC, AOE, and OOO). The abbreviations are explained as T—Tedlar (PVF film); P—PET film; K—Kynar (PVDF film); C—Coating of fluoropolymer; E—Polyethylene or EVA layer; f—Fluorine skin; A—Polyamide; and O—Polyolefin.

Earlier back sheet designs of the inner and outer layers that were extensively used were with Tedlar, a PVF material and the trademark of Dupont. Tedlar was synonymously called back sheet. After 2007, due to the sudden growth of solar PV systems, the demand increased for Tedlar, and Dupont was not able to supply the requirements of solar module manufacturers. To meet the growing demands and to reduce the price of the back sheet, the R&D efforts took place in two ways: fluoropolymer-based back sheets and nonfluoropolymer-based back sheets. PVF and PVDF are fluoropolymer-based materials with fluorine content that are kept as the outer layer of the back sheet. The core material is PET, which cannot withstand UV light and becomes brittle if exposed to light.

The most common back sheet defects found by the researchers were:

Prolonged exposure of the back sheet to UV light, high temperatures, and environmental stresses cause the back sheet to undergo discoloration. This is an early indicator of serious mechanical integrity issues, including delamination and cracking. Yellowing can compromise the back sheet's electrical insulating properties.

Macrocracks in the back sheet's outer layer and outer layer separation from the back sheet structure are called abrasion and delamination. The abrasion and delamination defects will lead to safety issues because they represent severe degradation of the back sheet's protective feature and expose the inner PET core layer to the elements.

Delamination and bubbling: Cracks in the outer layers of the back sheet responsible for delamination and bubbling. The defect has the potential to expose the core back sheet layers and compromise the structural integrity. Delamination can also result from hot spots (a bubble caused by the separation of the back sheet or encapsulant layers) or increased series resistance.

The back sheet serves as a moisture barrier and provides the necessary environmental protection. Hence, the lifetime of the back sheet under certain operating conditions directly impacts the performance and lifetime of the PV module. However, these materials gradually degrade and lose their performance due to environmental stresses, including high temperature, humidity, and UV radiation. Based on the results from observing a field-aged PV module, the cracking of the PET and delamination between polyvinyl fluoride (PVF) or polyvinylidene fluoride (PVDF) and polyethylene teraphthalate (PET) are commonly observed failure modes that also represent catastrophic failure. These failure modes can reduce the performance of the PV module and shorten its lifetime. Moreover, cracking of the back sheet, which is caused by a decrease in the tensile strength, results in the penetration of a large amount of moisture. It is the most catastrophic failure mode among the reported failure modes of the back sheet because the cracking of the back sheet allows both water vapor and liquid ingress into the PV module. It can significantly impact the performance and reliability of the PV module. Thus, the prediction of degradation patterns and the lifetime for the back sheet are critical to ensure that the PV module maintains its performance throughout its lifetime. Fig. 5.12 shows the chemical structures of different back sheet materials.

The degrading mechanisms can be prevented most effectively by an intact polymeric back sheet that forms an outer layer of solar panels and that is expected to provide protection for photovoltaic modules over the expected service life.

PET molecule contains a large number of ester groups in the main chain. It has good affinity with water, so, even a trace amount of water will lead to the degradation of the molecular main chain. PET aging properties change in hot and humid environments. The aging process is affected by three factors: the degree of crystallinity, water plasticization, and hydrolysis. These factors appear at different environments and play a leading role in different stages of performance of PET. The degree of crystallization is a dominant factor for early aging of the material. It increases the Young's modulus and maximizes the tensile stress, causing the material to become brittle, reducing the impact strength.

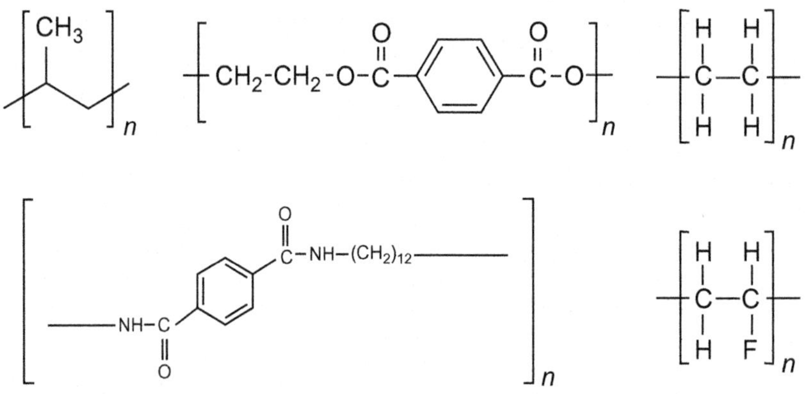

FIG. 5.12

The chemical structures of back sheet materials; (A) PP, (B) PET, (C) PVF, (D) PA, and (E) PVDF.

The water plasticization become a factor for the increase of material toughness causing the hydrolysis reaction to start. The hydrolysis reaction breaks the chains of PET macromolecules, and the molecular weight decreases, thereby causing the destruction of the mechanical properties. The temperature rise significantly accelerates the above-described processes. Water and heat cause reduction of physical and mechanical properties of PET material. In addition, UV radiation may cause the reduction of molecular weight of the PET, and so, the strength and elongation of PET are greatly decreased. UV radiation increases the degree of crystallinity causing the material to become brittle. Table 5.4 shows different configurations of the back sheet with different combinations of tri layers. PET with modifications is used as a core layer for almost all types of back sheets.

Table 5.5 gives the typical properties of a back sheet material.

Global photovoltaic back sheet market: Key players

Some of the players operating in the global photovoltaic back sheet market include DuPont, Isovoltaic, Coveme, Arkema, 3M, Toyo Aluminium, Taiflex, Krempel, Targray, Toray, Dunmore, Astenik, ZTT International, Madico, SFC, Hanwah Advanced Materials Corporation, Renewsys, Vishakha, Agfa-Gevaert NV, AluminiumFéron GmbH and Co. KG, Crown Advanced Material, Cybrid Technologies, Hangzhou First PV Material, Jingmao Technology, Krempel GmbH, Jolywood (Suzhou) Sunwatt, Royal DSM Group, ShingiUrja, and ToyalZhaoqing.

The following are the sample level tests conducted on back sheet material and the test coupons.

Moisture vapor transmission rate

Partial discharge 1000 or 1500 V

Out-gassing material test—post heat soak

Pressure cooker test

Surface morphology analysis at initial and post humidity freeze test

Shrinkage (at lamination conditions and during accelerated tests)

Interlayer adhesion (as received, postlamination, and post humidity freeze test)

UV exposure and weather O resistance

Table 5.4 Different configurations of the back sheet.

Back sheet structure	Tedlar (PVF)/PET/ Tedlar (PVF)	Kynar (PVDF)/PET/ Kynar (PVDF)	Kynar (PVDF)/PET/ Other Polymer film	Tedlar (PVF)/ PET/ Polyethylene
	TPT	**KPK**	**KPX**	**TPE**
Air side layer	Tedlar	Kynar	Kynar	Tedlar
Core layer	PET	PET	PET	PET
Cell side layer	Tedlar	Kynar	Polymer film	Polyethylene
	Kynar (PVDF)/PET/ polyethelyne	**Kynar (PVDF)/PET/ fluorine skin**	**Tedlar (PVF)/PET/ coating**	**Tedlar (PVF)/ PET/other polymer**
	KPE	**KPf**	**TPC**	**TPX**
Air side layer	Kynar	Kynar	Tedlar	Tedlar
Core layer	PET	PET	PET	PET
Cell side layer	Polyethylene	Fluorine Skin	Coating	Polymer film
	PET/PET/ polyethylene	**Coating/PET/ coating**	**Polyamide/ polyolefin/ polyethylene**	**Polyolefin/ polyolefin/ polyolefin**
	PPE	**CPC**	**AOE**	**OOO**
Air side layer	PET	Coating	Polyamide	Polyolefin
Core layer	PET	PET	Polyolefin	Polyolefin
Cell side layer	Polyethylene	Coating	Polyethylene	Polyolefin

Table 5.5 Typical properties of a back sheet material.

Sr. no.	Properties	Standard followed	Value	Unit
1	Interlayer peel strength	ASTM D1876	>5	N/cm
2	Peel strength with encapsulant	Relevant ASTM	>75	N/cm
3	Breakdown voltage	ASTM A-149	69.34	kV
4	Partial discharge	IEC 60664-1/IEC 61730	1500	VDC
5	Water vapor permeability	ASTM F1249 (38°C, 90% RH)	1.71	$g/m^2/day$
6	Damp heat exposure	IEC 61215 (85°C/85% RH)	1000	h
7	Maximum system voltage	IEC 61730-2; IEC 60664-1	1000	V
8	Thickness (total laminate)	ASTM D1593	320	μm
9	Tensile strength	ASTM D882	>100	N/mm^2
10	Shrinkage	ASTM D1204	>1.5	%

5.2.6 Solar ribbon

A crystalline silicon solar cell gives about $0.69\,V$ and a current density of about $40\,mA/cm^2$. The appliances require different voltages and currents for their operation. To get proper voltage and current, the solar cells (36, 60, or 72) are to be interconnected in series and parallel configurations. To connect the solar cells, an interconnect ribbon that is a conducting material is used. In a p-type Si crystalline solar cell, the top side contact is negative and the bottom contact is positive. To form a string with a series of cells, the top contact of the first cell is connected to the bottom contact of the neighboring solar cell as shown in Fig. 5.13. Thus, a string is formed by connecting 9, 10, or 12 cells in series. As the solar module contains 36, 48, 54, 60, 72, and 96 solar cells in series, they have to be arranged in different 6×6, 8×6, 9×6, 10×6, 12×6, and 12×8 matrix configurations, respectively. To connect the substrings and to bring the output of the strings to the junction box, a wider ribbon is used as a connector; this is called the busbar wire. The series and bussing connections are shown in Figs. 5.13 and 5.14.

Thus, there are two main types of PV ribbons: interconnect ribbon and busbar ribbons. The interconnect ribbon carries the generated current from all the PV cells to the busbar. Then the busbar ribbon connects two substrings and carries the accumulated current to the junction box and to the electrical distribution system.

The process of solar cell interconnection is called tabbing and stringing, and it is done by an automatic soldering equipment. The crystalline Si solar cell is a thickness of $180–200\,\mu m$ with three, four, five, and nine busbars. The front busbar is with silver metallization with continuous busbars, whereas the back side busbars are the discrete type with Ag-Al metallization. The busbar width decides the

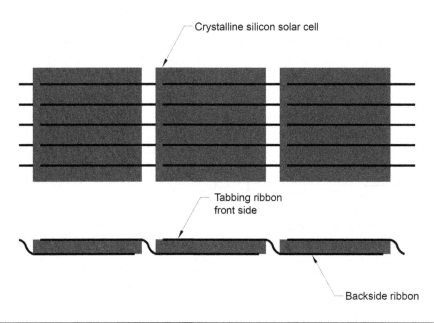

FIG. 5.13

Front and bottom side tabbing of a solar cell with a PV ribbon.

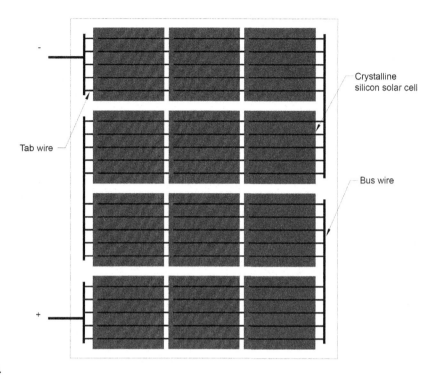

FIG. 5.14

Solar cell interconnection with a bussing ribbon.

width of the PV ribbon to be used. The number of PV ribbons to be placed on the solar cells will depend on the number of busbars of the solar cells. The solar module undergoes hot and cold temperatures and induces thermal stress on the solar interconnectors. The solar PV ribbon is one of the important components of the solar module for its reliable operation during its operational life. The following are the requirements of the interconnect ribbon:

The material should have high electrical conductivity to reduce the ohmic losses and to increase the generated power.

The material should have a better thermal expansion coefficient to match with silicon material to avoid thermomechanical-induced cracks during processing or its operational life.

The material should have a proper solder coating to enable an easy process of interconnection.

The contact processing should be easy to implement in manufacturing.

The interconnector ribbon should not create cracks in the solar cell due to thermal stress during the process of soldering.

The material should withstand thermal fatigue and should not break during its life of operation.

The solder material should be such that the solder joint should neither crack nor get disconnected during the operation of the solar module.

The interconnector ribbon design and its connection technology should not limit the amount of energy generated due to the thermomechanical reliability of the PV modules. These challenges include series resistance, shadowing losses, and induced thermomechanical stress in the solar cells. Series resistance losses are one of the major challenges associated with the manufacture of solar cells in the conventional form. These losses are created due to metallization for contact formation and the subsequent tabbing for current collection.

Another key challenge of conventional interconnection technology is shadowing losses. When cells are made wider, a thicker interconnection ribbon is required to conduct larger currents. The increase in the width of the interconnection ribbon cross-section increases the shadowing losses proportionally. The thickness of the ribbon strip is limited by built-up stresses in the soldered joint. The differences in the coefficients of thermal expansion between the ribbon interconnection materials and the silicon account for this stress accumulation. Furthermore, stress that occurs at the edge of the wafers due to the bending of the interconnection ribbon strip, which connects the front side with the rear of the neighboring wafer, impacts the reliability of the assembly. This situation entails that conventional interconnection technology makes a compromise between the width and thickness of the ribbon strip. However, the reduction of these losses is desirable to enhance solar cell efficiency.

A PV interconnect ribbon is 0.9–2 mm in width with a thickness in the range of 0.085–0.25 mm. The busbar ribbon is larger in size than the interconnect ribbon at 3–6 mm in width and 0.2–0.5 mm in thickness. The primary material of the PV ribbon is usually copper. Different grades of copper are used but it is important to have high conductivity to ensure the maximum efficiency of the solar panel.

The following factors play an important role in PV ribbon design and performance.

1. Base material: As far as the PV ribbon is concerned, copper is commonly used as the main material. The common alloys used are CDA 102 (oxygen-free copper) and CDA 110 (electrolytic tough pitch copper). These have to meet the requirements of ASTM/DIN standards. The type of metal selected directly affects its conductivity, and this in turn affects the power output of the solar PV module. With a high-purity oxygen-free copper core base, good electrical conductivity combined with guaranteed elongation and yield strength are achieved.

2. Solder coating: The solder composition and coating thickness are critical parameters that control the strength and reliability of the solder bond. The coating thickness is selected based on the interconnection process of the tabber and stringer followed by the manufacturer to meet the bonding requirements with the solar cell. It should be around 20 μm on all sides and the solder composition alloy decides the temperature of soldering. Earlier, SN62/Pb36/Ag2 solder was used to solder a solar cell. The 2% of silver in solder avoids the scavenging effect of silver metallization on the solar cell during solar cell interconnection. At present, Sn60/Pb40 solder-coated PV ribbons are extensively used, as they have passed all the tests. Some manufacturers prefer a lead-free solder coating on interconnectors.

3. Yield strength: The yield strength is the minimum amount of stress required to permanently deform the PV ribbon. This is an important specification because if the yield strength of the PV ribbon is too high, it can put large amounts of stress on the solar cells, leading to cell breakage during interconnection. For the solar cell stringing process, an interconnect with low yield strength will minimize the solar cell breakage by reducing the stress on the solar cell. Copper expands ($17 \times 10^{-6}/°C$) at a different rate than silicon ($2.6 \times 10^{-6}/°C$) during heating and cooling, thus creating stress at the joint. Hence, solar PV module manufacturers are incorporating thinner solar cells to reduce costs while enhancing the performance of solar PV modules. When selecting a PV ribbon, it is best to find a ribbon with a sufficiently low yield strength for the cell thickness. The thinner the cell, the lower the yield strength required. A low yield strength ribbon with an optimized cross-section is required to reduce cell breakage and maximize module power. For instance, it is best to find a PV ribbon with a yield strength of 80 MPa or less for solar cells that are 160–180 μm thick.

Using a ribbon with a low yield strength is an option to avoid or minimize solar cell stress during the interconnection process. The ribbon cross-section, soldering temperature, cell design, thickness, contact metallization, and soldering equipment are the parameters that influence the stress in the solar cell during the soldering operation.

4. Elongation: Elongation is a measure of the wire's ductility. A high elongation wire implies high ductility while enhancing the long-term solar PV module reliability. A high elongation wire is able to withstand the low and high temperature cycling in field conditions. The high elongation and low yield strength, which are required properties of a ribbon, help with the lowest solar cell breakage and higher solar PV module life.

5. Camber: The camber determines the wire's straightness. Interconnect and bus wires must be straight with little camber. Camber can be difficult to control when producing wires with low yield strength. The solar PV module manufacturer may experience problems during the stringing process, for example missed bonds, if there is no control of the camber. Exercising control in the rolling process, coating thickness, and winding parameters is required to minimize the camber. The straightness standard is min <5 mm/m.

6. Winding: Winding is a critical element to optimize the payoff of the wire. The size of the spool, the winding parameters, and the wire tension are to be controlled to minimize camber while eliminating tangle and preserving yield strength. The PV ribbon spools are to be packaged with corrosion inhibitor wraps or are to be vacuum-packed.

The PV ribbon is rolled from high purity copper round wire to high precision flat strip and, annealed to extra softness. The flat copper strip is plated with solder all-over in hot-dip-tinning process and precision level-winding done on the spools. The specifications of the PV ribbon are as follows:

- Copper percentage: $\geq 99.90\%$.
- Copper conductivity: $\geq 99\%$ IACS. IACS stands for International Annealed Copper Standard. The conductivity of the annealed copper (5.8001×10^7 S/m) is defined to be 100% IACS at 20°C.
- Tensile strength: ≤ 25 kgf/mm^2.
- Elongation: $\geq 25\%$ (copper width <3 mm) $\geq 15\%$ (copper width ≥ 3 mm).
- Bare copper flat wire thickness tolerances: ±10% of the nominal thickness.

Bare copper flat wire width tolerances: ±0.1 mm.

- Solder composition in wt.%: Sn60/Pb40; Sn63/Pb37; Sn62/Pb36/Ag2; Sn96.5/Ag3/Cu0.5.
- Melting point of solder: 183°C; 179°C; 217°C depending on the composition of solder.
- Thickness of single solder layer: 10–40 μm.

Tests to be conducted on solar PV ribbons online and offline

Dimensional testing—The width and thickness of the bare copper ribbon without the solder coating material is monitored continuously by inline noncontact and contact methods during the manufacturing process.

Tensile test—The tensile strength, yield strength, and elongation of the copper core without the solder coating have to be checked on the product.

Coating thickness test—The solder coating material thickness is tested with X-ray coating thickness equipment. Dedicated coating thickness measuring equipment is required to measure accurately the coating thickness and coating material and for verification of the correct amount of solder material for the soldering process.

Finished product surface test—A surface detection system that is an inline and noncontact type examines 100% of the PV ribbon surface to find surface defects.

Straightness test—The camber defines the straightness of the PV ribbon. The camber test on the soft annealed ribbon needs to be performed very carefully to avoid any error in the final results. A standardized SEMI test procedure allows accurate and comparable measurements.

Chemical composition of copper base material will be checked.

Electrical conductivity of copper base material and solder coating material will also required to be checked.

Peel test—Pull force to investigate the solder bond quality by making a solder joint with a solar cell.

Soldering tests—Soldering test to determine the soldering parameters for certain ribbon and cell specifications.

Metallographic specimen—Soft annealed PV ribbon grain structure is investigated using strain cycle fatigue analysis methodology.

Aging test—For simulation of shelf life, the solder ability of aged material is to be carried out.

Bending test—A bending test is done to check the performance/reliability of ribbons in solar PV modules using strain cycle fatigue analysis methodology.

5.2.7 Solar module junction box

Each solar module junction box (JB) has two output DC cables. One cable is positive (+), and the other is negative (−). The other ends of the cables are connected with lockable type connectors, which make the wiring connection of the solar PV array much simpler and faster.

The bypass diodes protect the solar cells in the event of partial shading of some solar cells. The solar module junction box seals the back of the solar module, where the interconnectors protrude, to avoid moisture entry.

As the junction box is an important component of the solar PV module, its reliability is a key factor for the long-term performance of the solar module. The junction box undergoes thermal, mechanical, and electrical energy stresses during its operation. It should protect all its inside components from the harsh environments of humidity, high and low temperatures, and rain and wind load conditions. So, it should meet all the design and operating requirements.

The junction box consists of a housing with a lid, electric connector terminals, bypass diodes, cables, and connectors. Fig. 5.15 shows the clampable-type junction box with cables and connectors.

Design requirements

The housing of the junction box should have high insulating properties and be able to withstand UV irradiance as well as hot and cold temperatures.

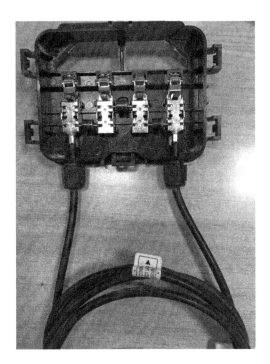

FIG. 5.15

Clampable-type junction box with cables and connectors.

The box should have sufficient space and should possess high thermal conductivity to dissipate the heat of the bypass diodes during their operation.

Moisture or water ingress in the junction box can cause electrical shorting and corrosion of the interconnects. So, the junction box has to be qualified for IP65 or IP67 ingress protection.

The base and cover of the housing should be made of HDPE/polyphenylene oxide (PPO)/Xyron and should meet flammability standards as per UL 94V-0. Junction boxes are typically constructed of Noryl (consisting of polyphenylene oxide and polystyrene) or Xyron (consisting of modified polyphenylene ether and polystyrene).

The solar PV module JB should have a rating of 5-V flammability, either by a material test or testing the JB design after water immersion and exposure of the JB as per IEC 60695-1-1.

– The solar PV module JB should have a minimum resistance to hot wire ignition as per IEC 60695-1-1.
– Should have suitable comparative tracking index.
– Should have suitable relative thermal index.
– Should have good electrical, thermal, and mechanical properties such as:
– Dielectric strength >49 kV/m.
– Volume resistivity > 10^{14} Ω m.

– Tensile strength > 59 MPa.
– Tensile strain > 80 MPa.
– Flame test characteristics as per UL 1704.
– UL testing certificate is required for JB.
– Contact resistance of the conductor should be lower than 0.5 ohms.
– Contact material—tin-plated beryllium copper.

A solar PV module junction box has to be certified according to IEC61215, IEC61730, and UL1703. The following are the typical specifications of the solar module junction box:

Rated voltage: 1000 V/1500 V.
Rated current: >1.25 times of I_{sc} of the solar cell.
Test voltage: 6 kV (1 min).
Operating temperature: −40°C to 90°C.
Pollution degree: 2. 1 if it is properly designed.
Protection class: Class II.
Flame class: UL94V-0.
Plug force: less than 50 N.
Withdrawal force: more than 50 N.

The total cable length should be as per the requirements of the customer from 90 to 120 cm. The junction box design has a significant impact on the thermal diode performance. When qualified without a solar module, the junction box has to meet DIN V VDE V 0126-5:2008 standard requirements. When qualified with a solar module, it has to meet the standard IEC 61215. The interconnect contacts should be solderable or clampable. It should have provisions to accommodate the required number of bypass diodes.

The junction box should be qualified to IEC 62790 standards. The junction box shall be certifiable to 1000 V/1500 V (IEC). It should be certified to IP65/IP67 requirements for ingress protection. The junction box will have a lid or cover to open the JB. But for sealed-type junction boxes, the cover will not be opened. The junction box will be compatible with a connected cable of size 4 mm^2.

There are different types of junction boxes such as unpotted, prepotted, and fully potted that are being used. Some have lids/covers that can be opened and diodes are accessible with interconnector connections that are clampable and solderable. Some are with prepotted diodes and the interconnector connections are clampable are solderable. Some junction boxes are with solderable connections with the completely potting type. In the clamping type of junction box, solar interconnectors coming out of the solar cell strings are attached to junction box terminals by mechanical clamping. There are no solder fumes or major cleanups in this method, but mechanical joints could get loose over time and poor electrical connections can cause arcing in the junction box, which could lead to fires. With the soldering and potting method, solar interconnectors coming out of the solar cell strings are soldered to the connector terminals in the junction box. The junction box will be filled with an adhesive to cover all the connections and bypass diodes while enabling the easy thermal transfer of heat. Junction boxes without pottants are usually vulnerable to water ingress and care should be taken to prevent that during the transportation, storage, and installation of the PV modules.

With the increased current and power of advanced solar module technology, the junction box has to be designed to handle the higher power, current, and higher temperatures while operating properly for its intended life of 25–30 years. The new type of small split junction boxes are developed and each one will have one diode. The split type of junction boxes are being used for glass/glass, glass-transparent back sheet-based bifacial and half-cut solar cell-based solar modules. For building integrated type solar modules also, split junction boxes are being used.

On bifacial solar modules, the junction boxes are to be positioned on the edge of the laminate or on three different locations of a side with split type ones. Edge junction boxes need to be mounted for bifacial solar PV modules, one each for the right side, middle, and left side of the bifacial solar module. This works effectively as one larger rectangular box. Module manufacturers are developing junction boxes for positioning along the absolute edge of the bifacial modules. The split junction boxes are mounted on the middle of the modules on the rear side for half-cut cell-based monofacial and bifacial solar modules.

Only materials and adhesives with proven long-term reliability are used to ensure the proper adhesion of the junction box to the module throughout the lifetime. Some adhesives may have reasonable adhesion strength in the beginning, but that could deteriorate during field exposure to heat, humidity, and UV. Connectors with proven field records should be preferred. Multicontact MC4 or compatible with MC4 connectors with socket and pin type lockable connectors are used.

Bypass diode

Bypass diodes are used in PV modules to prevent the application of high reverse voltage across the cells in the event of shading. These are the requirements:

It should have low forward voltage and fast switching characteristics. So, Schottky diodes, which are semiconductor-metal junction-based, are used for solar modules.
Bypass diodes should have a low leakage current.
The maximum repetitive reverse voltage (VRRM) of the bypass diode is directly linked with the number of cells bridged by the bypass diode.

Bypass diodes can fail in two modes: short-circuit mode and open-circuit mode.

For a PV module with three bypass diodes, a short-circuited diode causes the module to lose one-third of the power per failed diode. This immediately results in the failure of the PV module as the power drops below 80%. If a bypass diode fails in open circuit condition, it can pass no current and it does not affect the solar PV module power output. It is comparable to the associated solar cell substring not having a bypass diode. However, if the substring of cells with the open-failed diode is shaded, shaded cells could be forced into extreme reverse bias, causing cell breakdown, overheating, and—in the worst case—a fire hazard. Considering the potential safety issues, modules with open-failed diodes are also regarded as failed modules.

The failure mechanisms of bypass diodes in solar modules can be divided into two categories: catastrophic and wear out. Catastrophic failures include those due to arcing, electrostatic discharge, and thermal runaway. The wear out of bypass diodes is due to high temperature forward bias operation, high temperature reverse bias operation, and thermal cycling.

Arcing

If the electrical connection leads of the diode in the junction box are poor, it can lead to arcing in the event of partial shadowing of the solar cells due to the passage of high current. The arcing will destroy the diode.

Electrostatic discharge (ESD)

Schottky diodes are prone to failure by ESD. It is regarded as the most prominent failure mechanism for bypass diodes in PV modules during their attachment to the junction box and in flash testing. Some flash testers are also known to produce current surges that have the potential to damage the bypass diodes in PV modules [18]. Sometimes, ESD events cause latent failures in diodes, in which case the diode as a whole does not fail but small areas on the diode suffer degradation. If such bypass diodes end up getting installed in the PV modules, they are subject to premature failure in the field. Instruments such as a flash tester and insulation resistance testers could also generate current surges that can damage the diodes. In order to avoid ESD issue, the solar PV junction box manufacturers can implement ESD protective measures at factory level.

Thermal runaway

When the solar PV module is partially shaded, the diode corresponding to the shaded substring gets forward biased. The power dissipation in the diode in forward bias begins to increase the temperature of the diode. If shading is held for a sufficient time, the diode temperature stabilizes as it reaches thermal equilibrium. If the shading is suddenly removed, the diode returns to reverse bias. If the power dissipation in the reverse bias is greater than the heat dissipation capability of the junction box, the diode temperature begins to increase. The reverse current (and power dissipation) increases exponentially with temperature. This leads to even more power dissipation and this process continues until the diode is completely destroyed by overheating. This phenomenon is called thermal runaway. Also, it takes just one shading event under the right environmental conditions for thermal runaway to take place.

High temperature forward bias (HTFB) operation

When the module is partially shaded, the diode becomes forward biased. The ambient temperature and the current passing through the diode determine the operating temperature of the diode. Some modules regularly experience shading during their lifetime. Therefore, the diodes spend a significant amount of time in the forward bias state. Depending on the module mounting configuration, location, the extent of shading, and the current passing through the diode, the diode may have to operate at a high temperature for considerable time during its service life. Long-term, high-temperature operation is known to reduce the lifetime of the Schottky diodes, just like any other semiconductor components.

High temperature reverse bias (HTRB) operation

Bypass diodes operate in reverse bias in the normal, unshaded state of the PV module. The voltage generated by the substring of cells (12–14 V) gets applied in the reverse direction. The temperature of the diodes in this state is typically equal to the module temperature, as the reverse leakage currents are not sufficient to cause any significant self-heating. However, continuous operation of the diodes at reverse bias and moderately high temperatures is known to cause localized breakdown at the die level, which can eventually cause a complete failure of the device.

Thermal cycling (TC)

Diodes experience stress not only from operating at elevated temperatures for long amounts of time, but also from thermal shock caused by sudden or gradual changes in the
temperatures. A diode could experience a temperature difference close to 200°C just during a single day. The difference in the maximum and minimum temperatures (ΔT_J) experienced by the diode during a day is significantly more than that experienced by the module. This is because the increased current passing through the diode in the forward bias state causing increased heat dissipation.

Cable requirements for junction box

Conductor material: tinned copper wire.
Conductor size: 4 mm^2.
The cables extending from the module shall be rated for the appropriate system voltage, ampacity, wet locations, temperature, and sunlight resistance.
Insulation and jacket material: XLPO.

Connectors

Quick connectors are used to connect the PV modules in series and make the end connections for the PV array. These connectors are expected to provide protection to the electrical joints against moisture-induced corrosion and water ingress during the module's service life. Connectors that tend to get loose over time due to mechanical, vibrational, and thermal stresses increase the risk of overheating and the formation of arcs. Increased series resistance at the connectors due to corrosion will lead to power output loss by the solar PV array.

The connector used in the output circuit of a solar PV module shall be rated for the accurate voltage and current and be in line with the requirements of IEC 60130. Further, the connector shall comply with the standards with respect to flammability, comparative tracking index, and relative thermal index for the support of live parts.

A connector should have resistance to UV and the inclusion of water equivalent to IP55. The lockable-type male-female connectors are shown in Fig. 5.16.

The following are the tests connected on the junction box as part of the qualification test:

Flammability tests and weather-resistance test with xenon arc lamp/flash lamp for 500 h.
Glow wire test up to a temperature of 650°C and a gasket conditioning test.
Current carrying metal contacts should not show signs of corrosion after storing in 10% NH$_4$Cl solution.

FIG. 5.16

Lockable male and female connectors with socket and pin.

The following conditions are to be met for the ratings of the junction box and bypass diode:

1.25 Times solar module short circuit current (I_{sc}) at STC < fuse rating of the solar module ≤ maximum current rating of the junction box ≤ current rating of the bypass diode.

5.2.8 Frame

To get the proper voltage, the solar cells are to be connected in series with the required number of solar cells to form a solar cell circuit. The solar cell circuit is sandwiched between a glass and polymer sheet using an encapsulant such as EVA and laminated in a laminator. To get power to an appliance, the solar cell circuit with laminate has to be exposed to sunlight by mounting it outside. In order to mount the PV laminate and protect it from any moisture ingress, the frame is required to close the laminate on all four sides. Solar PV modules are fitted with aluminum frames that cover the laminate. The solar module frame plays an important role and its role lies in its mechanical characteristics.

The following are the functions and requirements of the frame:

– Frame should protect the laminate and its edges.
– Frame is used for handling of the solar module.
– Facilitates the solar module to be stored either in the horizontal or vertical position.
– Provides grounding holes to enable the solar PV module grounding/earthing.
– Mounting attachment points to facilitate the fixation of the module to the mounting structure.
– Frame will have water-draining holes to avoid accumulation of water in the hollow section of frame to cause freezing of water.
– Protects the internal components such as solar cells, glass, and back sheet of solar modules from thermal and mechanical loads.
– Protects the module during transportation.
– Provides resistance against mechanical loads such as wind load of 2400 Pa and snow load of 5400 Pa.

– Frame should have high mechanical strength to provide strength to the entire module and to withstand static and dynamic loads.
– Frame material should have good thermal as well as electrical conductivity.
– Frame should be corrosion-resistant.
– Frame should not have burrs at the edges.
– Frame should have a profile and height to accommodate the laminate such that a sufficient gap is there to accommodate the junction box.
– Frame should be easily attachable with laminate using silicone adhesive or double-sided adhesion tape.

Anodized aluminum is a superior material as a solar module frame. Coated solar module frames are prone to sustain scratches or damage and at the same time they are expensive. Solar module frames with closed-in cavities are to be avoided. Due to water collection and freezing, frames are likely to expand; this can deform or break the frame.

A flat tapered edge where the solar module frame meets the glass/cells is preferred as it reduces shading and allows water to run off. Also, it doesn't allow ice to build up in the winter. Ice can cause damage such as the separation of the frame from the solar PV module.

With tapered edges, the frame does prevent soiling, etc., which can lead to casting shadows on solar cells adjacent to the edge. If dust covers part of the solar PV modules, it can affect the solar output. Fig. 5.17 shows a typical frame with tapered edges.

The adhesives used to join the frame and glass together are important to avoid stress build up because the expansion and contraction ratios of the two parts of the solar module are different. The thermal expansion coefficient of Al 6063 is $2.34 \times 10^{-5}/°C$, whereas for glass the value is $9 \times 10^{-6}/°C$.

The solar panel frame must not have sharp edges and should be screwed together with specially designed screws. Otherwise, it should be firmly fixed with corner key elements. This type of tight fixing guarantees that the frame is electrically and mechanically conductive and can provide the required earthing.

FIG. 5.17

Frame with a tapered edge.

- Anodized aluminum with a suitable thickness of anodization coating increases the corrosion resistance.
- High tensile strength enhances the wind load impact and snow loading resistance to meet tests as per UL1703, IEC61215, and IEC61730 standards.
- The color of the frame is usually either silver or black, due to anodization.
- It is easy to handle anodized thinner aluminum frames, which are good for withstanding roof loads and are lighter for easy installation.

For solar PV modules to withstand a mechanical load, the frame properties determine the ability of the module to withstand wind and snow loads.

Aluminum has the following desirable properties as a frame material:

Weight—Aluminum is lighter in weight, having one-third the density compared to steel (2.69 g/cm^3).

Strength—Aluminum is strong with a tensile strength of 145–187 MPa, depending on the alloy and manufacturing process. Al extrusions with the right alloy and design are equally as strong as structural steel.

Elasticity—The Young's modulus for aluminum is 68.3 GPa, which is one-third that of steel. The aluminum extrusions should have three times the moment of inertia to meet the steel deflection requirement.

Formability—Aluminum has good formability, which is used in extruding and facilitating the shaping and bending of extruded parts. Aluminum can be milled, cast, or drawn.

Machining—Machining of aluminum is quite easy.

Corrosion resistance—When in contact with air, a thin layer of oxide is formed, which provides very good protection against corrosion even in extreme environments. This thin layer of oxide can be strengthened by surface treatments such as anodizing or powder coating.

Thermal conductivity—As compared to copper, the thermal conductivity of Al is very good.

Electrical conductivity—When compared to copper, aluminum has good electrical conductivity.

Linear expansion—In view of aluminum having a relatively high coefficient of linear expansion as compared to other metals, it is a good candidate for a frame. Differences in expansion can be accommodated at the design stage or in manufacturing.

Al 6063 is the alloy used for solar applications. It contains the following elements: Si, 0.2%–0.6% by weight; manganese (Mg), 0.45%–0.9%; iron, maximum 0.35%; and copper, chromium, zinc, and titanium each up to a maximum of 0.10%. It has good extrudability and can be solution heat-treated during hot working at the extrusion temperature. Solution heat treatment enables some of the alloying elements, such as Mg and Si, to go into solid solution and be maintained in a supersaturated state on quenching. This homogenous material is subsequently age-hardened to obtain the required mechanical properties.

The tensile strength decreases as the temperature increases while elongation before fracture usually increases. In contrast to steel, aluminum alloys do not become brittle at low temperatures. In fact, aluminum alloys increase in strength and ductility while the impact strength remains unchanged. As the temperature decreases below 0°C, the yield strength and tensile strength of aluminum alloys increase. As the alloy contains silicon, magnesium, and manganese, it shows good resistance to corrosion in seawater.

To get better mechanical properties and good structural rigidity, the alloy is tempered, which is a process of heat treatment to get a toughened alloy. The Al frame is tempered for a T5 or T6 temper. With a T5 temper, Al 6063 achieves an ultimate tensile strength of at least 140 MPa and a yield strength of at least 97 MPa. It has an elongation of 8%.

With a T6 temper, Al 6063 has an ultimate tensile strength of at least 190 MPa, a yield strength of at least 160 MPa, and an elongation of 10%.

Fig. 5.18 shows different profiles of frame. The side and bottom views also shown in the figure. Fig. 5.19 shows different corner key profiles used for solar module frames. The wall thickness of the frame and the height of the frame will decide the load-bearing capacity of the frame.

The extruded frame profiles are cut to different lengths and mounting and grounding holes are provided. The cut frames are electrochemically anodized in acidic and base chemical baths for about a 15 μm thick anodization.

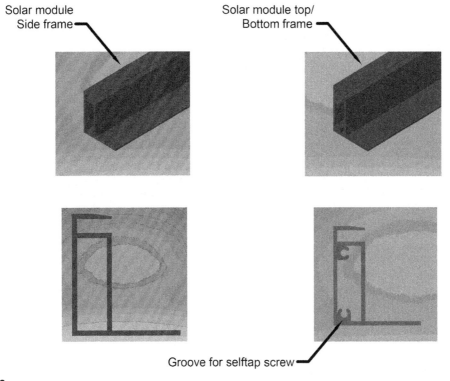

FIG. 5.18

Profiles of the frame. *Left bottom*: Profile of the frame used length side of the module; *Left top*: Three-dimensional view of length side frame profile. In the top of F type groove, the laminate is fitted. *Right bottom*: Profile of the frame used top and bottom side of the module. In the top F type groove, the laminate sits. Two grooves in the rectangular hollow space are used for joining the length and width side frames using self-tapping screw. *Right top*: Three-dimensional view of top and bottom side frame profile. Two grooves are seen for usage of self-tapping screws. At present, the most of the modules are using corner key to join the length and width sides of frames.

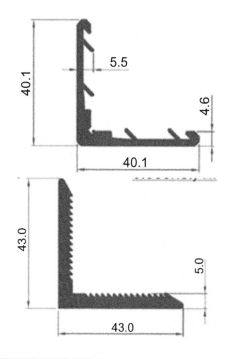

FIG. 5.19

Profiles of the corner key used to connecting the frames. *Top*: 40.1 × 40.1 mm size corner key used to join two frames at right angle interface. This corner gets locked inside the frame. *Bottom*: Saw tooth type of 5 mm thick 43 × 43 mm size corner key used to join the length side and width side frames.

Young's modulus, tensile strength, composition of alloy, visual inspection, dimensional check, and thickness of anodization are some of the tests conducted on the frame.

For the frame design, the size of the frame and material parameters are modeled in Solidworks or any other software. The load application is simulated and strain values are derived and with iteration process of changing parameters, the final design will be arrived.

5.2.9 Adhesive and sealant

The aluminum frame, which is fixed to the laminate, serve as an edge protection to avoid glass breakage and also shields the internal components of the solar module from environmental conditions. The junction box, which is mounted on the rear side of the modules, is a housing for current-carrying PV interconnectors and bypass diodes. An adhesive/sealant is required to fix the solar PV laminate to the frame and to mount and seal the junction box.

The following are the requirements of the adhesive for fixing the frame and mounting and sealing the junction box:

The adhesive should provide reliable bonding throughout a range of temperatures and conditions. It should be easily adaptable to solar module manufacturing and assembly processes while also providing increased production efficiency with fast curing.
It should have resistance to degradation from UV light and extreme temperature cycling.
It should have a very low moisture vapor transmission rate (MVTR) to act as a moisture barrier.
It should adhere properly to metal, glass, the back sheet, and fluorocarbons.

Historically, there have been two primary ways to mount and seal a junction box using adhesives: liquid adhesives such as silicone RTV (room temperature vulcanization), or adhesive tapes such as acrylic foam tape. New advances in both liquid and tape technologies are providing solar module manufacturers with a range of solutions for junction box mounting and sealing that can meet their needs. Both mounting and sealing methods have their pros and cons.

Liquid adhesives: Silicone-based adhesives have been used to fix the frame to the laminate and mount the junction box to the solar modules since the inception of the solar PV industry. They provide strong durable bonds with good resistance to temperature extremes, plus high moisture resistance. Moisture intrusion into the electrical connections within a junction box can cause tremendous damage. Liquid adhesives can be used with junction boxes for both potting/sealing and bonding on the rear side of the modules. Silicone RTV is a traditional potting and mounting adhesive, applied both manually or as part of an automated dispensing process. Silicone RTV has a good moisture vapor transmission rate (MVTR) plus the following characteristics:

It bonds to glass, metals, coated materials, and plastics.

- Able to withstand UV radiation and harsh weather conditions.
- Provides a flexible joint to withstand movement and temperature cycling.
- Provides higher long-term durability.
- Heat resistance from $-40°C$ to $150°C$.
- Wide operating temperature range from $-115°C$ to $316°C$.
- Has excellent electrical properties.
- Has good chemical resistance to humidity and water.
- No or low toxicity.

Depending on the temperature, these adhesives are incredibly flexible and have a wide operating temperature range. These adhesives are resistant to bad weather conditions, humidity, and mold and mildew, and also have excellent electrical properties. RTV silicones have a lifespan of 40 years with a high degree of elongation. It is easy to dispense these adhesives, even in cold temperatures. They are also volatile component compliant with properties such as excellent UV and thermal stability. RTV silicone is a silicone rubber that is generally supplied as a one-part system with a wide viscosity range. It is a combination of organic and inorganic compounds that makes it one of the most stable organic adhesive products and an ideal adhesive. RTV adhesives and sealants are high-temperature resistant and are flexible as compared to other industrial adhesives.

Table 5.6 gives the typical properties of a silicone adhesive.

RTV silicone adhesives and sealants utilize moisture in the atmosphere as a curing agent or a mechanism to form an adhesive bond or seal. When these sealants combine during cross-linking, a chemical byproduct is produced. Based on the materials that are being used, the byproduct can be acidic (that is, acetic acid), basic (amine), or neutral (oxime/alcohol). The curing agent will determine the final

Table 5.6 Typical properties of a silicone adhesive.

Typical properties of a silicone adhesive	
Chemical base	**One-component silicone**
Color (CQP1 001-1)	White
Cure mechanism	Moisture-curing
Cure type	Oxime
Density (uncured) (CQP 006-4)	1.32 kg/L
Nonsag properties (CQP 061-4/ISO 7390)	2 mm
Application temperature	5–40°C
Skin time (CQP 019-2) at 23°C/50% RH	23 min
Tack-free time 2 (CQP 019-1)	90 min
Shore A-hardness (CQP 023-1/ISO 868)	35
Tensile strength (CQP 036-1/ISO 37)	1.8 N/mm^2
Elongation at break (CQP 036-1/ISO 37)	450%
100% modulus (CQP 036-1/ISO 37)	0.7 N/mm^2
Thermal resistance (CQP 513-1) 4 h, short term 1 h	200–220°C
Service temperature	−40°C to 150°C
Shelf life (storage below 25°C) (CQP 016-1)	12 months
CQP, *corporate quality procedure.*	

properties of the silicone adhesive. The relative humidity and ambient temperature will also directly determine the cure rate, but RTV silicones tend to cure within a 72-h period while the adhesive can continue to firm up for up to 2 weeks after setting.

Acetoxy cure and neutral cure silicones

RTV silicones are two types such as acetoxy cure silicone and neutral cure (oxime) silicone, which are used abundantly. Acetoxy cure silicone has a faster curing rate and a short tack-free time, providing good quality adhesion while also being corrosive to metals. This type of RTV silicone releases acetic acid as byproduct, which gives smell of a vinegar. On the contrary, oxime, or neutral cure silicone, is noncorrosive and has resistant properties against oil and temperature. This type of silicone produces a neutral/nonacidic byproduct, which will not take longer time to cure and has a longer tack-free time. Oxime cure-based RTV adhesives are used for solar PV module applications.

As adhesives are liquids, they can be incorporated into dispensing stations to provide an inline manufacturing solution through automation, which can help to reduce material costs and processing time. Automated adhesive dispensing equipment eliminates the possibility of error in manual or semi-automatic processing. Processing systems in automated dispensing arrangements can minimize excessive adhesive application and waste or residue.

New advances in silicones for junction box potting and mounting are now offering fast-fixturing with fixture strength in just 2–3 min and handling strength in less than 20 min. Elasticity for these silicones has also been improved to provide more flexibility for mounted junction boxes. In addition to faster curing at room temperature, silicones provide a waterproof seal with strong adhesion to back sheet materials such as PVF and PET.

Liquid structural adhesives, elastomers, and thermally and electrically conductive adhesives and materials are undergoing continuous research as panel manufacturers investigate new technologies to meet customer warranties while also streamlining production time, effort, and costs. The following are the properties to be checked on the cured adhesive:

Cured elastomer hardness—It is defined as the final hardness of the cured rubber. The hardness will infer the adhesive's suitability to be used as a compression gasket and its capability to withstand thermal expansion or suppressing of the vibration.

Elongation—It is the percentage of elongation before the cured rubber snaps.

Tear—It is the force required to tear a sheet of cured elastomer after making a small cut.

Tensile strength—It is the force required to break the cured elastomer when subjected to tension.

Temperature resistance—It is the range of temperature in which the adhesive will retain its physical properties. The capability to withstand a wide range of temperature variations is linked to the choice of silicone polymer and the cross-linking system used. However, with increased temperature, resistance can be achieved with the addition of special fillers such as iron oxide.

Thermal conductivity—It is defined as the capacity to transmit heat by the elastomer. By adding exclusive fillers to the silicone polymers, it is possible to produce adhesives that are capable of dissipating heat.

Electrical conductivity—It is defined as measuring the electrical resistance of the adhesives. Silicones are by nature electrical insulators with high resistivity, but with the addition of conductive fillers, it is possible to create materials that will conduct or dissipate electricity.

Adhesive for potting junction box

The potting of the junction box with an adhesive by covering all the connections, solder joints, and bypass diodes improves the heat dissipation and insulation properties of the junction box. The sealant helps to avoid diode failure risk due to overheating, which ends up reducing the overall cost. The compounds used for potting of the solar module junction box help to decrease the operating temperature of the bypass diodes while increasing the safety and reliability of the solar module junction box.

The characteristics of the silicone potting compound include the following:

- Excellent fluidity and easy gap-filling.
- Excellent dielectric properties and flame retardancy.
- Good thermal conductivity.
- Excellent resistance to harsh weather.
- Suitable for automated mixing and dispensing.
- Excellent compatibility with EVA, ribbon, and J-box material.

Disadvantages of RTV silicones

While silicone RTV provides many advantages for mounting and sealing, there are also familiar draw-backs and some disadvantages:

(1) Curing times can run anywhere from 10 min to several hours to days, and this can slow down the panel manufacturing process.
(2) Silicone RTV use can be messy (even with dispensing equipment) and the time and labor devoted to clean up can be significant.
(3) Inconsistent application (even using dispensing equipment) can also cause quality, durability, and WVTR issues.

RTV silicones are susceptible to picking up dirt and have poor tear resistance, poor cohesive strength, and low tensile strength.

Double-sided adhesive tapes

Some solar module manufacturers find that their processes benefit more for mounting the frame and the junction boxes using adhesive tapes. In these cases, precut rolls of double-sided adhesive tape are used to eliminate the time-consuming application of liquid adhesives, the investment in dispensing technology, and, above all, the time required for the liquid adhesives to cure. In addition to slowing down production in their processes, the effects of discovering a problem might mean hundreds of panels could be rejected and scrapped.

There are three types of tapes used for solar frame fixing applications: acrylic foam tapes, polyethylene-based tapes, and polyurethane based tapes.

Polyethylene and polyurethane foam tapes. Polyethylene (PE) and polyurethane foam tapes have long been used by solar manufacturers for edge and frame sealing and for attaching junction boxes. These tapes are made in different grades and thicknesses. Typical thicknesses are 0.8, 1, and 1.55 mm with a thickness on the tolerance of ±20%, which is normal for a blown foam.

PE foam tapes are coated on both sides with adhesive using a transfer lamination process. The foam is corona-treated so the adhesives will key into the foam. The transfer lamination process gives a variance in the performance of the tape. The corona treatment consistency and quality of the foam along with the lamination process control affect tape performance. In the event of an uncontrolled process, the delamination of the adhesive can occur. PE tape is very useful in applications where gap filling is required and the bond is not subjected to a lot of stress. PE tapes are easily applied and can fit well into the typical solar panel manufacturing process. They are also cost-effective materials.

The compressive strength of PE foam tapes, however, is low. Solar cell rupture can occur with very little force and foam cells don't recover well from compression. The internal cohesive strength of foam is poor and tears can occur. Flex strength during elongation and maximum static load is also low. When subjected to repeated expansion and contraction as a result of high and low temperature exposure and cycling as well as being used with different materials such as glass, aluminum, and plastic, the foam will degrade and break down over time, leading to leakage and water absorption.

There have been issues with PE foam tapes in solar applications where the foams fail after 8–9 years of service life. Selecting the appropriate materials is a critical decision for solar manufacturers whose products are expected to last longer than 25 years.

Acrylic foam tape. High-strength acrylic foam tape provides an attractive alternative for adhesive in solar applications. The acrylic looks and feels like foam but in most cases, it is an acrylic with air bubbles and glass beads injected into it. It also gives the tape a viscoelastic effect that will allow it to stretch and retract to its original shape without breaking the bond. This provides the excellent expansion/contraction capabilities necessary for solar use without any adhesion loss. The acrylic foam tapes have tensile strength with good load-bearing characteristics, high shear, and peel adhesion characteristics; they are also resistant to plasticizer migration. Acrylic foam tapes also have excellent durability with solvent and moisture resistance.

Additionally, acrylic foam tapes can resist very high wind forces and snow loads. They are also more than capable of withstanding very high UV exposure for long periods without degrading or discoloring. Unlike PE foam tapes, they can withstand temperature extremes from −40°C to 160°C.

Acrylic foam tapes also provide moisture, dust, and air sealing for frame bonding, edge sealing, and junction box mounting. The tape can be precision die-cut for use as a gasket for a wide range of junction box sizes and shapes. It bonds well to polycarbonates, polypropylene, and other thermoplastics.

Eliminaion of adhesive curing time allows some manufacturers to keep the production line moving and there is no need to wait for liquid adhesives to cure. The selection of the appropriate junction box sealant technology will always depend on the specific manufacturing process. Tape offers new possibilities for solar panel manufacturers. However, like every option, it has its drawbacks, the largest being its higher cost.

If solar frame mounting and sealing is not proper, during rainy season, water seeps into the laminate and causes reduction of insulation resistance of the module. Because of lower insulation resistance of the module, the inverter connected to the module strings in solar plant considers this as a solar array string fault and gets switched off.

To reduce the processing time, solar module manufacturers earlier adapted double-sided adhesive tapes for fixing the frames and bonding the junction to the module. As the debonding of junction boxes and the detachment of frames has been observed in solar PV plants, the module manufacturers switched over to silicone adhesive.

The junction box adhesion strength and frame pulling strength are some of the tests to be conducted on solar modules as part of testing during the manufacturing process.

5.3 Manufacturing process of solar PV module and equipment

Solar cells generate electrical power by converting light energy using the photovoltaic effect. To effectively use the solar cells in different applications, the solar cells have to be connected in series to get the required voltage. To protect the solar cell circuit from the environmental conditions, the circuit is sandwiched between a glass and a back sheet and laminated. The laminate is hermetically sealed on all sides and framed with an aluminum frame for easy mounting. This final product is called a solar PV module.

The cross-sectional view of the flat plate solar PV module is shown in Fig. 5.1.

The solar module manufacturing process steps followed for manual and semiautomatic production plants are as follows:

- The glass is washed with warm demineralized water in a glass washer to remove dirt/contamination from the glass. At present, the antireflection-coated glasses are not being washed by water, as the available glasses are already washed after manufacture and packed in the box.
- The EVA and back sheets are cut to the required sizes using the foil cutter.
- The received solar cells are tested/verified for the electrical parameters.
- For lower wattage modules, the solar cells are cut using a laser scriber.
- The solar cells and the solar ribbon spools are loaded into the stringer. The solar cell tabbing and stringing will be done in an automatic stringer and 6–12 cell series strings will come out from the stringer.
- The glass is loaded on the conveyor belt and the cut EVA sheet is laid on the glass.
- The strings are picked up from automatic soldering station and placed on the glass-EVA layup. The strings are arranged in a matrix form using a semiautomatic layup station and manually, they are interconnected using thick and wider solar interconnectors. This is defined as bussing connection. Automatic bussing operations are also available.
- As an alternative process the strings are arranged in the matrix form as per the module drawing on a layup station. The strings are interconnected with wider interconnectors and the circuit is inspected. The soldered solar cell matrix is picked by a vacuum picker and placed on the glass-EVA layup.
- The alternative to the above process is the strings are picked by the arm robot and placed on the glass/EVA layup and the circuit matrix is formed by interconnecting the strings.
- The EVA and back sheets are laid on the solar cell matrix and the interconnectors are brought out to the rear side through the cut line of the EVA/back sheet and taped. Thus, the glass-EVA-solar cell matrix-EVA-back sheet layup is done.
- At this stage as part of prelamination, the solar cell circuit is checked for cell cracks and any other defects by electroluminescence equipment.
- The layups are loaded into the laminator through the input loading conveyor.
- With evacuation, heating, and pressure application operations, the layup is laminated in the laminator, unloaded by the output conveyor, and then cooled.
- The edges of the laminate are trimmed to remove the oozed-out excess EVA material along with the back sheet.
- The double-sided adhesive tape is fixed to the laminate or one-component silicone adhesive is used to fix the aluminum frames to the laminates using a framing machine.
- The junction box is mounted on the rear side of the module using RTV glue. If the junction box is the potting type, the adhesive will be applied, and humidity curing is allowed.
- The solar module is cleaned and visually inspected for any defects.
- The solar module is tested for its electrical parameters under a module tester.
- The solar modules are packed using various packing methods.

The process flow chart for solar module manufacturing for Al-BSF/PERC-based crystalline Si solar modules is shown in Fig. 5.20A

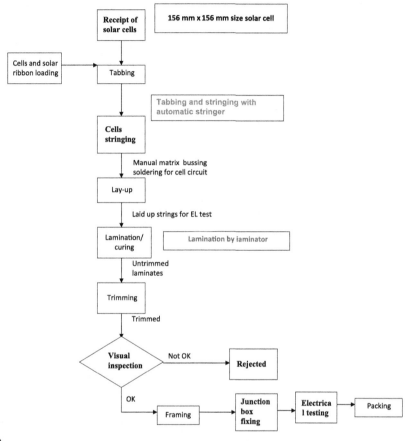

FIG. 5.20 (A)

Solar module manufacturing process flow chart.

At present, solar manufacturing plants have complete automation of all the processes. Fig. 5.20B shows the Jinchen Company's layout for a 200 MWp automatic solar PV module manufacturing plant.

Solar photovoltaic cell (PV) module assembly

There are two soldering process steps used to assemble a PV module. The first step is solar cell interconnection, called tabbing and stringing, and the second step is PV string assembly, called bussing. Initially, the cells are electrically connected using a tinned copper ribbon that is typically 1 mm wide. The solder-coated ribbon is either dipped or sprayed with flux. The flux is used to achieve wetting and for the removal of oxides from the ribbon surface. The tabbing ribbon provides the solder requirement to form a bond between the ribbon and the substrate.

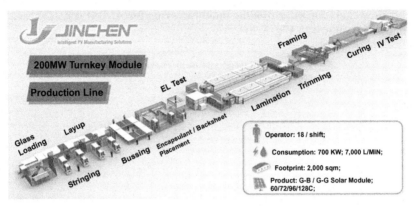

FIG. 5.20 (B)

200 MW automatic solar PV module manufacturing plant layout by Jinchen.

Courtesy: Jinchen.

Soldering is a metallurgical joining of two substrates using an interjoint material called solder, which can melt and solidify at a significantly lower temperature than the substrate. The process of soldering involves removal of oxide layers and contamination from the surfaces of the substrate and the wire to be connected using a flux. Next, the substrate, the connecting wire, and the solder are to be heated. The molten solder is then able to wet and spread on the surfaces of substrate and wire and join them. Thin layers of substrate & wire are dissolved in the molten solder and the molten alloy with the foreign atoms, subsequently form an intermetallic compound between the substrate and the wire.

The solar cells connected in series forming a column by reflow soldering is called a string, which is precisely placed onto a glass that is covered with EVA. From 6 to 10 rows of strings are then bussed/interconnected together using a 5 mm wide bus ribbon to create the collector busbar. Bussing together the columns of solar cells is manually performed using a soldering iron. The need to reduce solar PV module manufacturing costs is driving significant reductions in wafer and cell thicknesses. The soldering process has become critical in view of the thinner solar cells.

Thermal expansion of the copper and silicon elements occurs at a temperature of 300°C while following the manual soldering process. This differential temperature creates thermal stress that can result in the formation of microcracks that may not be detected using EL cameras during the module manufacturing process, resulting in a less-than-expected field lifespan. Time and temperature control becomes critical by the intermetallic layer requirement of within 1–2 μm that must be achieved during the soldering.

5.3.1 Solar cell tabbing and stringing

Tabbing

The solar cell has more than or equal to five busbars of width 0.9–1 mm or less with silver metallization as the front coating. Some solar cells have continuous pads on the front and rear and some will have discrete metallization pads, as shown in Fig. 5.21.

FIG. 5.21

(Top) Front side and (bottom) rear side of a solar cell with five busbars.

The discrete pads of the front busbars are joined with thin lines. For these contacts/busbars of solar cells, the solder-coated thin copper ribbon has to be connected. The process of contacting the solar cell busbars using tinned copper ribbons is called tabbing. Connecting the solar cells in a series of 6-12 in numbers is called stringing. The tabbing and stringing were done manually by the soldering process before the invention of automatic soldering station. The tabbed solar cell is shown in Fig. 5.22. The series interconnection of the tabbed cells more than two in number is called stringing. Fig. 5.23 shows a string formation by manual soldering. For semiautomatic and fully automatic manufacturing plants, tabbing and stringing are done in one automated step.

Visual inspection for any cracks, dark I-V measurements, solder joint strength and V_{oc} and I_{sc} measurements of the string were the quality checks followed for the manual tabbing and stringing process.

FIG. 5.22

Tabbing: the solar cell front contact soldered with a PV ribbon.

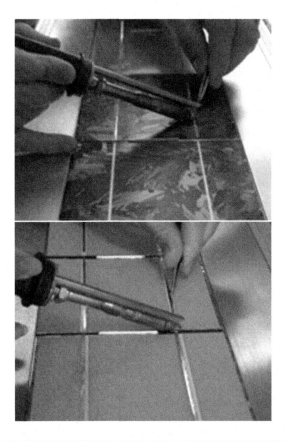

FIG. 5.23

String formation by manual soldering: (A) on front, and (B) on rear.

Automatic tabber and stringer

For the automatic interconnection of solar cells to form strings to meet a production demand of 1200–3600 cells/h, automatic tabbing and stringing equipment was developed and is in use.

All the equipment work on the principle of soldering.

The following are the different techniques of soldering being followed:

Hot air soldering—NPC, Japan; Ecoprogati, Italy.
Contact soldering—Somont, Germany, Team Tecknik, Germany.
Inductive heating-based Soldering—Komax.
Infrared heat-based soldering—Jinchen, Mandragon.

The tabber and stringer has four main remarkable elements: Cell quality control using artificial vision, advanced control of the soldering process, servodrives, and five busbar ribbon pool systems.

Vision control systems check the quality of the cells so that defective cells can be detected and rejected. Control of cell temperature measure ensures good soldering. Time control for correct self-diagnostics is required.

The main features of stringers are:

Contactless soldering so that no stress is generated in the cell.
Flux dispensing onto the busbar without contact with the cell or fluxing on the solar cells.
Anticamber systems to improve ribbon alignment.
Artificial vision system to find the cracks or defects in the solar cells.
The cell breakage rate should be less than 0.1%.
The uptime of the equipment is >95%.
The machine should provide many process formulae, where the same model or module can assume different process parameters, depending on the materials used.
It should have the flexibility to process different cell models or sizes with a very short change time.
It can be worked with cut cells up to one-half of 6″ cells.

The following are the steps in the tabbing and stringing of the solar cells:

The PV ribbon spools are to be mounted as per the required number of busbars of the cell.
The process recipe has to be set up in the HDMI of the stringer machine.
The packages containing 100 or 200 solar cells are to be unpacked and checked for any visual damage to the solar cells and the cell stack is manually loaded to cell magazine carrier.
The cell will be picked by the robot arm and keeps on the platform, where it is inspected by a camera for any defects such as edge chips, cracks, dimensional errors, etc. The rejected ones are kept in a separate bin while the visually acceptable cells are passed on to the conveyor belt, which contains the preheating plate.
For some machines, the cell busbars are fluxed with a spray from both sides.
For some machines such as Somont, Mandrogon, and Ecoprogati, the flux is not applied on the solar cell, whereas the PV ribbons are made to pass through the flux chambers for fluxing and dried before cutting and placement on the solar cell contact busbars.
The PV ribbon is drawn from the spools. It is precisely cut and precisely kept on the locations of the busbars of the cell. The interconnectors are aligned and with a slight pressure touch and an infrared heating element, the soldering process will join the solar cell and the PV ribbons. The next cell will

be soldered and like that, the defined number of solar cells will be connected in series to form the strings. The string is picked up by the robot arm, which has vacuum suction to pick up and flip the string such that the back side of the cell is facing upward.

Some stringer machines have cameras to inspect the defects in the strings. The strings are laid on the glass-EVA layup assembly.

The checkpoints:

The alignment of the ribbon with the contact busbar of the cell should be proper.

The flux should not spread around the contact busbar of the solar cell.

The preheating and temperature during soldering should be selected such that minimum thermal stresses will be there and there will be no damage to the solar cells.

The solder joint peel strength shall be checked for a 90 degree pull and it should be more than 5 N/mm. The peeled solder joint should be inspected to see whether the joint failure is the cohesion type or the adhesion type. When the joint strength is high, there will be an adhesion-type failure.

Cell interconnection is recognized as the most critical process with respect to module production yield. If the process is not carefully controlled, cell cracking and subsequent breakage may occur. Many manufacturers promise breakage rates below 0.3%–0.5% on their tabber-stringers, which applies to cells above 160–180 µm thickness that are free from initial cracks. In real production, this figure strongly depends on the materials, process parameters, and throughput.

Fig. 5.24 shows the automatic tabber and stringer and Figs. 5.25 and 5.26 depict the stringing process.

FIG. 5.24

Tabber and stringer of Jinchen.

Courtesy: Jinchen.

FIG. 5.25

Tabbing and stringing operation of an automatic stringer.

Courtesy: Jinchen.

FIG. 5.26

Vacuum picking up the string.

Courtesy: Jinchen.

Process control and defect detection

In order to reduce the damage of solar cells, consistently good intermetallic bonds are required between the ribbon and the solar cells.

In order to form an intermetallic layer of 1–2 μm, the precise control of time and temperature is required. The precise control of time and temperature will help in reducing the possibility of forming microcracks in the substrate. Damage to the crystalline silicon cells is checked by a peel test according to the soldering interconnect joint specifications.

FIG. 5.27

String repair station.

Induced thermomechanical stress in the solar cells is another challenge associated with the manufacture of solar cell modules in the conventional form. The process of interconnecting wafer-based silicon solar cells mostly involves the use of infrared reflow soldering.

The defective strings that are identified after stringing or after EL inspection are repaired by replacing the cracked cells. A string repair station is shown in Fig. 5.27.

The following are some of the manufacturers of tabber-stringer equipment:

TeamTechniks, Meyer Burger, Xcell Automation, Schmid, NPC, Mandrogan Assembly, Ecoprogati, P. Energy, Autowell Technology, M-10 Industries, Aaron, Hangzhou Confirm Ware Technology, Jinchen, and NingXia Automation.

5.3.2 Solar circuit layup

Glass loader machine

The front glass is usually tempered and antireflection coating is deposited on the sun side of the glass. The glass comes in pellets. The glasses are kept separate by means of a paper. Earlier, when glasses without antireflection coating were used, the glasses were cleaned with hot demineralized water and dried in the glass washer. At present, all manufacturers use ARC-coated glass and the glasses are thoroughly cleaned in the factory before packing. So, the glass washer is not being used. Sometimes, isopropyl alcohol is used to wipe the glass surface of the cell side to remove any dirt. It is recommended not to touch the glass with bare hands.

The dried glass is loaded onto the table for manual lay-up. Nowadays, automatic solar module manufacturing lines use a glass-loading system consisting of robotic arms, whose function is selecting the glass from the pallet and putting it on the conveyor belt.

The glass-loading system should consist of a glass-moving robot arm and a paper-picking robot arm. The glass-moving robot arm should move the glass to the transmission line and the paper-picking robot arm should remove the paper at the same time.

The machine should have two glass storage stations and one paper storage station.

The glass-loading and paper-picking robot arms are generally of cantilever type structure or gantry structure to facilitate easy handling of the glass and paper.

The machine should be compatible with different sizes of glasses and sizes of the glass pallet.

The suction/vacuum cup to pick up the glass is made of such a material that it will not make any print or contaminate the glass during loading. To avoid dust on glass, an air blower has to be equipped on the connected conveyor.

The machine should be equipped with a safety fence to ensure the safety of operation.

The machine should be fully automatic with all safety features of protection.

EVA cutting machine

The encapsulant material EVA/POE should be cut by an integral machine to ensure a flat and neat cutting edge. The machine should cut the EVA and lay the cut sheet on the glass with precise alignment. The EVA/back sheet cutting accuracy should be $\pm 2\,mm$ with a tolerance percentage of 95% and a diagonal error less than or equal to 1.5 mm.

The machine should automatically align and place the EVA/POE sheet without any wrinkles and the aligning scope should be less than or equal to $\pm 2\,mm$. The EVA film placing accuracy should be $\pm 1.5\,mm$, the back sheet placing accuracy should be $\pm 1.0\,mm$, and the diagonal accuracy should be $\pm 2\,mm$.

The safety shield and the dust shield should be equipped to the main part of the equipment. Fig. 5.28 shows the EVA cutting machine.

In a semiautomatic lay-up operation, the glass is laid on the conveyor table. The cut EVA is precisely laid on the glass sheet. The strings are picked by the vacuum pick up arm of the stringer and placed on the EVA with string polarities, as per the drawing. The strings are aligned and the bussing interconnectors are arranged as per the gap requirements of the module drawing. The bussing interconnectors are aligned and manually soldered to the strings. Fig. 5.29 shows the alignment of strings with the cell rear side facing upward. Fig. 5.30 shows the bussing connections of the strings and the tapes are fixed for the alignment of strings.

Function of automatic layup machine

The automatic layup machine with a long edge forward is one-to-one corresponding to the automatic string machine. The long edge of the cell string is paralleled with the long edge of the module. The machine with a single-set layup system can be applied for the layup of 96, 72, and 60 cells with machinery regulation.

FIG. 5.28

EVA/back sheet cutting machine.

Courtesy: Jinchen.

FIG. 5.29

Strings rear side connection.

The layup machine is connected to the autostringer, and connected to automation conveyor. The layup machine receives cell strings from the tabber and stringer and glass with EVA layup from the automation conveyor belt. The string pick up robot picks up the cell string and place them onto the glass-EVA layup according to the gap distance requirement. After completion of the activity of bussing interconnection either by soldering or welding, the layup station transfers the glass-EVA layup with the cell strings to the connecting automation conveyor belt.

FIG. 5.30

Bussing connections of strings.

It should have a busbar compatibility with 5BB or more to deal with different types of cells.
The string alignment and gaps between the strings and cell edge to glass, etc. are checked by CCD vision camera.
Layup accuracy: Repeat accuracy of strings ≤0.3 mm, strings locating accuracy ≤0.3 mm, strings distance ≤0.3 mm, dislocating accuracy ≤0.3 mm, and strings edge to glass edge distance accuracy ±<0.3 mm.
Vacuum suction-based pick-ups are used to pick up the strings.
Operation mode: manual/automatic.
Change over time for different types of module: ≤30 min.
Breakage rate during layup: ≤0.03%.
Material loading: Strings load from stringer or loaded manually, glass loaded from automation conveyor.

Automatic bus welding
- Cutting of automatic busbars, L-type production.
- Automatic welding of busbars.
- Automatic cutting and placing of insulation isolation strips.
- Conventional L-type manufacturing is produced by hot air welding and the pullout force is greater than 20 N.

- The welding of the lead wire and busbar adopts electromagnetic induction welding, and the pulling force is more than 8 N.
- No cleaning of welded joints; compatible widths of 5, 6, 7, and 8 mm busbar welding strips.

String bussing operations are done and the connectors are brought to one location on the front side, as shown in Fig. 5.31. It also shows the string interconnectors brought to the rear side through the cut of the EVA/back sheet.

In line electroluminescence inspection

The assembled solar string matrix with layup has to be subjected to EL inspection before lamination to find any defects or cracks in the solar cells. As it is not repairable once the module layup is laminated, so as part of prelamination, the layup undergoes EL testing.

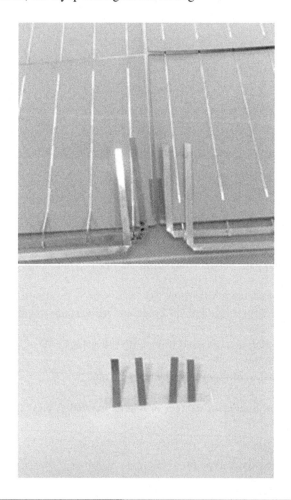

FIG. 5.31

String interconnectors after bussing: (A) on front side, and (B) on the rear side through the cut of the EVA/back sheet.

EL tester

This test has to be performed in the dark. The solar cell circuit in the dark is forward biased and 1–1.25 times I_{sc} is applied. With the application of forward bias, electron-hole pairs are generated and they recombine and emit radiation in infrared region of the sun spectrum. The emitted radiation is captured by the camera and a picture of the module with cells is created. The inactive parts and defective areas show lower or zero current and hence show partially dark or fully dark areas. The EL can find micro-cracks in the solar cells while also identifying solar cell shorting with interconnectors and defective zones.

Fig. 5.32 shows the EL equipment and Fig. 5.33A and B shows EL pictures of different laminates with and without defects.

5.3.3 Lamination

For EVA encapsulants, solar modules are typically laminated on flat-bed vacuum-bag laminators. Ideally, the laminator consists of two vacuum chambers separated by a flexible membrane. The chambers can be evacuated and ventilated separately, so adjustable pressure is applied to de-gas the laminate. There should be a cooling mechanism on the laminator or in an off-loading area with fans to cool the laminates. Conveyors on the postlamination side would be equipped with cooling fans to ensure cooling of the laminates below 60°C. Different laminator manufacturers provide various designs to achieve high production throughput and good quality control, such as large-area laminators, stack laminators, multiple-stage laminators, etc. However, they have the same working principle. The selection of packaging materials and the lamination process are crucial for the reliability of PV modules. Fig. 5.34 shows a laminator with laminates at the input conveyor ready to go for lamination.

A laminator consists of a heating plate, the heart of the system, mounted in the lower chamber, which enables the laminates to be heated quickly and homogenously. The laminator will have one

FIG. 5.32

EL machine used to find the defects in the solar cells.

Courtesy: Chaudary, Risen Energy.

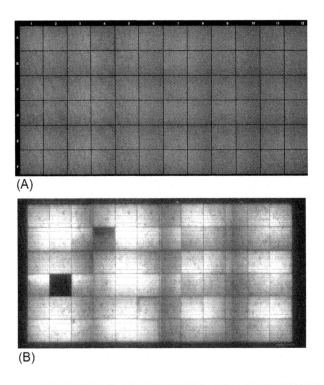

(A)

(B)

FIG. 5.33 (A)

EL image of a laminate without any defects. (B) EL image of a laminate. Dark cells are not working because they got shorted during stringing operation.

FIG. 5.34

Laminator with laminates at the input conveyor.

Courtesy: Jinchen.

incoming and one outgoing conveyor system for automatic loading and unloading of the laminates from the laminator. The control panel enables programming of the process sequence and comprehensive control of the lamination process. Some laminators have a unique heating system combining the compactness of a resistance heating element and the heating oil. Over the entire heating system, the temperature deviation is not more than 2°C with the setting temperature. The silicone membrane is stretched on the cover that separates the two chambers. There are mechanisms of lifting the laminate from the heating plate, so that direct heating of the laminate is avoided. Some laminate manufacturers use lift pin devices, which are regularly arranged in the grid. The laminator will have a vacuum pump to remove the air and volatile gases emitted during lamination from the chambers. A rotary vane pump can be used for this purpose.

The quality of encapsulation is primarily influenced by three steps in the module lamination process: the preheating step, the curing step, and the cooling step.

At the beginning of the lamination process, the module layup is automatically or manually placed in the laminator's lower chamber. The layup is gently heated from room temperature to approximately 60–80°C, while the lower and upper chambers are evacuated to approximately 1 mbar. The aim of this process is to remove the trapped air in the module layup. The volatile organic compounds (VOCs) are largely released from the encapsulant under the effect of heat. If VOCs are not properly removed during this first step of the lamination process, they may lead to bubbles in the final unit. The trapped air will not exit the layup completely if the evacuation is inadequate. This will result in air pockets being formed in the laminate [19].

In the second step, the laminate is heated through the heating plate and air is allowed to fill the upper chamber to a maximum atmospheric pressure of 1 bar. This presses down the flexible rubber membrane onto the layup module to ensure optimum adhesion between the encapsulant and the other layers of the layup module. This effectively results in a membrane pressure of 0.1 MPa (1 bar) applied to the layup of the module. The stress increases the transfer of heat from the heating plate to the layup and allows the layup to warm faster. Based on the composition of the EVA encapsulant, the temperature rises rapidly to the optimal heating temperature ranging from 140°C to 170°C. The main goal of the curing phase is to cure the EVA to achieve the necessary gel content. The curing reaction is determined by the temperature and the type of curing agent/coagent used in the EVA. EVA encapsulant curing begins at an initial temperature of approximately 110–130°C.

Due to either too low a lamination temperature or too short a curing time, there may be insufficient flow of the EVA encapsulant. This leads to incomplete wetting of the EVA on the surface of the adjacent components due to their complex surface geometries, particularly the cells and interconnections. This poor wetting can cause voids to form and the strength of adhesion between the EVA and the glass or back sheet will deteriorate. The curing reaction progresses so quickly in the case of a high lamination temperature that the EVA cross-links and becomes too viscous before the surfaces are fully wetted.

Upon cooling to room temperature, the residual thermal pressure in the system is proportional to the temperature of the lamination. The back sheet remains under pressure during the cooling of the module while the cells are subjected to compression [20]. The higher lamination temperatures increase the module's residual thermal stress and pose a threat to the cells and interconnections during field operations. The lower chamber will be vented to 1 bar after the curing step. The laminates are removed to the external table after the lamination process, where cooling fans are mounted under the table to cool the laminates to near room temperature. There are laminators that use a cooling press, which has water circulation, for applying the pressure on the laminate during the cooling process. The controlled pressure ensures that the laminates during cooling do not lose their shape. It minimizes stress points at the edges of the laminate that can cause delamination.

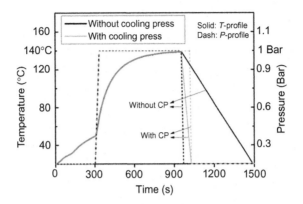

FIG. 5.35

The temperature-pressure-time (T-P-t) profile of a laminator with and without a cooling press.

Courtesy: Patrick Hofer-Noser (3S Solar Plus AG).

Temperatures and pressure profiles of the EVA encapsulation process with cooling in air and accelerated cooling under pressure are shown in Fig. 5.35.

The variations between the two cooling processes are primarily in the following two aspects: (i) a higher and regulated cooling rate of the module; and (ii) application of a controlled pressure of 0.1 MPa (1 bar) throughout the cooling process. The EVA is a substance that is almost amorphous with low crystallinity. The difference in cooling rates will definitely affect the resulting crystallinity and other properties of the material. The cooling step also affects the encapsulation performance of the modules in the following aspects.

The residual normal stress in the Si solar cell in the encapsulated module after cooling can be reduced by as much as 22%–27%, depending on the EVA gel content [21], by applying pressure during cooling and using a higher cooling rate. The peeling strength between the EVA and the glass can be increased by more than 10% [21] under pressure cooling. In the case of air cooling, the temperature of the module may remain for 1–5 minutes at or above the starting temperature of the curing reaction during the cooling stage. The gel content of the EVA may develop further during this time. Consequently, the cooling step also affects the quality of the EVA gel content, which is more pronounced for undercured EVA.

After module lamination, the manufacturers generally perform a series of technical control evaluations to ensure the production quality of the PV modules. From those controls, they obtain the following quality factors that are used to judge the quality of the manufactured series of modules:

Gel content, adhesion strength of glass and back sheet, visual inspection to check void formation, cell breakage, cell/interconnection swimming, glass breakage and discoloration, and postlamination EL testing.

The laminate is inspected for air bubbles, misalignment of solar cells, PV ribbons, proper curing of EVA, voids, and shrinkage in the EVA under illumination. The laminate undergoes EL testing to find out any microcracks are any other defects. If this all checks out, the laminate is taken for edge trimming.

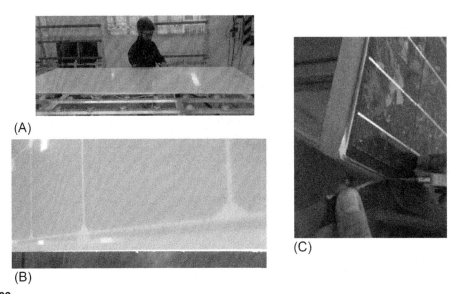

FIG. 5.36

Edge trimming of a laminate. (A) Edge trimming with hot knife, (B) edge trimmed laminate, and (C) edge trimming with knife.

5.3.4 **Edge trimming**

Postlaminate inspection, the oozed out EVA along with the back sheet are trimmed using a knife or a hot knife. Automatic edge trimming is done with a hot knife without damaging the glass. Fig. 5.36 shows the manual edge trimming of a laminate. Care has to be taken to avoid glass edge breakage during manual trimming of the laminates.

5.3.5 **Framing**

A single-component silicone adhesive is used to fix the frame to the laminate. Some manufacturers use double-sided tape to fix the frame with the laminate.

The grooves of frames and the corner key are applied with adhesive. Then, the frames are aligned to the laminate and fixed using pressure application in the framing machine. Fig. 5.37 shows fixing of the frame to the laminate by a framing machine. The frames are firmly fixed with screws or with corner keys.

If the frame is not fixed properly with the laminate, water enters the module and reduces the insulation resistance of the module. In the solar PV plant, during cleaning of the modules or during the rainy season, the inverter recognizes the low insulation resistance of the module as earth fault in the string and gets switched off automatically. Fig. 5.38 shows leakage in the sealant, and a low amount of sealant causing water entry into the module. Hence, the frame joining strength has to be checked for every lot of adhesive received. A wet leakage test has to be performed on modules on a sampling basis.

FIG. 5.37

Fixing of the frame to the laminate by a frame fixing machine.

Courtesy: Chaudary, Risen Energy.

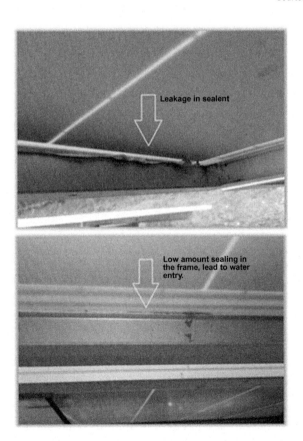

FIG. 5.38

(Top) Leakage in sealant due to water entry into the module, and (bottom) low amount of sealant in the frame leads to water entry into the module.

FIG. 5.39

Fixing of the junction box to the laminate.

Courtesy: Chaudary, Risen Energy.

5.3.6 **Junction box fixing**

The junction box is fixed on the rear side of the module, that is, on the back sheet, using a double-sided adhesive tape or silicone adhesive. Presently all the manufacturers are using silicone adhesive for junction box fixing. Presently all the manufacturers are using silicone adhesive for junction box fixing. The solar interconnectors, which are coming out of the solar cell strings as shown in Fig. 5.31B, are connected to the terminals of the junction box using a clamping technique or a soldering method. Fig. 5.39 shows the fixing of the junction box to the laminate.

For the potted type of junction box, the interconnect terminals coming from the solar cell strings are soldered and the junction box is filled with adhesive for sealing purposes. It will be cured in the controlled humidity room. If the adhesion strength is not good, for small handling loads, the junction box debonds and hangs from the modules during field operation of the solar PV plant. The water/moisture directly attacks the interconnectors and causes corrosion. This corrosion causes cuts in the interconnector and it becomes loose. This causes DC arcing and in turn burning of the junction box. A junction box with corroded terminals due to moisture ingress is shown in Fig. 5.40. A burnt junction box due to DC arcing caused by moisture due to the use of an improper sealant for fixing is shown in Fig. 5.41.

A junction box pull strength check will be conducted on some samples in the beginning of process finalization.

5.3.7 **Electrical testing of solar PV module**

Solar PV modules will be 100% tested for current-voltage (*I-V*) characteristics to know the electrical output of the module at standard test conditions (STC). To test the modules, a standard procedure has to be followed by all solar PV module manufacturers. Under STC conditions, the intensity of solar irradiance is $1000\,\text{W/m}^2$, the solar module operating temperature is 25°C, and the light spectrum is AM 1.5G. The light will be simulated with a sun simulator. A laboratory-calibrated reference solar module with a similar type of construction and cell has to be used as the reference solar PV module to set the intensity of the sun simulator and the temperature of the solar PV module.

FIG. 5.40

Terminal corrosion of the junction box due to moisture penetration.

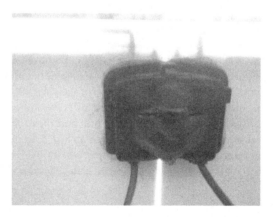

FIG. 5.41

Burning of the junction box due to DC arcing caused by a loose connection of terminals.

There are three types of solar simulators: a constant light simulator, a pulsed light simulator, and a pulsed light and decaying simulator. The constant light simulator continuously produces light and, during measurement, the temperature of the module will go up. Hence, it has to be controlled. Pulsed light simulators provide light in the form of short or long pulses in a definite interval of time. A pulsed light and decaying simulator can measure the *I-V* characteristics at different solar irradiation levels. The IEC 904-9 standard describes the requirements for solar sun simulators.

Table 5.7 Standard parameters for solar sun simulator classes.

Characteristics	Class A	Class B	Class C
Spectral match (ratio of the actual percentage of total irradiance to the required percentage specified for each wavelength range)	0.75–1.25	0.6–1.25	0.4–2.0
Nonuniformity of irradiance	$<\pm2\%$	$<\pm5\%$	$<\pm10\%$
Temporal instability, short-term, STI	$<\pm0.5\%$	$<\pm2\%$	$<\pm10\%$
Temporal instability, long-term, LTI	$<\pm2\%$	$<\pm5\%$	$<\pm10\%$

The performance of sun simulators is classified as Class A, Class B, and Class C. This is based on three aspects: positional nonuniformity, spectral match, and short- and long-term Temporal instability. The various parameters for a sun simulator are given in Table 5.7.

Requirements of solar sun simulators

A solar simulator is a complete test system that includes reflectors, lenses, shutters, power supplies, control electronics, and a high-intensity light source that simulates the solar spectrum.

A xenon lamp provides a light source that is closest to that of the solar spectrum, and it is typically in the visible spectrum from 400 to 700 nm. Xenon short-arc lamps that provide power levels of 150–1600 W are ideal for solar sun simulators, especially for a large-area solar PV module that requires higher overall power.

Solar sun simulators measure the *I-V* characteristics of solar modules and cells during a single flash of a xenon gas discharge lamp or an LED-based system. LED-based systems are able to offer longer exposure times, which is required for high-efficiency solar panels. This allows solar module manufacturers to improve the solar module efficiency classification and yields. The spectrum is filtered to comply with the Class A requirements of the standard IEC60904-4. The spectral characteristics of sun simulator are measured with a calibrated real-time spectrometer. Also, the irradiance nonuniformity is measured following the guidelines and specifications of the standard.

Measurement setup

The basic measurement setup is shown in Fig. 5.42. The current voltage characteristics of a solar module/cell are measured using a xenon flash as an irradiation source. During a single flash, the electronics unit records the voltage and current signals while the module is swept from short circuit into open circuit. A four-wire measurement principle is essential in order not to measure any voltage losses due to parasitic resistances in cables and connectors.

The irradiance signal is recorded using a monitor cell and the voltage and current signals are corrected accordingly, following the procedure in the standard IEC60891.

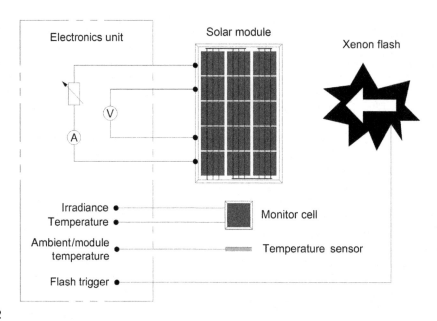

FIG. 5.42

Measurement set up for *I-V* testing of a solar PV module.

Courtesy: Endeas Oy.

Two temperatures are measured in order to perform the necessary correction calculations to the desired target temperature, typically 25°C. The module is assumed to be at ambient temperature and monitoring the cell temperature is used to correct the irradiance signal.

Measurement principle

Fig. 5.42 shows the current-voltage curve measurement setup for a solar PV module. By generating a flash, the measurement is carried out. When the tail of the flash pulse has reached the preset level, typically $1200 \, \text{W/m}^2$, an electronic load, a field effect transistor, is triggered to sweep the voltage linearly from zero to open circuit. Because of the decaying flash pulse, the maximum power point will be measured very close to the typical target irradiance level, $1000 \, \text{W/m}^2$, minimizing irradiance correction effects. High-efficiency solar cells exhibit capacitance. To overcome the capacitance effect, the signal is reverse swept from V_{oc} to zero voltage and the curve will be obtained. The recorded signals, voltage, current, and irradiance, are transferred to a PC for both the forward and reverse sweeps and the data is processed and the *I-V* parameters are obtained. High-efficiency solar cells such as the passivated emitter and rear contact (PERC) cell, the heterojunction technology (HJT) cell, and the interdigitated back-contact (IBC) cell are currently gaining in production volume and market share. Due to well-passivated surfaces and high effective minority carrier lifetimes, these architectures store a considerable amount of charge mainly in the quasineutral regions of the device [22]. PV modules with high

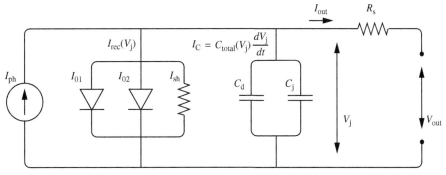

FIG. 5.43

Equivalent circuit of high-efficiency solar cell with inclusion of junction capacitance and diffusion capacitance [22].

Courtesy: Jaakko Hyvärinen, Endeas Oy.

efficiency often have a high internal capacitance, which requires a relatively long measurement time. PERC cells and modules require a 30-millisecond measurement duration, while heterojunction (HJT) modules may require more than 1 second [22].

If sweep time is too short considering the dynamic effects of the PV module, the resulting P_{max} is underestimated (forward sweep) or overestimated (reverse sweep). The capacitance compensation method is based on a physical model of a solar cell. It uses measurement data of both forward and backward sweeps to find the voltage-dependent capacitance, and removes its dynamic effects on the I-V curve. Fig. 5.43 shows the equivalent circuit of high-efficiency solar cell [22]. It consists of two diodes representing the Shockley diffusion direct current (I_{o1}) and the space charge recombination current (I_{o2}), lumped shunt resistance R_{sh}, lumped series resistance R_s, junction capacitance C_j, diffusion capacitance C_d, and the photogenerated current under illumination I_{ph}. The diodes are modeled with the saturation currents I_{o1} and I_{o2}. rec: total recombination current, c: capacitive current, out: output current, I_{o1}: current of diffusion of diode, I_{o2}: current of recombination diode current, sh: shunt current, total: total capacitance; j: junction voltage; out: output voltage. The internal capacitance together with the series, parallel, and diode differential resistance form an RC circuit that introduces a transient time constant into the measurement process [23]. The time constant determines the quasistatic condition. Its magnitude depends on various parameters such as the operating point (voltage and current), temperature, irradiance level, minority carrier lifetime etc.

Endeas developed capacitance compensation method (CAC) to correctly measure the I-V curve of solar module by taking care of noise, series resistance, and capacitance effect by double sweep method [24]. Capacitance compensation method is based on a physical model of a solar cell. The smart electronic load in Endeas QuickSun 600 Solar module tester can generate both a forward sweep (module voltage is swept from I_{SC} to V_{OC}) and a reverse sweep (module voltage is swept from V_{OC} to I_{SC}) during a single flash. The two-diode model with capacitance is used for the optimal extraction of the junction capacitance. Internal capacitance of solar cells is higher at high cell voltages. At low voltages capacitance is low and voltage can be changed rapidly without creating dynamic effects. When the applied

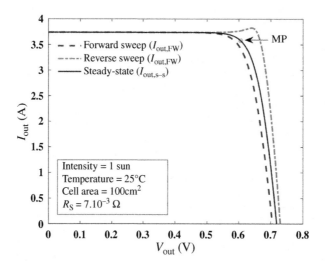

FIG. 5.44

I-V curves of high-efficiency solar cell with forward scan from I_{sc} to V_{oc} and reverse scan from V_{oc} to I_{sc} and capacitance compensated corrected one [22].

Courtesy: Jaakko Hyvärinen, Endeas Oy.

voltage nears maximum power point the voltage ramp should be slow enough to minimize effects of the capacitance. Optimal way to perform the voltage sweep is then to change voltage rapidly at low voltages and change it slowly at high voltages. This two part sweep is called double slope indicating the two different voltage ramps. Typically when measuring a PV module, the applied voltage is increased linearly from zero to open circuit voltage (V_{OC}). The time it takes for the voltage to reach V_{OC} is the sweep time (or measurement time). Fig. 5.44 shows forward scan from short circuit current I_{sc} to open circuit voltage V_{oc}; reverse scan from V_{oc} to I_{sc}. A hysteresis can be observed demonstrating underestimation of the real I-V curve in the forward direction and overestimation in the reverse direction. A forward sweep can underestimate module power if internal capacitance of the module is high and measurement duration is too short. Accordingly, reverse sweep can overestimate module power. The two-diode model with capacitance is used for the optimal extraction of the junction capacitance.

The reference module that will have a similar type of solar cell and the same spectral response will be calibrated at a standard test agency. The calibrated reference solar PV module will be considered as the primary reference module, and secondary reference solar modules have to be made and used in regular production.

The reference module has to be tested and the intensity of the sun simulator has to be adjusted to get the I_{sc} of the module. The temperature of the solar module will be reflected in the measurement of V_{oc}. Once the specified parameters of the calibrated module are obtained, the test modules will be tested.

If the temperature of the solar module is not exactly 25°C during measurement, the correction to the parameters is to be applied based on the temperature coefficients of the parameters, as an input. The accuracy of the measurement depends on various factors such as the reference solar module, the cell operating temperature, etc.

5.4 Special type of solar PV modules and manufacturing process

5.4.1 Half-cut cell solar modules

The tariff rates for the energy generated by solar PV power plants are falling and grid parity has been achieved in different countries. The solar cell and module manufacturers and researchers are putting efforts into finding new techniques to improve the solar cell and module efficiency. The driving force behind new techniques in solar module technology is to increase the solar module power. There are different techniques being followed to increase the module power. One of the methods is using half-cut solar cells in the solar module. Though the concept was tried earlier, it has become popular in recent times due to the manufacturing trend to increase module power.

When solar cells are connected in series to form a string, the string voltage is the sum of the single cell voltages while the string current remains equal to the single cell current. In view of this, dividing a whole cell into two half-cut cells causes a one-half decrease in current. Thus, the ohmic loss is reduced. The dependence of the ohmic loss on the current is described by the following equation:

$$\text{Powerloss } P = I^2 R;$$

where I is the current in Amp and R is the resistance in ohms.

The main contribution to ohmic loss is the squared current, and the resistance has a linear influence on power. It is more advantageous to reduce the current to get a lower power loss.

Fig. 4.26A shows the *I-V* characteristics of a full multicrystalline solar cell of size 156.75×156.75 mm with n/p-type Al-BSF technology.

The I_{sc} and I_{mp} are directly proportional to the incident radiation and the area of the solar cell. V_{oc} and V_{mp} are dependent on temperature, and vary logarithmically with the incident solar radiance. The fill factor depends on R_s and R_{sh}.

The resistance of the solar cell comprises the resistance of the cell grid lines and busbar metallization, semiconductor bulk resistance, back contact resistance, and the resistance offered by the emitter. The resistance offered by the PV ribbon that is being used to interconnect the solar cells also affects the power loss in the solar module. The PV ribbon resistance depends on the resistivity of the material, the area of the cross-section of the ribbon, the length of the ribbon, etc.

When the solar cell is cut into two half pieces, the I_{sc} will become half and there will be no change in V_{oc}. The maximum power point voltage and current values depend upon the various resistances and current leakage paths of the half-cut solar cell.

The solar module technology is advancing with the use of higher-efficiency solar cells, which produce higher currents. The higher current causes a higher voltage drop due to the resistances offered by the cell and the interconnector.

Due to the lower series resistance of the solar cell and lower current, the voltage drop of the half-cut cell will become low and hence, the loss of power will be lower. If the solar cell is cut properly, it maintains the same shunt resistance. While cutting the solar cell with a laser scriber, there is a possibility of generating microcracks that cause the increase of leakage current at the edges and reduce the shunt resistance of the solar cell.

The series resistance and the shunt resistance are estimated from the *I-V* curve taken in the dark.

Design of half-cut solar cell module

A solar cell module consisting of 60 full solar cells contains three strings and each string has 20 solar cells connected across with a diode. There are three diodes and these three strings are connected in series, as shown in Fig. 5.45A.

The full cell is cut into two halves and a 20 half-cell string is formed. Two such strings are paralleled end to end to form a 20Sx2P (20 cells in series and two cells in parallel). Three such strings are connected in series and a diode is connected across the 20Sx2P string, as shown in Fig. 5.45B. The same configuration can be shown in a different way, as shown in Fig. 5.45C.

The strings A_1 and A_2, B_1 and B_2, and C_1 and C_2 are paralleled end to end and three diodes are connected across each substring. This is shown as the curved type in Fig. 5.45C. Without disturbing the parallel connections of the half-cut cell strings A_2, B_2, and C_2 can be flipped to the bottom side, as shown in Fig. 5.46A.

After flipping, the curved figure can be brought to a straight string; it appears as shown in Fig. 5.46B. Thus, finally the half-cut cell module configuration looks like that in Fig. 5.46C.

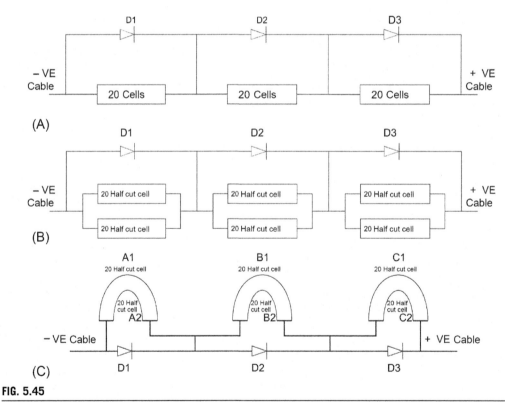

FIG. 5.45

Design of half-cut solar module—explanation: (A) substring arrangement in a 60 full cell solar PV module, (B) substring arrangement in a 120 half-cut cell solar PV module, and (C) substring arrangement in a 120 half-cut cell solar PV module in a different form.

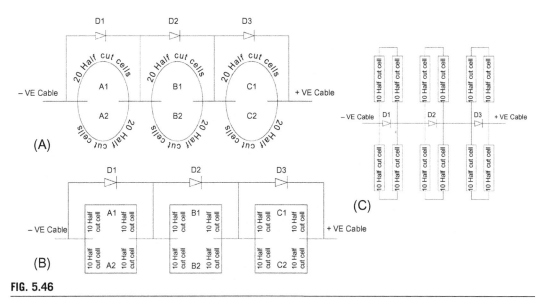

FIG. 5.46

Arrangement of strings in half-cut solar module. (A) Arrangement of half-cut cell substrings in a solar PV module in a different form, (B) arrangement of half-cut cell substrings in a solar PV module on the top and bottom side, and (C) actual arrangement of half-cut cell substrings in a solar PV module.

In the upper side, two 10 half-cut cell substrings connected in series form a string (A_1) with 20 half-cut cells. The string A_2 is formed in the lower side by connecting two half-cut cell substrings of 10 cells. Strings A_1 and A_2 are paralleled at the positive and negative sides end to end. Similarly, the B_1 and B_2 and C_1 and C_2 strings are paralleled in the middle of the panel. The individual junction boxes catering to each diode are fixed between the upper and lower side circuits on the rear side of the solar modules. With this design, there are three cuts in the EVA and back sheet to take out the string interconnectors to connect to bypass diodes and output cables.

Finally, the 60 full cell-based 120 half-cut solar module looks like that shown in Fig. 5.47.

Upon cutting a solar cell into two halves, each half-cut cell will produce one-half current and its resistance will be one-fourth. This lower resistance helps to reduce electrical losses and in turn improves solar module efficiency. There will be two-fold connections and with that, changed internal resistance. The reduced electrical losses will assist in a 3% increase in solar module power. Hence, the solar module efficiency increases from 18% to 18.5%. In terms of solar module wattage, it will increase from 300 to 309 W, approximately.

The laser is used to put a groove in the cell, but not to cut it. Then, it's snapped into two half solar cells.

Further, there is an extra expense of connecting twice as many cells in a solar PV module and doing double soldering in view of the twice half-cut cells. In the case of manual soldering, the labor cost is high. In view of machines, the soldering costs are lower with improved reliability.

Disadvantages of half-cut solar cell configurations are: higher cost, twice the potential for soldering defects, and internal defects due to the slicing of the cell.

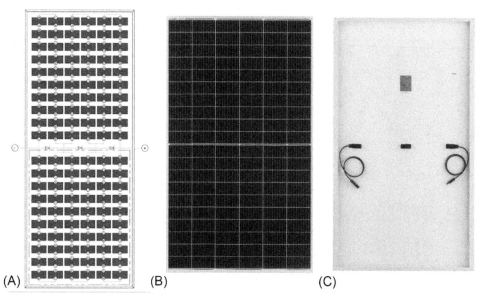

FIG. 5.47

120 Half-cut cell-based solar module. (A) Schematic of half-cut cell module, (B) front side of the half-cut cell module, and (C) rear side of the half-cut cell module showing three junction boxes mounted on the middle of the module.

Courtesy of images (B) and (C): Chaudary, Risen Energy.

Standard solar modules have a single junction box mounted on the rear side of the panel near the top portion. Half-cut-based solar modules with 120/144/156 half-cut solar cells require three junction boxes as per design and these are mounted on the center of the solar module. The split junction boxes mounted near the edges house the diode and connect to output cables, but the middle junction box is connected to the middle bypass diode only.

When one or a few of the solar cells get shaded, the other cells that aren't shaded, direct their generated energy into the shaded cells. This results in heating of the shaded cells and this is called hotspot phenomenon. This hotspot phenomena can potentially damage the solar module, if the hotspot condition persists for a long time. The shaded solar cell has half the area to radiate heat as a full cell, but the decreased amount of heat produced is less damaging to the solar PV module and there is an improvement in resistance due to the hotspot.

Two half-cut pieces have the same voltage as the full cell, but the current, which is a function of the surface area, gets divided accordingly by half. Consequently, the internal electrical losses, which are in the second order of the flowing current, are reduced to one-fourth for a half cell compared to full square solar cells. There is an additional process step to slice the solar cells into two halves. This is a mechanical process that might induce additional breakage losses and microcracks. On top, half cells also reduce the throughput of the manufacturing capacity of the solar cell interconnection tools by half.

In a half-cell solar PV module, there is an increase of module power about by 1.48% in the fill factor due to reduced electrical power losses in solar cell connectors. Also, about 3% of the solar module current is increased due to the optical gains from solar cell spacing. Parallel substrings allow the solar module to save up to half of the string's power under partial shading conditions.

There is a process difference in manufacturing the full-sized and half-sized solar cell-based solar PV modules. An additional laser cutting step is needed to make a half-sized solar cell, which introduces additional cost. Cutting solar cells in half is a harmful process using a laser. The major problems are shunts induced by laser cutting and additional edge recombination. Further mismatch losses may be introduced by cutting the solar cells into half. This could be one possible reason for the difference between the theoretical prediction and actual measurement results.

Half cells create additional gaps in the solar module, which result in additional length of the ribbons as well as cost. For the ribbons connecting the half-cut solar cells in the solar module, the total length increase is approximately 15 mm per full-size 5 busbar cell. The bussing ribbon length used by a half-cell solar module also increases. The additional gaps between the half-cut solar cells in the solar module lead to an increase in the total solar module area by 0.7%. This change in area does not affect the solar module power output, but it reduces the solar module efficiency. Despite the described problems, the gain in terms of power/efficiency is high. Also, this half-cell solar module has the potential for cost reduction of PV modules, especially while using high-efficiency solar cells.

The half-cut solar cell module is nothing but two nos. of 60 half cell circuits are paralleled and the three bypass diodes are shared commonly. Fig. 5.48 top one shows the I-V curves at taken at STC for 120 half-cut solar cell module without any shadow, whereas the bottom I-V curve is of the same module, where the lower half of the circuit is fully shadowed. There is a reduction of 1.11 V in V_{oc} of the shadowed condition of the module and the I_{sc} of the shadowed module shown in Fig. 5.48 bottom is 4.42 A, which is half of the I_{sc} of the top curve and the power is 124.38 W. It shows that one of the circuits is shadowed completely, the other half circuit provides the half of the current operating with the same voltage of the module in unshaded condition. Fig. 5.49 top and bottom I-V curves are of the 120 half-cut cell module in partially shadowed condition. For the upper half circuit, one column of substring (10 half cells) is shadowed. The bottom curve is of the module where both upper and lower half one substring (10 cells each) is shadowed. Both the substrings are part of the same bypass diode. The top curve V_{oc} is same as that of unshaded module. The bypass diode is activated and the power is 165.2 W. In the bottom I-V curve, the bypass diode is activated and gives a power of 127.25 W operating at 15.82 V.

In Fig. 5.50, the top figure, the complete substring (20 half cells) is fully shadowed. This lower half shadowed substring shares the diode with the upper substring. The diode got activated and shows two peaks with and providing a power of 154.3 W at operating voltage of 19.2 V. The bottom curve of the module is with two substrings of lower half circuit belonged to two bypass diodes are shaded. So, two bypass diodes are activated, showing two kinks producing 159.22 W. The V_{mp} is shifted toward V_{oc}.

In Fig. 5.51A, the top curve, a half cell shadowed, whereas in bottom figure, two half cells of different strings are shaded. In both the conditions, the provided power is around 180 W with operating voltage of around 32 V, which is same as that of unshaded module. In Fig. 5.51B, in lower half circuit, horizontally one cell of each column is shadowed. This means in all substrings of lower half circuit, one half cell is shaded. It is equivalent to lower half circuit is in off condition. The I-V curve does not show any kink. The current is reduced to half and the power is 130 W, which is around half of the unshaded module.

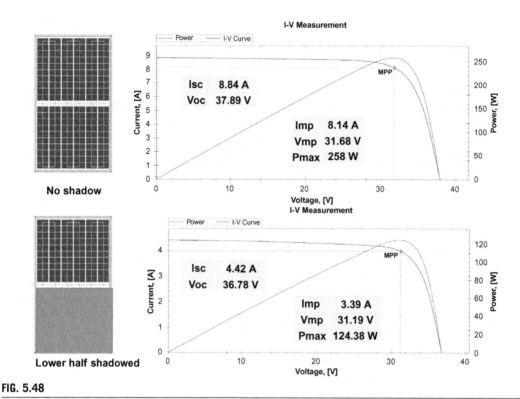

FIG. 5.48

I-V curves of 120 half-cut solar cell module at STC. (*Top*) Without any shadow, (*bottom*) lower half of the circuit is fully shadowed.

This discussion shows that, if at least one cell in all of the substrings of one-half circuit is shaded, the module operates in the same V_{mp} and gives half the power that of unshaded module. If one or two cells are in any half circuit are shaded, the module gives lower power and operates at different voltage compared to unshaded module. The half-cut solar modules are connected in series to form a string and number of strings are connected in parallel to a central inverter in a solar PV plant. If one module is partially shaded, the concerned bypass diode will activate and provide lower power operating at different voltage compared to other module. It behaves like a shaded full cell module. As the shaded module is a part of many strings connected to inverter, it pulls the voltage of other modules that are connected to inverter.

It is observed that in a half-cut cell solar module, the diodes are difficult to trigger so as to eliminate the hotspot. This leads to the hotspot heating the solar cell continuously. The shaded string and unshaded string both are in parallel and work in the optimal maximum power point is incorrect. There is 100% bypassing of the partial shaded half-cut solar module output, as observed in the curves of Figs. 5.49–5.51. Until and unless 90% of the area of a half-cut solar PV module is shaded, the bypass diodes will not turn on. This would not help to prevent the hotspot problem.

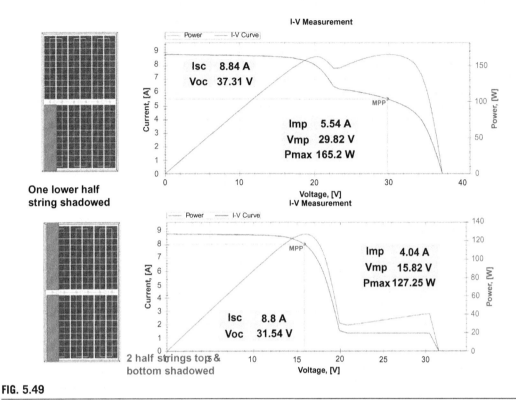

FIG. 5.49

I-V curves of 120 half-cut solar cell module at STC with different shadings. (*Top*) One column of substring (10 half cells) is shadowed in upper half circuit, (*bottom*) both the upper and lower half circuits, one substring (10 cells each) is shadowed.

There is a growing demand for half-cut solar PV modules. The multicrystalline, mono-PERC, HJT, and bifacial solar cells are being used to manufacture half-cut solar PV modules.

5.4.2 Glass to glass and bifacial solar modules

Bifacial solar cells absorb and convert light captured on the front and back sides of the solar module. The rear side of the solar module absorbs the reflected or diffused light from the ground or surface on which the module is mounted and contributes to increased power. The reflected or diffused radiation coming from the ground surface is called the albedo.

Based on ground surface conditions the front side of the solar PV module will intercept most of the sunlight, the back of the solar module still can intercept sunlight to the extent of less than 10% or more as compared to the front glass. The bifacial interception is shown in Fig. 5.52.

Bifacial solar PV modules have several designs of the broadly framed or frameless type. A typical design uses glass-glass while some manufacturers use transparent back sheet as rear cover. Bifacial solar modules use either multi- or monocrystalline silicon solar cells. In both the cell-based designed

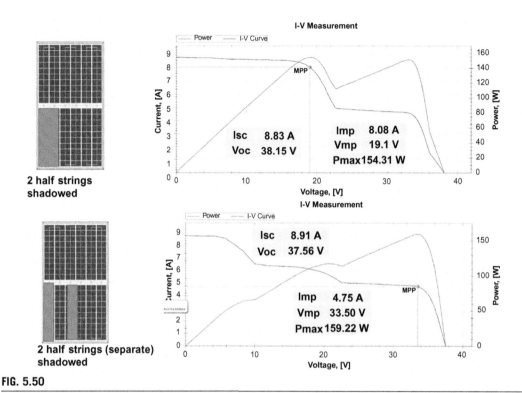

2 half strings shadowed

2 half strings (separate) shadowed

FIG. 5.50

I-V curves of 120 half-cut solar cell module at STC with different shadings. (*Top*) Complete substring (20 half cells) is fully shadowed in lower half circuit, (*bottom*) Two 10 half-cut cell strings of lower half circuit belong to two different bypass diodes are shadowed.

bifacial solar modules, power is generated from both sides. Bifacial solar PV modules have contacts/busbars on both the front and back sides of the solar cells.

The bifacial concept is superior to the standard monofacial solar PV module, where the photovoltaic effect is restricted to sunlight absorbed by the front surface only. Even diffused or reflected light, to which the solar cell's rear side is normally exposed, can generate charge carriers. A bifacial approach simply fully exploits all possibilities. The power gain for a bifacial solar module is between 5% and 30%, depending on various aspects such as the bifacial solar PV module design, the albedo, the module mounting arrangements, and other factors. Bifacial solar module design encompasses three commercial solar cells—HJT, PERT, and PERC. HJT, PERT, and PERC solar cell technology demonstrates different levels of bifaciality. HJT and PERT solar cells have a bifaciality greater than 90% while PERC and IBC solar cells have a bifaciality of about 80% and 70%, respectively.

A typical PERC structure that uses Al-BSF (local) is not considered bifacial. In order to convert the cell into a bifacial device, the aluminum BSF has to be interchanged with a grid. The structures of monofacial and bifacial solar cells are given in Fig. 5.53.

Conventional solar modules use an opaque back sheet. In comparison, bifacial solar modules either have a transparent or reflective back sheet or have a glass-glass combination. In view of the frameless design of the solar PV module, there is less of a possibility for potential induced degradation.

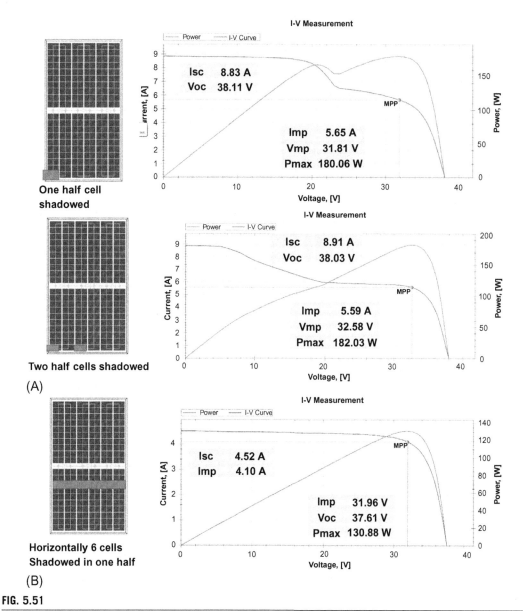

FIG. 5.51

(A) *I-V* curves of 120 half-cut solar cell module at STC with different shadings. (*Top*) One half-cut cell is shadowed in lower circuit, (*bottom*) two half cells belong to two different bypass diodes are shadowed. (B) *I-V* curve of 120 half-cut solar cell module at STC with different shadings. In lower half circuit, horizontally one cell of each column is shadowed.

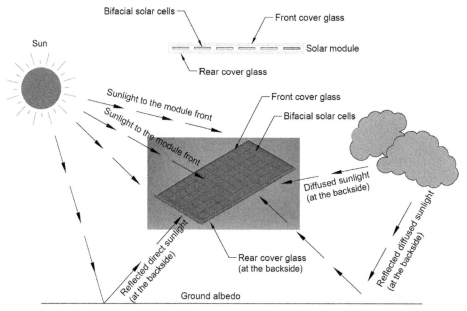

FIG. 5.52

Principle of a bifacial solar PV module.

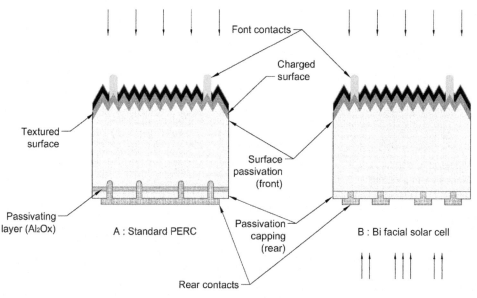

FIG. 5.53

Structure of monofacial and bifacial solar cells.

Advantages of bifacial solar modules

- Increased energy generation. Unlike PV systems deployed with monofacial modules, bifacial PV systems can convert light that shines off the back of the module into electricity. This additional back-side production increases energy generation over the life of the system. Ongoing research and side-by-side testing suggest that a bifacial PV system could generate 5%–30% more energy than an equivalent monofacial solar PV module, depending on the installation of the solar modules. Moreover, the manufacturers' linear performance warranties for bifacial PV modules for 30 years are some of the best in the industry.
- In a bifacial solar module made from a double glass combination, it becomes rigid and helps in reducing mechanical stress on the solar cell during transportation. Also, double glass helps in handling and ease of installation as well as from harsh environments such as high wind or snow. Double glass prevents the ingress of water, which in turn reduces annual degradation.
- If the bifacial solar PV module is frameless, eliminating the aluminum frame effectively reduces opportunities for occurring potential induced degradation.

While designing a bifacial solar module, the replacement of the opaque back sheet with transparent material makes it bifacial. The transparent material such as glass opens up the back side of the solar PV module for the penetration of sunlight. With the availability of less expensive solar glass, glass-glass modules are becoming a clear winner. The double glass structure enables extending the warranties of the solar module beyond 30 years. For commercialization of bifacial solar PV module, three issues need to be addressed: bankability, standardization, and power measurement/solar simulation. There is significant progress for these parameters.

There is considerable growth in the deployment of bifacial solar PV modules, which is helping to improve bankability. In addition, several solar PV module manufacturers have set up large MWp-level facilities to increase bifacial solar PV module production and subsequent deployment. IEC standards are under approval, which will help to measure the *I-V* characteristics of bifacial solar PV modules. A new type of solar sun simulator has been developed exclusively for bifacial solar PV modules. This is now helping solar PV module manufacturers accurately measure bifacial solar module power.

Basics of bifacial technology

As mentioned above, a bifacial PV is nothing but making a solar cell light-sensitive on both sides. All major advanced cell architectures are bifacial by nature. But the optimization needed to make the cell's rear side receptive for sunlight absorption is very minor. It is primarily about printing a rear metallization pattern that is similar to the sunny side. The bifacial concept requires some changes at the solar module level. The major effort is to replace the conventional opaque back sheets with either a glass or a transparent back sheet. An important attribute of the bifacial cell and module is the so-called bifaciality factor, which is the ratio of the front side to the rear side efficiency. The bifacial modules also offer the opportunity for vertical installations. Such installations can be considered for nontraditional applications such as sound barriers on highways and railways.

Key drivers for development of bifacial solar module technology

- Solar power is competing with other power generation technologies, and the tariff rates for generation have come down. To bid tenders and to win power purchase agreements, long-term system yields are becoming increasingly important for developers. Therefore, it has become inevitable to improve the energy yield.
- Reduce the levelized cost of electricity (LCOE) of solar photovoltaic energy.
- Many recent technology developments such as PERC and multibusbar cells in the PV industry are directly or indirectly supporting the move to bifacial modules.

  The move away from the aluminum-BSF cell structure to advanced cell concepts such as PERC is a main driver, as these devices are bifacial by nature.
- Another factor that backs the move to a glass-glass configuration is thinner glass that is getting cheaper.
- With thin glass getting cheaper and the products being mature, glass-glass modules are becoming increasingly competitive, even more as they allow module manufacturers to offer longer warranties.

  The development of transparent back sheets is actually complementing bifacial technology, as replacing the opaque back sheet with a transparent material such as glass on the rear side is the prerequisite for bifacial technology at the module level.
- n-Type wafers are currently more expensive while bifacial gain is much higher than that for p-type PERC, which is why companies are starting to look into this technology.

  The glass-glass configuration with the usage of polyolefin as an encapsulant helps in avoiding the acid formation that originates from the EVA, causing degradation. So, several companies have already started to use polyolefins instead of EVA for the encapsulation of bifacial modules.

  Interconnection: As for the core interconnection process of a bifacial module, only a slight optimization is required, especially regarding the heating and cooling of the cells. That's because bifacial cells without aluminum BSF behave differently during thermal treatment. Off-the-shelf solutions are available from all leading stringing equipment suppliers.

  Bifacial modules are the perfect application for half-cut cells. The gain due to bifaciality mainly reflects in increased currents, which also increases the losses. Thus, half-cell configuration is better for interconnecting bifacial cells. In this way, not only is the power rating increased by at least 5 W per module, but also the cell-to-module (CTM) power ratio of the module will be improved to 98%.

  Junction box: Though not critical, replacing or moving the junction box is recommended so that it doesn't cover any cell rear sides in bifacial modules. New split type compact junction box designs with a single bypass diode that can be placed at the corners or middle region of the module, are available commercially from several vendors.

  Testing modules and standardization: Among the main challenges with bifacial PV are how to test and finally how to label the module. Because there are no standards, every company follows a different approach. Some companies are flashing the module from the front side with an intensity of 1.1 suns and covering the module from the rear side. But, some solar module manufacturers are measuring both sides of the module at standard testing conditions (STC) by covering the other side with a nonreflective sheet and reporting both values.

  Transparent back sheet: While the majority of PV panel producers are adapting glass-glass structures for their bifacial modules, some manufacturers are promoting panels with transparent back sheets on the rear side.

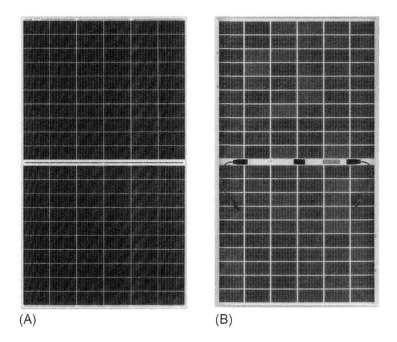

(A)   (B)

FIG. 5.54

Bifacial solar PV module.

Courtesy: Choudary, Risen Energy.

Fabrication of bifacial solar module

As per the IEC 61215 and IEC 61730 latest standards, the thickness of the superstrate is 3.2 mm, whereas different manufacturers are considering both and front side glasses as 2 or 2.5 mm thickness. The actual thickness to be decided based on the mechanical and wind load requirement. If both of the glasses are tempered, it gives better mechanical strength to bifacial modules.

A wind load simulation study has to be done to analyze the ability of glass-glass solar PV modules to withstand a 2400 Pa wind load and a 5400 Pa snow load. Based on the wind load study the thickness of the glass can be finalized. A laminator with a cooling press is required for proper lamination of the solar module, which ensures the long-term reliability of the module.

Polyolefin-based encapsulants give better performance and increase the durability of the solar module. Special junction boxes are to be mounted on the rear or front side of the solar module at edges or in the middle. Automation has to be used to properly align the glass during layup to avoid misalignment. Fig. 5.54 shows the bifacial solar PV module.

PV companies working on different bifacial cell technologies

p-PERC: JA Solar, LONGi,SolarWorld, Trina

p-PERT: NSP, SolAround

p-PERT (multi): Shanxi Lu' an
n-PERT: Adani, HT-SAAE, Jollywood, LG Linyang, PVGS, Trina, Yingli
HJT: Hanergy, Jinergy, Sunpreme, Panasonic

As every advanced cell concept can be tweaked to bifacial, the number of companies implementing the technology is quickly increasing.

5.4.3 PERC solar modules

The Al-BSF-based solar cell and module technology has dominated the solar PV industry for more than 15 years. There is no visibility for further improvements in the technology as the efficiency has saturated. With the increased growth of the solar industry and reduced tariff rates for solar PV plants, there is an interest among solar module manufacturers and researchers to increase the efficiency and power of the solar cell module. The continued efforts for improved solar cell efficiency have led the industry to adopt PERC solar cells.

What is PERC?

PERC stands for passivation emitter rear contact cell. To improve the current flow, the back surface is deposited with insulating film, which is called the passivating layer. The removal of defects due to passivation enables higher energy from the solar cell.

Fig. 5.55 shows the structure of a PERC and an Al-BSF cell. Compared to Al-BSF technology, the following are the changes in the PERC solar cell. The cell designs are very similar, except for the rear surface. There is no change in the front surface of the solar cell. For both types of solar cells, the front surface is passivated using a thin layer of silicon nitride, which serves as both a passivation layer and an antireflective coating for the cell. The rear surface of a PERC cell is improved by the addition of two steps in the cell manufacturing process—rear side passivation and rear contact opening. The passivation layer of Al_2O_3 and the cap layer of SiN are deposited on the back side of the solar cell using

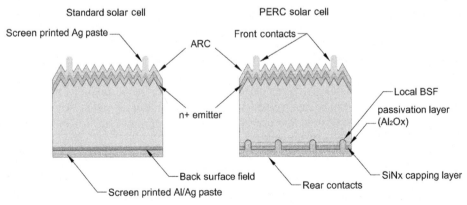

FIG. 5.55

Structure of a PERC and an Al-BSF cell.

plasma enhanced chemical vapor deposition (PECVD) technique. To make the contacts, the passivation layers of $Al_2O_3 + SiN$ layers are opened at some locations by laser ablation and a special aluminum paste is used by the screenprinting technique.

The appearance of a PERC solar cell is similar to that of Al BSF technology. The same process of interconnector connection by a tabber and stringer and lamination by a laminator is applicable for a PERC cell in solar module fabrication.

Normally, a solar module shows light-induced degradation, year-wise degradation in the electrical parameters, and potential induced degradation. The light-induced degradation is attributed to the presence of boron-oxygen complexes in p-type Si getting activated during light exposure and causing degradation in the module. This type of mechanism is less apparent for multicrystalline cells because they have a low concentration of oxygen. For monocrystalline cells, the wafers contain a higher oxygen concentration and the LID impact is more pronounced. The conventional Al-BSF module shows 1%–2% as LID, and the actual value depends on the quality of silicon material used and can vary between manufacturers. PERC cells manufactured using monocrystalline technology are particularly sensitive to LID as compared with Al-BSF cells. So, in data sheets, the degradation in power at the end of the first year is shown as 2%–3%.

PID is caused by the voltage difference that occurs between solar cells and the grounded module frame. The primary mechanism for PID in conventional Al-BSF modules is the migration of ions from the front glass into the front junction of the solar cell that leads to shunting and power loss. There is no evidence in the literature that PERC cells are more susceptible to PID than Al-BSF cells. Any PID test developed for a conventional Al-BSF module can be applied to a PERC module with similar results. The current standard within the industry is IEC 62804.

The year-wise annual degradation of the module is related to a number of failure mechanisms related to environmental factors such as water ingress, temperature stresses, mechanical stresses, and UV light. Based on an extensive study, a linear loss of 0.5%–0.8% per year has been assumed for the majority of conventional Al-BSF modules. In general, these mechanisms are not related specifically to cell technology. Furthermore, PERC and Al-BSF cells are similar enough that the same module materials and manufacturing are used for both. It is reasonable to expect that PERC modules would display a yearly degradation comparable to Al-BSF modules.

Recently, multiple research groups have shown that both multicrystalline and monocrystalline PERC modules can degrade severely (up to 15%) when exposed to both light and temperatures above 50°C. This degradation is associated with light and temperature, and it is called light and elevated temperature induced degradation (LeTID). Industrial LID suppression techniques and tools cannot be applied directly to LeTID as the mechanism associated with each is different. However, some convincing suppression strategies have been presented in the literature [25, 26] for LeTID: (a) regeneration using illumination and high temperatures, (b) careful wafer material selection, and (c) reduced peak temperatures during the firing process. Fig. 5.56 depicts a PERC device cross-section and summarizes the known degradation mechanisms.

There are three degradation mechanisms in PERC solar cells: (a) hydrogen-induced degradation (HID), (b) light-induced degradation (LID), and (c) depassivation of rear dielectrics.

This type of light and temperature induced degradation was found in multicrystalline Si PERC cells [25] and also in mono-PERC cells [26]. The reason for this type of degradation is based on the high hydrogen content in the device. The back side passivation takes place by concentrated hydrogen dielectrics. The bonding of the released hydrogen passivates the defect states. Due to solar irradiance

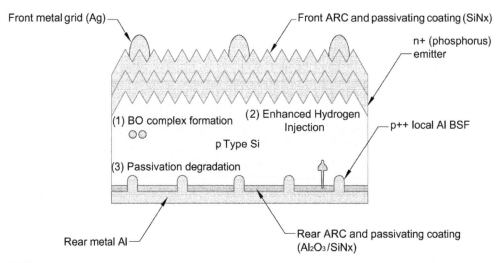

FIG. 5.56

Degradation mechanisms in a PERC solar cell. Light-enhanced temperature-induced degradation can be attributed to (1) boron-oxygen complex formation, (2) excess injection of hydrogen, and (3) the degradation of passivation.

and ambient temperature, these bonds break at a faster rate to free up weakly bonded hydrogen. This phenomena lead toward degradation. To mitigate HID issues, hydrogen-poor dielectric film has to be used for passivation and low temperatures are to be followed for contact firing.

For the LID mechanism, boron-oxygen complexes are responsible. So, the Si wafer has to be selected such that it has a low concentration of oxygen atoms if it is the p-type. N-type Si wafers can be used to get freedom from LID.

It has been observed that degradation in PERC solar cells is based on the depassivation effect of the rear side dielectrics [27]. This degradation effect was earlier observed on the front side of IBC solar cells.

The PERC solar module is driven by monocrystalline technology. Fig. 5.57 shows the achieved wattages of solar modules by different companies in 2019. At present the power rating for 72 cells is around 430 W.

5.5 Reliability and tests conducted on solar PV modules

The solar PV testing requirements are determined by various IEC standards such as IEC 61215:2016 and IEC 61730 Parts 1 and 2: 2016. IEC 61215 defines design, qualification and type approval tests to be conducted on the modules. IEC 61730 Part 1 provides guidelines for module construction and Part 2 defines safety tests that are to be conducted on solar modules. Some of the tests of the standards are described.

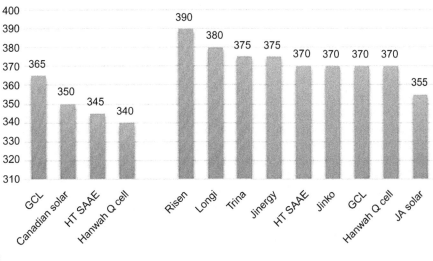

FIG. 5.57

PERC module wattages achieved by different companies [28].

Courtesy for data: Shravan Chunduri, Tiayang News.

- UV exposure.
- Thermal cycling test.
- Humidity freeze test.
- Damp heat test.
- Dynamic mechanical load.
- Damp heat under bias.
- Visual inspection.
- Maximum power (P_{max}).
- Insulation resistance.
- Wet leakage current test.
- Temperature coefficient estimation.
- Nominal operating cell temperature (NOCT) measurement.
- Outdoor performance.
- Hot-spot endurance.
- Bypass diode thermal test.
- UV irradiance test preconditioning.
- Robustness of terminations.
- Mechanical load test.
- Hail impact.

Visual inspection, an *I-V* flash test, a Hipot test, and EL tests are conducted on 100% of solar modules as part of the finished product during manufacturing of solar modules. To assess the quality and reliability of solar modules, they will be tested according to the IEC standards IEC 61215 and IEC 61730 Parts 1 and 2 as well as PID test, ammonia tolerance test, salt spray test, and a sand and dust test. Some companies conduct accelerated tests on the modules with 2–3 times the IEC standards to get more confidence. Some of the tests are described below.

Thermal cycling

PV modules are constructed with several materials, including silicon, copper, glass, and various polymers with varying coefficients of thermal expansion. These materials expand and contract at different rates due to fluctuations in temperature and irradiance. The material interfaces are mechanically stressed due to differences in their coefficients of thermal expansion. The solar PV copper ribbons experience thermal fatigue and the mechanical stress stimulates the solder joint stress, causing cell cracking and increase in series resistance causing decreased performance at high irradiance conditions.

IEC 61215 specifies 200 thermal cycles. PV modules are subjected to extreme temperature swings from −40°C to +85°C with dwell time at temperature extremes as shown in Fig. 5.58. During the temperature ramps maximum power current (I_{mp}) is applied to the modules. This additional current injection drives extremely localized thermal stresses if localized points of increased electrical resistance develop. Fig. 5.58 shows the temperature profile and duration of dwell time for thermal cycling test.

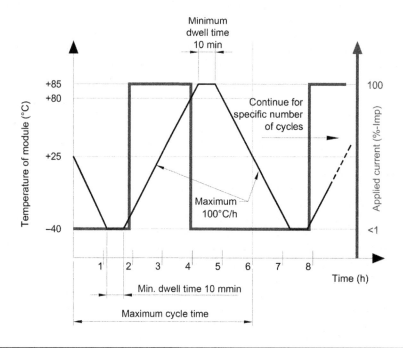

FIG. 5.58

Thermal cycling test—temperature profile.

DNV GL conducts 800 thermal cycles on modules as part of accelerated testing. The 800 cycles are divided into four stages each with 200 cycles. Visual, *I-V* characterization, EL, and wet leakage tests are conducted as pre and post testing of 200 cycles.

Humidity-freeze

PV modules exhibit many interfaces and are constructed with back sheets and encapsulants. The presence of moisture can cause interfacial adhesion to weaken over time. When absorbed moisture in the module package freezes, ice crystals can cause additional damage to the interfaces, and possibly result in delamination. The purpose of humidity freeze test is to investigate the module's ability to withstand high temperature and humidity followed by subzero temperatures.

PV modules are subjected to hot (85°C) and humid (85% RH) conditions for 21 h to cause the module package to be saturated with water. The module temperature is then reduced to −40°C to cause all moisture to freeze. As per IEC standard, this cycle is repeated 10 times. High-temperature glass corrosion results when alkali is removed from the glass surface. The freezing of the moisture propagates the corrosion effect deeper into the glass. Fig. 5.59 shows the requirements of temperature profile and dwell time in humidity freeze test.

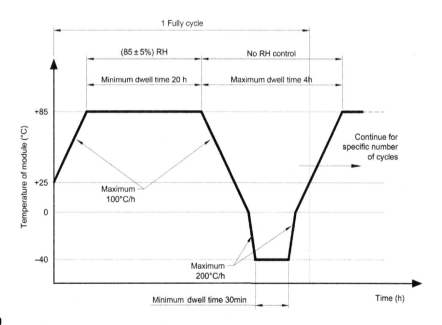

FIG. 5.59

Humidity freeze test—temperature profile.

UV exposure test

Solar PV modules are constructed with various polymers such as encapsulants, back sheets, and adhesive. High-energy UV photons can break polymer chains, resulting in decreased optical transmission and degraded mechanical properties. A reduction in optical transmission leads to a reduced output current of the solar PV module. Degraded mechanical properties can result in delamination of glass or cracks in the solar cells. So, the purpose of UV test is to determine whether solar PV module materials are susceptible to UV degradation. This test is conducted before the thermal cycle and the humidity freeze tests.

Solar PV module is subjected to a total solar irradiance of 15 kWh/m^2 (280–400 nm) while ensuring 33% in the UVB region (280–320 nm) and maintaining a temperature of 60 ± 5°C.

Dynamic mechanical load test

The thickness of silicon solar cells is reducing and thinner cells are more prone to fracture, during events of extreme mechanical loading. The microcracks may have no effect on performance of the module in the beginning, but after subsequent thermal cycling, the crack may propagate and cause failures in the module.

The mechanical stress caused by wind and snow loading impact the glass, frame, encapsulant, and solar cells. This type of regular deflections or stress cause performance loss due to solder joint fatigue, microcrack development and propagation, and cell corrosion. The dynamic mechanical load test sequence evaluates a module's ability to withstand cyclic mechanical deflection as an accelerated test for wind and snow loads.

The module is installed as per the manufacturer's installation manual is subjected to 1000 cycles of alternating loading at 1000 Pa. During the test, continuity of the module's electrical circuit and leakage current to the module frame is monitored. The modules are then subjected to thermal cycling stress for 50 cycles (from –40°C to +85°C) to cause microcrack propagation. The cell cracks do not impact module performance substantially until the cell cracks propagate through the metallization. After thermal cycling, the modules are subjected to 10 humidity freeze cycles to fully realize the potential power loss. The high humidity followed by freezing temperatures in this stress test causes cracks to propagate through the cell metallization.

Potential induced degradation test

For the evacuation of power from solar PV modules, they are connected in series to achieve a system voltage of 1000/1500 V DC, as the case may be. In addition, the solar PV module frame needs to be earthed or grounded as per several standards and laws. Hence, this grounding will give rise to a voltage between frames (which are grounded) and solar cells of the solar PV modules.

This voltage drives charge and mobile ions either toward or away from the cells. Low-iron soda-lime glass, the typical superstrate of PV modules, contains sodium ions. The mobile sodium ions can migrate toward the cells and degrade their properties. This degradation mechanism has been correlated to the defect density in the solar cells' silicon nitride (SiN) antireflective coating. Hence, it is described as potential induced degradation (PID). In addition, static charge can build up on the cells' front surfaces, resulting in an additional parasitic electric field inside the solar cells. This effect is described as polarization. Solar PV modules are subjected to the steady-state temperature of 85°C and humidity of 85% RH conditions while concurrently biased at a positive or negative 1000 V DC with respect to the solar PV module frames.

Hipot test: The Hipot test or dielectric withstand test is centered around testing the insulation level of the solar PV module. This test simulates the conditions that solar PV modules will undergo during operation.

A DC high voltage between the positive and negative terminals of the solar PV module is applied to test the solar PV modules, as the solar PV modules are connected in series to achieve a system voltage of 1000/1500 V DC. This applied voltage gives rise to leakage current (μA), which is measured between various locations on the frame that can come in contact with operating personnel. It is assumed that the insulation of the solar PV module can withstand high voltage for a short duration and the solar PV module can function without any risk of shock hazards in its lifetime of operation.

Ground continuity test: This test is done in conjunction with the Hipot test. This test determines that ground connections of solar PV modules (with metallic frames) are done adequately as per the relevant standards or electrical rules. By applying a voltage between the frame of the solar PV module and the ground, the current has to be checked. It is required that the solar PV modules in the production line should undergo ground continuity tests.

Visual inspection test

The purpose of this visual inspection is to find the major visual defects defined below by checking the solar PV module in a lighted area with 1000 lux light levels.

(a) Broken, cracked, or torn external surfaces, including superstrates, substrates, frames and junction boxes.
(b) Bent or misaligned external surfaces, including superstrates, substrates, frames, and junction boxes to the extent that the installation and/or operation of the module would be impaired.
(c) A crack in a cell, the propagation of which could remove more than 10% of that cell's area from the electrical circuit of the module.
(d) Bubbles or delamination forming a continuous path between any part of the electrical circuit and the edge of the module.
(e) A loss of mechanical integrity to the extent that the installation and/or operation of the module would be impaired.

　　The sum of the area of all bubbles shall not exceed 1% of the total module area. Evidence of any molten or burned encapsulant, backsheet, front sheet, diode, or active PV component. Voids in or visible corrosion of any of the layers of the active circuitry of the module extending over more than 10% of any cell. Broken interconnections, joints, or terminals are also the rejection criteria.

This is repeated multiple times throughout all the test sequences and is conducted more than any other test.

Maximum power (P_{max}) test

This test is performed before and after each of the environmental tests. This test is carried out using a sun simulator.

It is a standard practice among solar PV testing laboratories around the world to perform tests on solar PV modules at STC, which is defined as a solar irradiance of $1000 \, W/m^2$, a solar cell temperature of 25°C, and an air mass of 1.5 (as prescribed in IEC 60125 and IEC 60904-3).

The detailed requirements for solar sun simulators can be referred to in IEC 60904-9:2007. In general, the solar sun simulators in testing laboratories have a spectrum close to AM 1.5 and these sun simulators are typically located indoors. The manufacturers of solar sun simulators offer systems with the highest rating, AAA, where the first A represents the spectrum quality, the second A represents the uniformity of solar irradiance on the test area, and the third A represents the temporal stability of the solar irradiance (as per IEC 60904-9:2007).

The referred IEC standards specify the accuracy of the voltage, the current, the temperature, and the solar irradiance measurements.

It may be noted that as per IEC 61215, the repeatability for the power measurement is $\pm 1\%$.

Insulation resistance test

This test is conducted to find out the electrical insulation between its current-carrying parts and the metallic frame. This test is formulated to give stress to the insulation of the solar PV module when high DC voltage is applied between the positive and negative terminals. This test helps to determine the insulation resistance of the solar PV module, which is measured in $M\Omega$. By using a dielectric strength tester, 1000 V DC plus twice the maximum system voltage is applied for 1 minute. The test pass criteria is that no breakdown or surface tracking should be observed. The measured resistance using a megger should not be less than $40 \, M\Omega/m^2$ of solar PV module if twice the system voltage is applied for 2 minutes.

Wet leakage current test

This test evaluates the insulation resistance of the solar PV module. There could be possible moisture entry into the solar PV module due to fog, rain, dew, or snow. This can cause corrosion, ground faults, and, at times, electric shock. This test simulates what will happen if anyone accidentally touches a wet solar PV module. This test is identical to that of the insulation resistance and Hipot tests. Hence testing or safety organizations have called for this wet leakage test.

The test is conducted with the immersion of the solar PV module in a suitable water tank, which needs to cover the full surface of the solar PV module except for cable entries. The terminals (+ve and −ve) are short-circuited and high voltage is applied for 2 minutes between the shorted terminals and the water bath solutions based on the system voltage. The measured insulation resistance must exceed $40 \, M\Omega/m^2$ of solar PV module ($>0.1 \, m^2$). Make sure that the mating connectors are well inside the water bath and any fault in the connector can be found out if the test fails.

Temperature coefficients test

This test is conducted to find out the temperature coefficient of the current (I_{sc}), the voltage (V_{oc}), and the P_{max} from the measurements of the solar PV module. By using a solar sun simulator and a temperature controlled heating plate, the *I-V* parameters are measured, at solar irradiance of $1000\,W/m^2$ and temperature coefficients for current, voltage, and power are determined.

As per IEC 60891, one can calculate the temperature coefficient over a range of solar irradiance and this is valid if there is a linearity over a particular solar irradiance range.

The solar sun simulator makes a current-voltage measurement (open circuit voltage, short circuit current, and max power are measured during a 5°C interval from 25°C to 55°C). Plotting of the three parameters is done as a function of temperature for each dataset. The temperature coefficients for current, voltage, and maximum power are derived from the slopes of the least square fit straight lines for the three parameters.

For the calculation of the simulated energy yield of solar PV modules in temperate climates, the various temperature coefficients are used.

Nominal operating cell temperature (NCOT/NMOT) test

The nominal operating cell temperature as defined for a ground-mounted solar PV module is obtained at the following standard conditions:

Solar irradiance: $800\,W/m^2$.
Ambient temp: 20°C.
Tilt angle from horizontal: 37 ± 5 degrees.
Wind speed: 1 m/s.
Solar PV module connected to load in NMOT measurement and it is open condition in NOCT measurement. The NOCT is used to estimate the temperature of the module at different irradiance and different ambient temperature conditions. NOCT depends on thermo optical properties such as absorptance and hemispherical emittance of the module package materials.

The NOCT is determined by an outdoor test method. The setup required is a data logger, a pyranometer (for solar irradiance measurement), temperature sensors (ambient temperature), thermocouples (cell temperature), and an anemometer (wind speed and wind direction). Under maximum power conditions, electric energy is withdrawn from the module, therefore less thermal energy dissipates throughout the module than under open-circuit conditions and so, NMOT will be <NOCT.

Outdoor exposure test

This test assesses the ability of the solar PV module to operate in outdoor conditions, when it is exposed to an outdoor environment. This testing process is limited to the outdoor exposure of a solar PV module to a cumulative $60\,kWh/m^2$ of solar irradiance over consecutive days. The pass criteria for this test as per IEC 61215 is that the P_{max} should not be reduced less than 5% of the original value prior to the exposure test. Outdoor exposure and NMOT determination tests are performed simultaneously on the same module.

Hotspot endurance test

The purpose of this test is to determine the ability of the solar PV module to withstand localized heating due to hotspots arising out of partial shadowing, soiling, cracked or mismatched solar cells, and interconnection failures.

When a module is partially shaded, the shaded cells are usually forced into reverse bias. If bypass diodes are not used to protect the cells in the module, the current flowing in reverse bias can cause extreme heating, leading to dissipation of heat. The localized temperature can exceed 150°C or even 300°C for minutes or hours, causing permanent damage within the module package. Serious hot spot phenomena can be as dramatic as outright burns of all the layers, cracking or even breakage of the glass. The withstandability of the test depends on shunt resistance and the reverse characteristics and the reverse breakdown voltage of the solar cell. Four test cells are selected in the module. They are one lowest shunt resistance cell at module edge, two lowest shunt resistance cells, and highest shunt resistance cell. Determination of worst-case shading can be done by shading the cell. The worst-case shading condition for each selected cell is maintained for 1 hour and the shaded cell temperature is monitored. If temperature of shadowed cell is still increasing after 1 h, the total exposure time will be 5 h.

Bypass diode test

This test is designed to determine the thermal behavior of the solar PV module under hotspot conditions. Bypass diode is a very important aspect of module design. It is a critical component determining the thermal behavior of the module under hot-spot conditions and therefore also directly affecting reliability in the field.

The method requires attaching a thermocouple to the bypass diode body and heating the solar module to 75°C±5°C. Then, a current is applied for 1 h, which is equal to the I_{sc} at STC.

Robustness of terminations test

This test is designed to find out the robustness of the solar PV module terminals. In this test, the terminations are subjected to a stress test, which is designed to simulate the normal assembly and handling by a number of cycles of tensile strength, bending, and torque tests, as per IEC 60068-2-21.

Damp-heat test

This test is designed to evaluate the how the solar PV module is able to withstand long period exposure to humidity. This test uses high temperature and high humidity to evaluate module construction, such as lamination process and the edge sealing from ingress of moisture and material quality. The various layers in a typical crystalline-Si PV module need to stay securely adhered for decades in the field. The high-temperature/high-humidity conditions occur regularly in many parts of the world, the damp heat testing sequence is effective at uncovering degradation and failure modes associated with long-term exposure even in moderate climates.

In this test, solar PV module is held at a temperature of $85 \pm 2°C$ and relative humidity of $85 \pm 5\%$ for 1000 hours in a chamber continuously. This test evaluates the lamination process and the edge sealing from the ingress of moisture.

Mechanical load test

This test is designed to find out the static loading capability of a solar PV module when subjected to wind or snow loading conditions.

This load test is conducted following the damp heat test, which has undergone stress in severe harsh environmental conditions.

This test is done using the fixing points of the solar PV module on the mounting structure with the application of a mechanical load equivalent to 2400 Pa for 1 h on the back side of the solar PV module. For qualification for heavy snow loading, the last cycle of this load test is increased to 5400 Pa loading.

The pass criteria is that no major visual defects and no intermittent open circuits are detected during the course of the test. In addition, the $P_{\max}$ and insulation resistance are measured after this load test.

Hail stone test

This test is designed to determine the capability of the solar PV module to withstand the impact of hailstones. The test equipment is a launcher capable of propelling various weights of ice balls (25 mm/7.53 g) at a velocity of 23 m/s so that the ice balls can hit the solar PV module at 11 specified impact locations with a ± 10 mm distance variation.

The time elapsed between the removal of the ice ball from cold storage and the impact on the solar module shall not exceed 60 s.

In addition to the IEC 61215 and IEC 61730 tests and other tests such as PID for system voltage requirements (IEC 62804), an ammonia resistance test (IEC 62716), a dust and sand resistance test (IEC 60068), a salt spray test (IEC 60701), and some accelerated tests on solar PV modules should be conducted to demonstrate the ability of the designed solar PV module to work for 25/30 years of life guaranteed by solar PV module manufacturers.

5.6 Manufacturers of crystalline silicon solar PV modules

Table 5.8 shows the top solar module suppliers and their shipments in 2018 and 2019.

Table 5.8 Top 10 PV module suppliers in the world.

S. no.	Name of the company	2018 Shipment (GW)	2019 Shipment (GW)
1	JinkoSolar (China/Malaysia)	11.4	14.2
2	Trina Solar (China/Thailand)	8.1	9.7
3	JA Solar (China/Malaysia)	8.8	10.3
4	Canadian Solar (Canada/China/Vietnam)	7.1	8.5
5	Hanwha Q CELLS (Korea/China/Malaysia)	5.5	7.3
6	GCL System Integration Technology (GCLSI) (China/Vietnam)	4.1	4.8
7	LONGi Green Energy Technology (China)	7.2	9.0
8	Shunfeng PV International Ltd	3.3	4.0
9	Risen Energy (China)	4.8	7.0
10	First Solar (United States/Malaysia)	2.7	5.5

There are companies with more than 10 GW/year of production capacity.

5.7 Technology trends of solar PV modules

There have been many advancements in solar module technology:

Half-cut solar cell-based module technology.
Glass-glass-based modules.
Monocrystalline PERC solar modules.
Bifacial solar modules.
Solar modules with multinumber busbars and smart wire technology.
HJT solar cell-based modules.
Shingling-based solar modules.
Higher wattage modules with larger size solar cells.

Advanced module technologies represented by the bifacial module, the shingled-cell module, the half-cell module, and the multibusbar module (MBB) work together with high-efficiency solar cell technology to improve the reliability of the PV system and the performance of power generation as well as to minimize the levelized electricity costs (LCOE).

In a double glass solar PV module, the glass replaces the back sheet of the solar module. The double glass module is highly reliable, weather-resistant, and fire-resistant. Sandwiched with bifacial PERC, HJT, or PERT cells, the glass-glass solar panel is a bifacial system with some increases in cost. It can produce additional power of 0%–25%. But the disadvantage with double glass module is its weight, trapping of acetic acid in the glass, more breakage rate due to usage of heat strengthened glass. The rear glass of bifacial modules is being replaced with transparent back sheet with frame to overcome the issues encountered with glass-glass modules.

With the new test standards and test equipment and the recognition of the value of bifacial power generation, the bifacial module has now increased its market share by more than 50%.

In a shingled cell module, the cut cells are bonded to each other in a shingled manner. In the area, more than 12% of the solar cells can be packed compared to a similar standard PV module area. These modules provide higher output power, fewer hotspots, and lower losses. Seraphim Energy shingled modules were tested on the ground in a large grid-connected plant in January 2018 while DZS Solar produced a new generation of high-efficiency shingled modules with 335 Wp module power during the same year, using 60 cells and >20% efficiency.

Multibusbar (MBB) technology reduces the covering area by the busbar and at the same time increases the solar cell efficiency and solar module output. This also helps in the optimal utilization of the ribbon area. In addition, MBB cells help in consuming less paste while there is less damage in the cell level. The main key technologies are the MBB cell metallization and the interconnection process.

The busbar-free technique, which is based on MBB, has a round copper wire coated with a low-temperature alloy. Round wire soldering usage will increase the reuse rate of light causing increase in optical gain. If more busbars are used, they can be made thinner and finer making cost saving in consumption of silver paste. Reduction in electrical resistance due to denser interconnection causing lower voltage drops, thereby improvement in the power. The effect of microcracks is reduced, as the more busbars are concentrated and the failure area will be smaller than the conventional microcracks.

The solar cell size is increasing from 156.75×156.75 mm to 210×210 mm due to increase in wafer size. Mono crystalline Si wafers are available in full square. With the increase of cell size, the current will increase and in-turn the power will increase. The modules are coming up with half-cut, triple-cut cells, multibusbars with the usage of larger size M10/G12 wafer-based p-type mono-PERC cells. The power of the module is increasing and 600 W+ modules were introduced by different companies at SNEC, Shanghai 2020.

JA Solar introduced 810 W model called Jumbo, the panel has quadruple layouts of 47 cells and dimensions of 2.220×1.757 mm. This panel utilizes a triple-cut cell design with 11 busbars on 210 mm wafers. It uses mono-PERC technology and the module efficiency is 20.5%.

Jinko Solar exhibited 610 W solar module of efficiency 22.31% with the usage of 78 number N-Type Bifacial solar cells. The module uses n-type ToPCON cell of 24.79% efficiency with tiling ribbon with MBB technology and the cell size is 182×182 mm. It uses cell with HOT tunneling layer passivated contact and advanced metallization technologies.

Tongwei introduced 760–780 W solar module of efficiency 21.9% and weighing 39 kg. The module size is 2357×1612 mm using 210×210 mm size mono-PERC solar cell with shingling technology.

Trina Solar exhibited 660 W solar module of efficiency 21.20%. It uses 210×210 mm size mono-PERC cell using multibusbar technology. It uses cut cells with non-MBB, nondestructive cutting, high-density encapsulation.

Risen energy introduced 615 W solar module of efficiency 21.2%. It uses mono-PERC half-cut cells with 12 busbar configuration. Suntech exhibited 605 W solar module of efficiency 21.3%. It uses mono-PERC triple-cut cells with multibusbar configuration. Astronergy exhibited 605 W solar module of efficiency 20.8%. Jollywood introduced 615 W solar module of efficiency 22.1% with the usage of TOPCon/Bifacial 11BB solar cells. ZNshine introduced 600 W solar module of efficiency 21% with the usage of 150 cell. DMEGC introduced 600 W solar module of efficiency 21.98%. Haitai Solar introduced 600 W solar module of efficiency 21.7%. DZS Solar introduced 635 W Shingled solar module of efficiency 21.1% with the usage of G12 large wafer Bifacial solar cells. Longi and Eging came up with 540 W modules with the usage of Ga doped solar wafer-based mono-PERC solar cells.

HJT solar modules

HJT is an age-old technology. The cell was developed by Sanyo and later taken by Panasonic. The junction is formed with amorphous Si material. It uses a high-quality n-type wafer as the base material.

It has a low temperature-based simple process for manufacturing. It has a low temperature coefficient of power due to the formation of a heterojunction with the a:Si material. It has a lower annual degradation in power. There is no light-induced degradation in power due to the use of n-type wafers.

The key patents of HJT technology expired in 2010, which attracted the interest of researchers in the development of high-efficiency cells. In mass production, a 22.5% efficient cell is available on the market, whereas at the R&D level, the expected efficiency of the cell is 25.6%.

Panasonic, CTI, Kaneka, Sharp, Havel Solar, NSP Solar, Tesla, Solar Tech, Inventec, Sunpreme, GS Solar, CIE Power, Hanergy, J Energy, GCL, Tanguei, ENN, and Eco Solar are working with HJT solar cell technology. The Intel and Enel groups are involved with the development of HCT cells.

HJT module manufacturing

A standard tabber and stringer are used with a low-temperature process for soldering the interconnector to the solar cell. An electrically conductive adhesive can also be used for the interconnection of solar cells. SWCT, which is an advanced interconnection process with the use of circular wires to realize the solar cell interconnection, can also be implemented with this type of solar cell. The solar PV systems that use high-efficiency HJT cells have better LCOE in the module and system levels.

The capex required to manufacture HJT solar cells is 2–3 times that used to manufacture PERC-based solar cells. The throughput is rather low compared to the CPIA technology roadmap. It is expected that HJT cells will outperform ToPCon and PERC solar cells. The capacity forecast for n-type technologies are HJT, ToPCon, IBC, and PERT. There are no LID and PID effects with HJT solar cells. These solar cells can be used for panels with half-cut cell technology. The cell structure is bifacial and can be modified for bifacial solar cells. The HJT bifacial solar cell and the front and rear side illuminations show almost the same efficiency.

References

[1] Mercom India Research Report.
[2] Wood Mackanzie Report, https://www.woodmac.com/press-releases/global-solar-pv-installations-to-reach-record-high-in-2019/.
[3] IEC 61730-1:2016, Photovoltaic (PV) Module Safety Qualification—Part 1: Requirements for Construction, 2016.
[4] S.M. Sze, Physics of Semiconductor Devices, second ed., Wiley, NewYork, 1981.
[5] AN3432 Application Note of ST Electronics "How to Choose a Bypass Diode for a Silicon Panel Junction Box", www.st.com.
[6] Fraunhoffer Institute Web Site, https://www.cell-to-module.com/software/.

[7] H. Hanifi, et al., Investigation of Cell-to-Module (CTM) Ratios of PV Modules by Analysis of Loss and Gain Mechanisms, 2016. www.pv-tech.org.

[8] M. Rubin, Optical properties of soda lime silica glasses, Sol. Energy Mater. 12 (4) (1985) 275–288.

[9] Borosil Solar Glass Data Sheet.

[10] F. Kaule, W. Wang, S. Schoenfelder, Modeling and testing the mechanical strength of solar cells, Sol. Energy Mater. Sol. Cells 120 (Part A) (2014) 441–447.

[11] H.K. Raut, V.A. Ganesh, A.S. Nair, S. Ramakrishna, Anti-reflective coatings: a critical, in-depth review, Energy Environ. Sci. 4 (2011) 3779–3804.

[12] IEA-PVPS 2014, Performance and Reliability of Photovoltaic Systems Subtask 3.2: Review of Failures of Photovoltaic Modules IEA PVPS Task 13 External Final Report IEA-PVPS, 2014.

[13] L.S. Bruckman, N.R. Wheeler, J. Ma, E. Wang, C.K. Wang, I. Chou, J. Sun, R.H. French, Statistical and domain analytics applied to PV module lifetime and degradation science, IEEE Access 1 (2013) 384–403.

[14] S. Pingel, O. Frank, M. Winkler et al., Potential induced degradation of solar cells and panels, in: 2010 35th IEEE Photovoltaic Specialists Conference, June 2010, Honolulu, HI, pp. 2817–2822.

[15] M.C.C. de Oliveira, A.S.A.D. Cardoso, M.M. Viana, V.d.F.C. Lins, The causes and effects of degradation of encapsulant ethylene vinyl acetate copolymer (EVA) in crystalline silicon photovoltaic modules: a review, Renew. Sustain. Energy Rev. 81 (Part 2) (2018) 2299–2317.

[16] G. La Roche, K. Bogus, Assessment of new components for space solar arrays, in: Proceedings of the Fifth European Space Power Conference (ESPC): Spain, 21-25 September, Organised by European Space Agency [ESA] (ESA SP; 416), 1998, p. 609.

[17] M. Kempe, Overview of scientific issues involved in selection of polymers for PV applications, in: Conference Record of 37th IEEE Photovoltaic Specialists Conference, 2011, pp. 85–90.

[18] K. Whitfield, ESD surge characterization of Schottky diodes, in: Paper Presented at the PV Module Reliability Workshop, Denver, CO, USA, 2013.

[19] H.-Y. Li, L.-E. Perret-Aebi, R. Théron, C. Ballif, Y. Luo, R.F.M. Lange, Optical transmission as a fast and non-destructive tool for determination of ethylene-co-vinyl acetate curing state in photovoltaic modules, Prog. Photovoltaics Res. Appl. 21 (2) (2013) 187–194.

[20] U. Eitner, M. Pander, S. Kajari-Schröder, M. Köntges, H. Altenbach, Thermomechanics of PV modules including the viscoelasticity of EVA, in: Proc.26th EUPVSEC, 2011, pp. 3267–3269.

[21] H.-Y. Li, L.-E. Perret-Aebi, V. Chapuis, C. Ballif, Y. Luo, The effect of cooling press on the encapsulation properties of crystalline photovoltaic modules: residual stress and adhesion, Prog. Photovoltaics Res. Appl. 23 (2) (2015) 160–169.

[22] H. Vahlman, Application Note 20180829HJV accurate flash testing of high-efficiency solar cells and modules using capacitance compensation, 2018. https://www.endeas.fi/high-efficiency-measurement/.

[23] M. Herman, M. Jankovec, M. Topič, Optimal I-V curve scan time of solar cells and modules in light of irradiance level, Int. J. Photo Energy (2012) 151452. 2012.

[24] H. Vahlman, J. Lipping, J. Hyvärinen, A. Tolvanen, S. Hyvärinen, Capacitive effects in high-efficiency solar cells during I-V curve measurement: considerations on error of correction and extraction of minority carrier lifetime, in: 35th EU PVSEC Photovoltaic Solar Energy Conference and Exhibition, September 24–28, Brussels, Belgium, 2018.

[25] K. Ramspeck, et al., Light induced degradation of rear passivated mc-Si solar cells, in: 27th European Photovoltaic Solar Energy Conference and Exhibition, 2012, pp. 861–865.

[26] F. Fertig, et al., Mass production of p-type Cz silicon solar cells approaching average stable conversion efficiencies of 22%, Energy Procedia 124 (2017) 338–345.

[27] A. Herguth, et al., A detailed study on light-induced degradation of Cz-Si PERC-type solar cells: evidence of rear surface-related degradation, IEEE Photovolt. 8 (5) (2018). 1190–120.

[28] S. Chunduri, M. Schmela, Advanced module technologies-2019 edition, Taiyang News.

Solar PV systems and applications

6.1 Introduction

Solar photovoltaic systems are very flexible. They can be smaller, such as 0.3 Wp for powering a flash-light, or larger than a football stadium, powering several villages or a city. Solar PV power systems can power everything from a watch to an entire city or town. They can be used as specific solar power suppliers for a variety of standalone or specific purpose applications.

There are numerous applications of solar PV systems, some of which are mentioned below.

6.2 Different applications of solar PV systems

6.2.1 Lighting

With the invention of LED (light-emitting diode) technology as a low-power lighting source, PV systems have found ideal application in remote or mobile lighting systems. PV systems combined with battery storage facilities are mostly used to provide lighting for billboards, highway information signs, public-use facilities, parking lots, vacation cabins, and lighting for trains.

(a) **Solar lanterns:**

Solar-powered lanterns are portable devices, for example, with three 1 W LED lamps (the power drive is for 2 W light output). The solar PV module capacity is 3.3 Wp with a sealed lead acid battery of 6.0 V, 4.5 Ah capacity. It has a built-in solar controller with an ON/OFF switch. Also, it has a socket for mobile charging (optional feature). Fig. 6.1A and B shows images of solar lanterns used with CFL and LED lamps.

(b) **Solar power home lighting system:**

A solar-powered home lighting system is a standalone power supply system that can be used to light homes. A typical system comprises two 3 W LED ceiling- or wall-mounted lamps with cables, a 10 Wp solar PV module, a battery box with a 12 V, 12 Ah sealed maintenance-free battery, and a built-in solar charge controller. There is an ease of connection using suitable male/female connectors for all light loads and solar PV modules. It also has a mobile charging facility. The housings for the battery box and the luminaries are from ABS and polycarbonate materials. Fig. 6.2 shows a solar home lighting system with a 10/40 W solar panel, a 12 V, 12/40 Ah lead acid battery supported with two 3 W LED luminaries with one table or with one 12 V DC, 10 W color TV.

Solar PV Power. https://doi.org/10.1016/B978-0-12-817626-9.00006-X

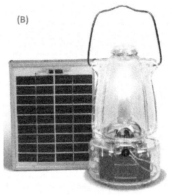

FIG. 6.1

(A) Solar lantern with a CFL lamp, and (B) solar lantern with LED.

FIG. 6.2

Solar home lighting system with (A) 10 Wp solar module, 12 V/12 Ah lead acid battery with two 3 W LED luminaries; (B) 40 Wp solar module, 12 V/40 Ah lead acid battery with two 3 W LED luminaries, one table fan; and (C) 40 Wp solar module, 12 V/40 Ah lead acid battery with two 3 W LED luminaries, one 12 V DC, 10 W TV.

(c) Solar PV powered street light system:

Solar PV powered street lighting system consists of a solar PV module (37/75/80 Wp or 2 × 125 Wp); LED lamps (6/15/18/35 or 40 W); a battery (tubular flooded lead-acid or sealed maintenance free (VRLA gel type) with a capacity of 12 V, 40/60/75/180 Ah; a pole of 4 m or higher; a built-in charge controller; and other controlling functions such as a dusk-dawn or fixed timer, automatic operation, etc. Fig. 6.3 shows a solar street lighting system with and without integration of an LED-based lighting system with a solar panel.

6.2.2 Water pumping

(a) Solar PV powered DC surface pump:

This system uses a DC surface pump (1 HP or 846 W) directly connected to a solar PV array of say 900 Wp and an interfacing solar controller.

(b) Solar PV powered DC submersible pump:

This system uses a DC submersible pump directly connected to a solar PV array of say 900 Wp and an interfacing solar controller to maximize the solar array power to the pump. Fig. 6.4A shows the assembled solar water pumps with different capacities from Mercury Electronic Corporation. Fig. 6.4B shows different parts of water pump in exploded view.

(c) Solar PV powered AC submersible pump:

This system uses an AC submersible pump (starting with 0.5 HP to several HP, typically 3/5 HP) connected via a specialty inverter (variable voltage and frequency) to the solar PV array of say 1400 Wp or higher. Fig. 6.5 shows a solar water pumping system with surface pumps and controllers developed by Mercury Electronic Corporation.

6.2.3 Water purification

(a) Solar PV powered water purification system:

This system uses a water purifier fitted with primary (25 μm, nylon) filters and secondary filters (10 μm, nylon), which are effective against viruses, bacteria, protozoa, worm eggs, etc. It also has a

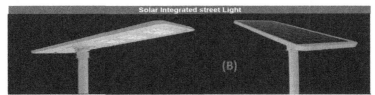

FIG. 6.3

Solar street lighting system (A) solar modules and lamps arranged separately, and (B) LED-based lighting system integrated with solar panels.

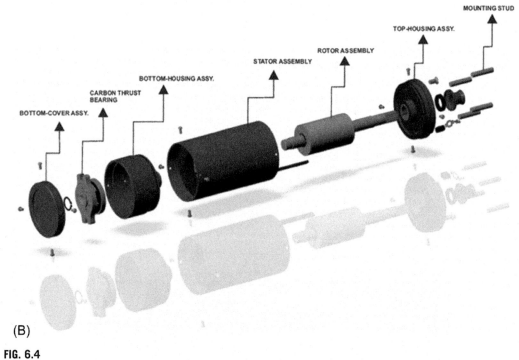

FIG. 6.4

Solar water pumps. (A) Pumps with different capacities and (B) different parts of water pump with exploded view.

Courtesy: MEC Electronic Corporation.

FIG. 6.5

Solar water pumps and controllers. (A) Controllers for water pumps and (B) surface water pumps of different capacities.

Courtesy: MEC Electronic Corporation.

20 W UV lamp and is combined with a 75 Wp (12 V) solar PV module and a 12 V, 40–65 Ah sealed lead-acid battery. The purified water has a daily capacity of 2000 L with a 100 L tank capacity with a normal flow rate of 5–6 L/min. Fig. 6.6 shows the installed PV powered water purification system.

6.2.4 **Refrigeration**

(a) Solar PV powered vaccine refrigerator:

Vaccine refrigerators, needed to maintain a temperature of 2–8°C for vaccines in rural villages and community health centers, demand a continuous power supply, and solar power is the right candidate. Vaccine refrigerators (typically 40 L storage and 2 L ice-making capacity), as per the World Health Organization (WHO) standard, should not consume more than a specified energy or a specific solar array capacity. The solar array and the battery bank are external to the vaccine refrigerator while the solar controller is built-in.

(b) Solar PV powered refrigerators can also be used for recreational vehicles as well as for commercial and residential usage. DC powered refrigerators such as those powered by solar PV modules are becoming cost effective and can be used for a variety of applications, as they avoid DC-AC conversion losses and the extra cost for an inverter while improving reliability with the latest technologies. These units come in either 12 or 24 V DC and are 4–5 times more efficient than similar AC units. Hence, they require smaller solar PV systems.

FIG. 6.6

PV powered water purification system.

6.2.5 Electric fences

Solar PV powered electric fences:

This system uses a high-voltage pulse generator to energize metallic wires in electric fences. This in turn is powered by a solar PV array-charged battery bank with a controller. The energizer is capable of charging long lengths of fencing, as it draws minimal current under normal operating conditions.

6.2.6 **Battery charging**

(a) Solar PV powered battery charging system:

Solar PV modules can be effective battery chargers for rechargeable batteries. During low-power use conditions, the solar PV modules can charge the batteries to a 100% charge or can supply the loads directly while charging the battery during deep discharge conditions. Fig. 6.7 shows the concept of an energy storage system with renewable energy power plants such as a PV array or a wind turbine farm. The battery is placed in a container. Batteries store energy generated by the solar array and the wind and grid in different conditions.

(b) Charging vehicle batteries: PV systems may be used to directly charge vehicle batteries, or to provide a "trickle charge" to maintain a high battery state of charge on little-used vehicles such as firefighting equipment, snow removal equipment, and agricultural machines such as tractors or harvesters. Direct charging is useful for boats and recreational vehicles. Solar stations may be dedicated to charging electric vehicles also.

6.2.7 **Telecommunications**

(a) Powering a microwave repeater station:

Solar PV power is an ideal power supply source for remotely and isolated microwave repeater stations. In view of the remoteness of the location of repeater stations and the high carrying cost of diesel, solar provides a cost-effective power supply solution.

(b) Powering of a base transceiver station (BTS):

These base transceiver stations (BTS) require a dedicated and continuous power supply with high reliability. Solar PV power systems play an important role for rural and remote BTS systems while becoming cost-effective. Considering the cost of conventional power or using a DG set for remotely located BTS systems, solar power provides in situ power with an optimized cost of power supply solutions as well as higher reliability.

FIG. 6.7

Concept of an energy storage system with renewable energy power plants such as a PV array or a wind turbine farm.

FIG. 6.8

Solar PV system to support power to telecom towers (A) solar array, (B) battery bank, (C) charge controller, and (D) solar module mounting system.

Fig. 6.8 shows pictures that provide a typical example of such a solution. It is a PV system to support power to telecom towers, and it contains its major components.

6.2.8 Oil and gas platforms

(a) Solar-powered gas detection system:

Unmanned oil and gas platforms in high seas use solar PV power systems for powering gas detection systems and associated other loads while avoiding the high cost of power using undersea cables from onshore or a DG set. It has proven to be a cost-effective, reliable, and effective power supply system for use in classifieds area of oil and gas platforms. This has various critical parameters for design, including 7 days of storage using either VRLA or a flooded lead acid battery while the solar PV module is mounted on the helideck and the solar module mounting structure is designed for 280 kmph. Typical energy/power requirements vary on the number of loads and the duty cycle. Fig. 6.9 shows a clean power energy concept with an oil pump using solar panels at sunset. Fig. 6.10 shows the PV system in a control system on a platform of the power and energy business industry. Solar panels are mounted at the flare bridge at the offshore oil and gas wellhead remote platform to charge the batteries of the electrical and control systems on the platform.

(b) Solar-powered data telemetry system:

Telemetry and SCADA systems of unmanned oil and gas platforms are critical systems that provide communication access from the shore. Hence, these telemetry and SCADA systems need a continuous power supply. A solar PV power system with 7 days storage and adequate solar array power installed on the helideck is a favored cost-effective and reliable power source candidate as compared to other sources. Typical energy/power requirements vary on the number of loads and

FIG. 6.9

Clean power energy concept, and oil pump with solar panels at sunset.

FIG. 6.10

Solar panels at the flare bridge at the offshore oil and gas wellhead remote platform to charge the batteries of the electrical and control systems on the platform.

FIG. 6.11

Solar modules provide electric power for monitoring equipment on a pipeline in Texas.

the duty cycle. This is an independent solar power system. Fig. 6.11 shows solar panels providing electric power for monitoring equipment on a pipeline in Texas.

(c) Solar-powered navigational aids:

In view of the fact that these unmanned oil and gas platforms are located in the middle of high sea and pose obstruction to ships, boats, etc., it is mandatory that oil and gas platforms be provided with navigational aids, typically four marine lanterns (Morse U code, 15 s) at the four corners as well as one foghorn (Morse U code, 30 s). These lanterns and foghorns are powered by a solar PV power system with a 7-day battery backup and an adequate solar array. This is a dedicated and isolated solar PV power system for navigational aids. Fig. 6.12 shows an electrical and instrument technician inspection of the lighting of the navigation aid system at an oil and gas wellhead remote platform.

6.2.9 **Cathodic protection**

Pipeline corrosion protection

Naturally, metallic pipeline structures experience corrosion when exposed to soil and water due to electrolytic action. This is because the metals lose ions when exposed to an electrolyte. To prevent corrosion, a voltage may be applied to the metallic structures so that the loss of ions from the metal will not occur. This method of protection is called cathodic protection. Only a small DC voltage is necessary to protect the metals.

Corrosion of the metal surface can be avoided by using impressed current from a DC source, and solar-powered cathodic protection systems found most attractive and reliable for pipeline (water, steel, and fuel pipelines) corrosion protection. The heart of the system is the cathodic

FIG. 6.12

PV modules for lighting of the navigation aid system at an oil and gas wellhead remote platform.

protection controller powered by a solar array and a battery bank. Cathodic protection is used on pipes, tanks, wellheads, wharves, bridges, and buildings.

6.2.10 Space

(a) Powering satellites:

Solar arrays in satellites deliver power to spacecraft for space applications.

To power satellites orbiting in space, solar PV panels supply power during sunlight conditions for all the electrical loads and charge the battery. During an eclipse period, the onboard battery supplies power. Crystalline silicon solar cell-based panels were used earlier to power satellites. At present, space solar arrays use III–V compound-based multijunction solar cells. Each solar cell has germanium, gallium indium arsenide, and gallium indium phosphide junction layers monolithically grown on a Ge wafer. At 28°C and with one solar constant intensity with AM0 spectrum, the efficiency of the solar cell is 30%. The manufacturing processes of space solar cells and space solar panels are entirely different compared to the terrestrial solar fabrication process. Fig. 6.13A shows solar array powering a space station. Fig. 6.13B shows a satellite with a solar array in deployed condition in orbit.

(b) Space-based solar power generation:

This is the concept of using solar energy available in outer space, then converting and transmitting the energy directly to a receiver station on Earth or other planets. This energy source would be able to convert solar energy on a 24/7 basis in outer space. The power from the sun can be converted into energy and could be received by a rectenna (an antenna for receiving energy on the Earth's surface as shown in Fig. 6.13B from outer space, and then the power can be distributed using a standard power system network).

To create a solar power satellite, a solar panel is required on the spacecraft. It should be very large area as per the requirements of power. After the spacecraft is launched into space, it will convert solar energy into electrical energy and then transmit that to Earth by converting it into microwave energy.

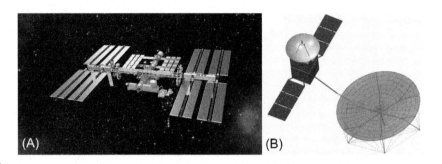

FIG. 6.13

Space applications of solar PV power: (A) a solar array powering the space station; (B) the concept of a solar power satellite: solar array in deployed condition; and the satellite is transmitting the generated power to antennae at Earth.

The microwave energy is received by a large antennae system on the ground. It then needs to be converted to electrical energy. Fig. 6.13B shows the solar power satellite transmitting power to Earth.

6.2.11 Transport

(a) Electric mobility:

Electric mobility is connected with all street vehicles, which are powered by electric motors. An electric vehicle's (EV) electric motors are combined with a storage battery such as Li-ion (due to energy density and charge/discharge characteristics) and charged from grid power using a proper charging device.

The charging of the EV storage battery can be from grid power or solar power sources. Fig. 6.14 shows the concept of EV charging with an electric charging station using alternative fuel sources to save green plants on the planet. Fig. 6.15 shows an EV charging station charging a four-wheeler.

(b) Solar-powered car:

A solar-powered car is a specially designed car used for transport on the road. The source of power is from a solar array that is integrated with the body of the car and also backed by a battery. The use of highly efficient solar cells will provide a greater traveling distance for the car, as more power is packed on the roof of the car with limited space.

6.2.12 Other applications

(a) Powering vegetable vending trucks:

This is a very novel solar-powered application, powering vegetable vending trucks. In order to address the challenges faced by horticultural producers, solar-powered vending trucks help to increase the shelf life of fruits and vegetable by two days or more.

Solar PV modules help in maintaining the required humidity and maintain a temperature of 14°C.

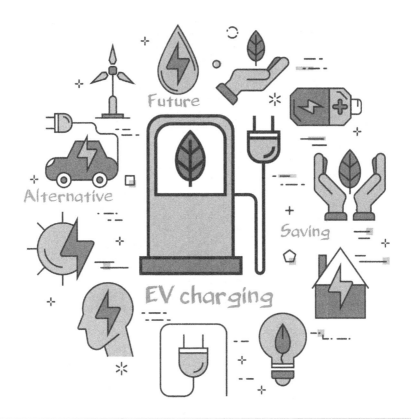

FIG. 6.14

Concept of an EV charging station with alternative fuel sources.

(b) **Powering water-dispensing machines**:

This application deals with powering water-dispensing machines with solar PV power. These water-dispensing systems are standalone water-dispensing units. Each unit has a water storage tank with an optional dispensing pump, a dispensing tap, a dispensing controller, and a solar PV power system comprising a solar PV module, an inverter, and a battery bank. The solar-powered units help to increase the effectiveness of the present water purification system, making it possible to locate these units in rural, remote, and far-off locations.

(c) **Solar-powered wheelchair**:

The wheelchair is an important device for elderly or sick people. It needs a reliable power supply. Hence, powering with a solar PV array makes sense.

(d) **Solar-powered drones**:

Drones are becoming important for many applications such as photography as well as checking the integrity of solar PV modules by scanning to find any defects in solar cells through imaging techniques using infrared cameras. Powering drones with solar power makes them operate longer, although they have a limited battery capacity.

FIG. 6.15

EV charging station charging a four-wheeler.

(e) **Mobile solar power generating system**:

   The solar-powered remote area containerized solution provides in-situ power for remote villages and for various mobile power applications, as and where it is required, and makes solar a better candidate as remote area power supply system. Fig. 6.16 shows a mobile power station with a containerized solution. The solar array is mounted above the container. The inverter and battery bank are kept in the container. If required, even the diesel generator can be kept in the container and the space in the container can be used for office purposes.

(f) **Solar-powered billboard**:

   LED lights powered by solar power can be used for signs and billboards. Solar power to light billboards is a commercially cheaper power supply option as compared to expensive commercial grids, utility power, or locations where there is difficulty in getting access to electrical energy. Also, in view of the isolated and remote locations, it becomes more important to power LED billboards with solar PV power and a battery back-up. A solar power system, which is an integral part of signs or billboards, saves long trenches and cabling work. Dusk-to-dawn timers can be more effective when integrated with these solar-powered signs to optimize the solar PV system design and also the cost.

   Signs and signals: Devices such as navigational beacons and audible warning signs such as sirens, highway warning signs, railroad signals, aircraft warning beacons, buoys, and lighthouses are generally remote or even impossible to connect to the utility grid. PV systems provide reliable power for these critical applications.

FIG. 6.16

Mobile power station with a containerized solution.

(g) Solar power for disaster management:

Using solar power for disaster management is a fitting solution in view of the fact that severe damage is inflicted to a grid power network during natural disasters. During storms such as Hurricane Maria, which was responsible for large amounts of devastation in Puerto Rico, solar power with a battery backup came in handy to immediately establish a power supply system. During hurricanes or storms, local needs for temporary power are in the range of up to a few hundred kilowatts or less.

A portable and foldable 5 kWp solar array with about 150 kWh storage can be operational in a half-hour or less. Other options for 30–40 kWp power is, containerized solar PV power system solutions, which can also provide temporary shelter to people.

(h) Solar-powered signaling system:

Solar-powered road traffic signaling systems are becoming popular, as these are standalone with a storage backup and are reliable for powering critical road traffic signaling systems.

Solar-powered rail signaling systems, including railroad crossing signals, are ideal and are reliable in view of the criticality requirement.

(i) Solar power for scientific experiments:

In view of climate change investigations, it is becoming more relevant to use solar power sources to provide power for scientific experiments. For example, Antarctica is located at the remote South Pole, where it is very cold with 24 h of darkness in the winter months. Scientists have created zero emission research stations using solar and hybrid power sources with suitable storage devices. A notable solar system is the Princess Elisabeth.

(j) Solar PV powered boat:

A solar energy-powered boat is a novel idea to reduce or eliminate the use of conventional liquid fuels such as diesel or gasoline. The solar PV modules, which are fitted on the top of the deck, are exposed to open sun, helping to charge the batteries. In turn, the batteries power the electric motors to drive the boats in rivers, canals, or high seas. Thus, solar energy is converted efficiently to kinetic energy without causing pollution. Fig. 6.17 shows a solar-powered boat that is mounted with rigid solar panels. Lightweight flexible panels also can be used to power the boat, along with a battery. On railway coaches, flexible solar modules are also used to power fans and light the coach with the use of a proper battery and charge controller.

(k) **Solar-powered refrigerator**:

Solar PV-powered refrigerators are good and efficient devices that operate on directly converted solar energy.

The refrigerators may run using solar PV modules or solar thermal power. These refrigerators are mainly designed to keep vaccines and provide ice packs for remote areas, mostly health centers.

New developments such as direct solar power driven vaccine refrigerators without any battery are becoming popular.

(l) **Solar-powered ATM**:

Currency dispensing is becoming a necessity in remote and isolated areas. A solar-powered ATM is the best solution for these locations. The DC power-driven ATMs are solely powered by solar PV modules and backed by a battery bank.

(m) **Solar-powered telephone booth**:

It is important to provide phone booths that are standalone and powered by a solar PV module with a backup storage battery. In remote locations, telephone booths are powered using solar. These are reliable and effective communication devices for rural and remote areas where there is no mobile coverage.

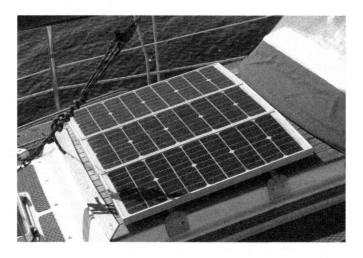

FIG. 6.17

Solar-powered boat.

6.2.13 **Grid connected**

In a typical grid-connected solar PV-powered residential system, solar PV modules are directly connected to the inverter. The DC electricity generated by the solar PV modules/array is converted into AC power by inverter. The inverter is in turn connected to the household LT bus, which powers the connected loads or may feed excess power to the grid.

As the building is connected to the grid, the inverter provides all interfaces for all power, either supplies power to all the loads or exports the power into the grid, with all required protection as per IEC/VDE/UL or other standards.

(a) Solar PV-powered residential grid-connected system (without battery storage)

A solar PV-powered residential grid-connected system (without battery storage) typically consists of solar PV modules, grid connect inverter and module mounting system with associated cables, a net meter unit, and other components.

These systems are directly connected to the grid at the LT bus (400/415 V AC, single or three-phase system).

The solar PV array generates DC power while the sun shines and in turn the DC power is converted into AC power by the grid-connected Inverter. Fig. 6.18 shows a schematic of a typical grid-connected solar PV system without battery storage.

(b) Solar PV-powered residential grid-connected system (with battery storage)

A solar PV-powered residential grid-connected system (with battery storage) typically has these major components: a solar PV module, a hybrid inverter, a battery, and a module mounting system with associated cables, a net meter unit, and other components.

These systems use hybrid inverters that can be directly connected to the grid at the LT bus (400/415 V AC, single or three-phase system) while simultaneously charging the battery bank with surplus power from the solar PV module to feed power into the grid, based on set of parameters. Fig. 6.19 shows a schematic of a typical grid-connected solar PV system with battery storage.

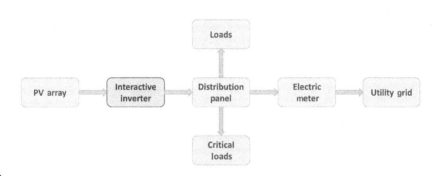

FIG. 6.18

Schematic of a typical-grid connected solar PV system without battery

FIG. 6.19

Schematic of a typical grid-connected solar PV system with battery storage.

6.2.14 Hybrid power

In a hybrid solar PV power system, solar PV power forms a constituent power source while there can be more than one renewable-based power source such as wind power, hydro power, diesel, biogas-based power, etc. The addition of other power-generating sources supplies continuous energy to the loads.

(a) Hybrid solar PV power system (with battery storage)

Hybrid solar PV power systems are becoming 100% reliable energy supply systems. They employ more than one renewable energy power supply source such as solar PV modules, a wind turbine battery charger, etc., while combining one or more storage devices. Fig. 6.20 shows a schematic of a connected solar and wind-based hybrid PV system. The principles of the solar cell, the Li-ion battery, and the generators are also shown in the figure. The generated energy from the solar and wind farm is stored in the battery but also supplied to the loads. Fig. 6.21 is the hybrid system with solar and a wind farm.

6.2.15 Microgrids

(a) Microgrids (also known as hybrid solar power systems) provide continuous power to remote and rural households, which are away from the grid and have no other local power source. The microgrids use solar PV power as one source in addition to another source such as a wind power generator, a biogas generator, or a DG set, backed by storage battery backup for continuous power supply. A typical village power system uses a kWp to MWp level solar array along with a DG set and a battery bank, associated control electronics, and an inverter. Examples include the Bihar/UP/ Tamil Nadu rural village electrification and India's first fully solar powered village, Irumbai in Tamil Nadu.
(b) Microgrids to power remote islands are becoming a reality, replacing expensive power from diesel generators. One such example is the Sabang power supply system on Palawan Island in the Philippines.

The Sabang Renewable Corporation (SREC) hybrid power plant and microgrid is located in Barangay Cabayugan, Palawan, the Philippines. It supplies electricity to a community of around 700 consumers consisting of 10% small and large commercial consumers with the remaining 90% a mix of low- to high-income residential consumers.

Green power for your home infographic

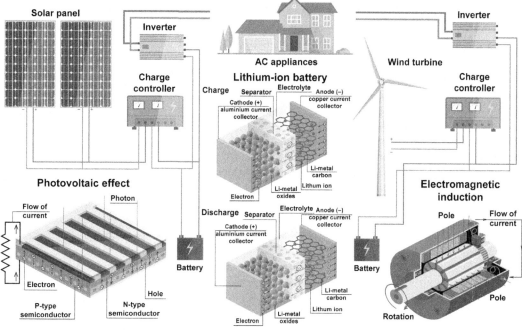

FIG. 6.20

Schematic of a typical solar and wind-based hybrid PV system connected with a battery.

FIG. 6.21

Solar array and wind farm as part of a hybrid PV system.

The renewable power in this hybrid power plant is generated by 5400 polycrystalline solar PV modules with a total capacity of 1.4 MWp. The excess solar power generated during sunny days is stored in Li-ion batteries with a total capacity of 2.4 MWh, and discharged to the load in the evenings. At night and during cloudy days, four diesel generators with a total capacity of 1.3 MW ensure that the power demand of the community can be satisfied on a continuous basis, which is a requirement of the Philippine government for off-grid electrification by qualified third parties.

While the total energy demand in the first year was around 2.5 MWh with a peak load of 550 kW, the plant is designed to meet the demand for the next 20 years, which is estimated to reach more than 5 MWh p.a. with a peak demand of 1 MW.

The electricity is supplied to consumers by means of a 13.8 kV three-phase distribution grid with a total length of 13 km. Each consumer has an EDMI smart meter that transmits data through the cloud (using a 3/4G wireless connection). These data are received in real time at the SREC operations center (and at any other place of administration beyond the site) and used for monthly billing and some analytics to understand consumption patterns.

The total project cost was $8.8 million, of which approximately 30% was equity financed with the remaining 70% paid through a loan facility of the Development Bank of the Philippines. WEnergy Global, the project developer, holds a 40% stake in SREC while its local partners hold the remaining 60% of the shares.

Given the fact that SREC is the first project of its kind in the Philippines and Southeast Asia, it took approximately 7 years to obtain all government approvals and permits because the rules and regulations for proper project evaluation were not clear for a hybrid power system. However, this project clearly demonstrated that hybrid-powered microgrids can significantly reduce government subsidies (i.e., more than $10–15 million in 20 years) by achieving its electrification targets, if a hybrid power system is compared to a conventional fully diesel-powered microgrid. Learning from this experience, the Philippines Senate Commission for Energy is currently working closely with several stakeholders, including WEnergy Global, to draft a new bill that aims to remove regulatory barriers and reduce the number of agencies involved in the approval process in order to shorten the approval time for such projects. Figs. 6.22 and 6.23 show the images of the solar PV plant with battery for energy storage and diesel generator as part of microgrid system installed by SREC in the Philippines.

6.2.16 Solar roads—Highways

Solar cells are used to make roads and produce electricity. The solar road built by China consists of a protective surface layer made of transparent concrete, which can reportedly handle 10 times the pressure of standard asphalt. Beneath that is a middle layer of solar laminates, which generate the road's electricity, above a waterproof insulation layer to prevent any dampness from the ground below. It's estimated that roughly 40,000 cars will be able to drive over the solar highway every day with two lanes to choose from in addition to an emergency lane to help ease congestion. France also constructed a solar road.

6.2.17 Solar-powered battery charging system

A solar energy storage system stores the excess solar energy generated by a solar array during sunshine hours into the battery and provides back-up power whenever required or during a blackout. This is perfect for homeowners, although for the majority of businesses that operate during daylight hours, a common grid-fed solar system is still the most economical choice.

FIG. 6.22

Solar PV-based microgrid system of the Sabang Renewable Corporation, Philippines. (A) Aerial view of the 1.4 MW solar PV plant, and (B) solar array in the PV field.

Courtesy: Sabang Renewable Corporation, Philippines.

FIG. 6.23

Solar PV-based microgrid system of the Sabang Renewable Corporation, Philippines. (A) Li-ion battery, (B) inverter/controller, (C) diesel generator, and (D) transmission line.

Courtesy: Sabang Renewable Corporation, Philippines.

The change in the electricity tariff rate structure depending on the time of day, utility grid outages, maintaining and upgrading the grid, and changes in regulatory policies is creating new applications and opportunities for PV systems with integrated storage. To manage the impact of solar and wind energy sources on regional utility grids, utility-scale storage will likely become a design requirement for renewables integration.

One of the most developed markets for grid-connected energy storage is commercial retail peak-demand reduction. Electricity peak demand charges are increasing at an even steeper rate than energy prices.

There is an increasing interest in home storage systems, and the most common driving factor is the desire for backup power. As storms and other climate-related disasters have increased in intensity in recent years, more people are interested in securing their own energy to avoid hardships.

Solar energy storage systems enable the public to store solar energy and use it when they are at home during the evening when the cost of electricity is typically at the peak rate.

Battery systems can be used as an alternative to diesel generators during grid outages.

This energy storage system would act as an additional power supply if the customer is unable to get the required capacity of contract demand from distributing agencies due to certain limitations.

Battery-based systems can be charged when utility prices are low and discharged during peak demand when energy prices are high, causing reduced demand charges.

Battery-based systems can automatically discharge power every time with load spikes.

A battery-based system provides the functionality of an uninterrupted power supply to operate with a solar PV plant, even during a grid outage.

It also enables advanced energy management such as peak shaving as well as energy independence by increased reliability with the increased consumption of self-generated solar power, reducing the proportion of grid power used each month.

It also reduces power consumption from the grid (reduced demand).

Several applications of energy storage are frequency regulation, renewable integration, peak shaving, microgrids, and black start capability.

Storage can provide backup power to homes and businesses, or it can cut peak-demand charges. Energy storage can be used to provide firm peak capacity to the grid and also provide frequency regulation and other services that offer value to the grid.

Typical customer requirements are ramp control, or reducing the volatility of the intermittent output from PV generation, and various grid-support applications such as frequency response, frequency regulation, voltage support, and responsive reserves. They are also interested in improving grid stability and reducing the wear and tear on conventional generators by using the energy storage systems' ability to deliver power quickly and accurately, to help with frequency or voltage fluctuations that high levels of renewable penetration may be causing.

Solar-plus-storage systems add value to traditional solar arrays by enabling benefits such as emergency power, peak-demand reduction, improved power quality, ramp-rate control, load shifting, and grid services.

Smart energy storage systems (SESS) are a magnificent addition to home energy needs that combines the use of green energy, the convenience of back-up power, and the strength of a high-end storage system supporting the uninterrupted running of home appliances. An SESS is a compact and easy-to-install system with a $LiFePO_4$ battery, a fast charging rate, and a low discharging rate. An SESS is able to run a load of fridge/TV/kitchen appliances and a mixer/air conditioner for more than 2 h in case of a power cut.

Bidirectional inverters with the latest technology of batteries can achieve ancillary services such as:

- Load leveling.
- Peak load shaving.
- Energy shift and time of day charging.
- Frequency regulation.
- Reactive power control and others.
- Load management.
- Uninterrupted supply.

Technical considerations for grid applications of battery energy storage systems

An electric energy time-shift involves using energy that is available for cheaper prices during off-time periods to charge the battery so that the stored energy can be used or sold at a later time when the price or costs are high. The excess energy production from wind and photovoltaic systems, which would otherwise be curtailed for some time, can be stored in the battery and used later. The following are the applications of energy storage.

As the energy storage provides peak generation capacity for electric power systems, it could be used to defer or reduce the need to buy new central station generation capacity.

Load following: Provide fast-responding resource to match the generation to the fluctuation load.

Reserve capacity: Provide reserve capacity should normal supply resources unexpectedly become unavailable.

The energy storage provides voltage support to service works by maintaining voltage levels by injecting or absorbing reactive power.

Congestion relief: Avoid congestion-related costs and charges associated with inadequate transmission facilities—grid system.

Upgrade deferral: Defer or avoid the need for transmission or distribution system upgrades—grid system.

Supplies back-up power to protection, communication, and control equipment of the substation.

Time of use cost management: Store off-peak energy to supply customer loads when peak time of use rates apply.

By demand charge management, the stored energy is used to reduce the instantaneous power draw from the grid and in turn reduces the peak demand charges.

It provides service reliability by riding through the power outage.

The energy storage protects the onsite load from voltage spikes, dips, and snags, thus improving the power quality.

Time shift: Store low-value energy generated by renewables at off-peak times until it has a higher financial value—renewable integration.

Capacity firming: Mitigate variability by discharging stored energy when renewables are not producing full power—renewable integration.

6.2.18 EV charging system

In China, the United States, India, and other countries around the world, hundreds of thousands of electric vehicle chargers are being installed in homes, businesses, parking garages, shopping centers, and other locations. The number of chargers is projected to grow rapidly in the years ahead as the electric vehicle stock grows.

The EV charging industry is currently a highly dynamic sector with a wide range of approaches. The industry is emerging from infancy as electrification, mobility as a service, and vehicle autonomy interact to produce some of the most far-reaching changes in the transportation sector in a century.

Electric mobility is expanding at a rapid pace. In 2018, the global electric car fleet exceeded 5.1 million, up 2 million from the previous year and almost doubling the number of new electric car sales. China remains the world's largest electric car market, followed by Europe and the United States. Norway is the global leader in terms of electric car market share.

Policies play a critical role. Leading countries in electric mobility use a variety of measures such as fuel economy standards coupled with incentives for zero- and low-emission vehicles, economic instruments that help bridge the cost gap between electric and conventional vehicles, and support for the deployment of charging infrastructure. Increasingly, policy support is being extended to address the strategic importance of the battery technology value chain.

Technology advances are delivering substantial cost cuts. Key enablers are the developments in battery chemistry and the expansion of production capacity in manufacturing plants. Other solutions include the redesign of vehicle manufacturing platforms using simpler and innovative design architecture, and the application of big data to right-size batteries.

The private sector response to public policy signals confirms the escalating momentum for the electrification of transport. In particular, recent announcements by vehicle manufacturers are ambitious regarding intentions to electrify the car and bus markets. Battery manufacturing is also undergoing important transitions, including major investments to expand production. Utilities, charging point operators, charging hardware manufacturers, and other power sector stakeholders are also boosting investment in the charging infrastructure.

Off-grid solar photovoltaic systems

7.1 Introduction

Off-grid solar photovoltaic (PV) systems are also known as standalone solar PV power systems. These store electrical energy generated by solar PV modules in storage devices such as batteries. Energy stored in batteries can be used when there is a demand for power supply or at night when there is no sunshine. More precisely, these systems are used in areas with no electricity supply, a shortage of energy, or remote locations/islands where there is no accessibility to a grid.

The following are some of the examples of various types of standalone or off-grid solar PV power systems:

(a) Solar PV-powered water pumping system.
(b) Solar PV-powered home power system.
(c) Solar PV-powered cathodic protection system.
(d) Solar PV-powered telecom systems.

7.2 System components of off-grid solar PV systems

An off-grid solar PV system provides electricity in situations where utility power is not available. Solar PV modules are connected in series to form strings and these strings are combined together in a string combiner box (array junction box), which is located near the array. The generated DC electricity from the PV panels flows through the string combiner box and from there to a controller in the inverter system. The controller regulates the DC power to the batteries. The inverter converts the DC power from battery to AC electricity and the AC electricity goes to the main electrical panel. From the main electrical panel, the electricity is used by the loads connected to the main panel. These systems allow the solar power to be stored in batteries, then power can be supplied to the loads during the grid outage or whenever required.

Fig. 7.1 gives the schematic of an off-grid solar PV system. The broad solar system components of a typical solar PV standalone or an off-grid solar PV system are as follows:

Solar PV module.
Solar charge controller.
Battery.

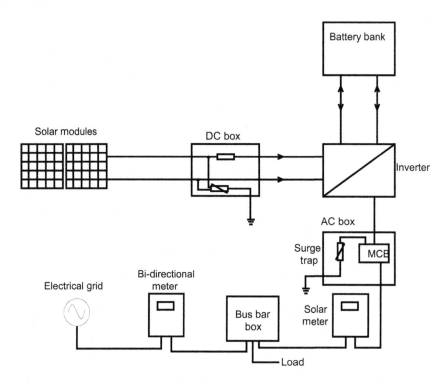

FIG. 7.1

Schematic of an off-grid solar PV system.

Off-grid/hybrid inverter.
Module mounting structures.
Array junction box.
AC distribution board.
DC and AC cables.

Description of solar PV system components:

7.2.1 Solar PV module

Solar PV modules are available in various sizes with different technologies, different wattages, and different voltages of operation. The solar modules are available with a nominal voltage of 12, 24, and 30 V. Solar PV modules with high conversion efficiency and voltage from different suppliers with advanced technology are available.

The current and voltage values of the module vary depending on the solar irradiance incident on the module and the operating temperature of the module. The current-voltage and power-voltage curves for different irradiance conditions for a 405 W module maintained at 25°C are shown in Fig. 7.2. A module of 405 W gives power of 324 W at 800 W/cm^2 intensity at 25°C. This is because the I_{sc} of the solar module varies proportionally according to the intensity of the radiation falling on the module. The V_{oc} logarithmically depends on the incident radiation. So, in Fig. 7.2, the V_{oc} of

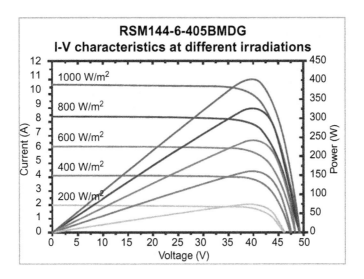

FIG. 7.2

Current-voltage and power-voltage curves for different irradiance conditions of a 405 W module.

Courtesy: Mr. Chaudary of Risen Energy.

the module is slightly reducing with the decrease of the intensity of the radiation. The current-voltage curves for different module temperatures at $1000\,W/cm^2$ irradiance condition for a 405 W module are shown in Fig. 7.3. The voltage and the power reduces with an increase of temperature due to a negative temperature coefficient of voltage.

The V_{oc} of an n/p monocrystalline PERC solar cell is about 675 mV, whereas 640 mV is the V_{oc} for a polycrystalline n/p BSF technology-based cell. The V_{oc} doesn't change with the size of the cell. The V_{oc} of the cell reduces in the range 1.8–2.2 mV per 1°C rise of temperature depending on the cell technology. The V_{mp} of the cell is around 80% to 90% of the V_{oc} of the solar cell.

The I_{sc} of an n/p monocrystalline PERC solar cell is around $40\,mA/cm^2$, whereas $38\,mA/cm^2$ is the I_{sc} for polycrystalline an n/p BSF technology-based cell. The I_{mp} of the cell is around 92% of the I_{sc} of the solar cell.

The V_{oc} of a 72 solar cell module will be $72 \times V_{oc}$ of a single solar cell.

The V_{mp} of a 72 solar cell module will be about 80% of the module V_{oc}. The resistive losses due to interconnection of the cells will cause the V_{mp} of the module to be less than $72 \times V_{mp}$ of one solar cell.

The I_{sc} of the solar module will be 1%–2% lower than the I_{sc} of the solar cell due to cell to module conversion losses. The I_{mp} of the module will be around 95% of the I_{sc} of the module. The values of I_{mp} and V_{mp} depends on the series and shunt resistance values of the solar cell/module.

The lower wattage modules are made by cutting the cells to the appropriate size based on the requirement of current. The number of cells in series is selected based on the requirement of the V_{oc}, V_{mp}, and P_{max} of the module. The bypass diode is connected across the group of $\leq$24 solar cells. Table 7.1 gives the electrical parameters of typical mono-PERC and polycrystalline Si solar modules. 36 cell solar modules with nominal voltage of 12 V were used earlier for lead-acid battery charging. For a 72 cell module, 24 V is considered as the nominal voltage.

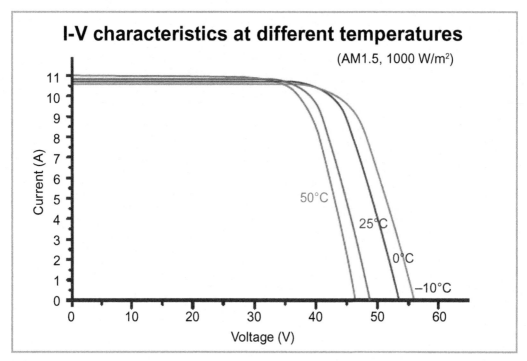

FIG. 7.3

Current-voltage curves for a 405 W module at different temperatures.

Courtesy: Mr. Chaudary of Risen Energy.

7.2.2 Solar charge controller

A solar charge controller is used to charge the battery by regulating and controlling the output from the solar PV array; it also protects the battery from being overcharged or overdischarged. Overcharging of the battery creates release of hydrogen and oxygen gases from the electrolyte, which could cause explosion and failure. If the battery is allowed an excessive discharge of current, the battery charge will be drained. So, the life of the battery will be reduced and cause premature failure of the battery.

Solar charge controllers redirect or switch off all or part of the array to reduce the current flow to the battery when it is becoming full. If the battery is discharged below a specified voltage, which is a low voltage preset point, the disconnection of some or all the loads takes place. To protect the battery from overcharging, the charge controller will have a high voltage disconnect (HVD) point. So, the controller will have set points such as low voltage disconnect (LVD) and HVD. The controller voltage must be compatible with the nominal system voltage and it must be capable of handling the maximum current produced by the PV array.

Table 7.1 Parameters of solar PV modules [1,2].

Parameter	72 cell mono PERC	144 half cut cell mono PERC [2]	36 cell multi	36 multicut cells	76 multicut cells
Peak power watts, P_{max} (Wp)	380	405	150	100	100
Maximum power voltage, V_{mp} (V)	39.89	40.55	17.72	17.46	35.6
Maximum power current, I_{mp} (A)	9.54	10.0	8.47	5.73	2.81
Open circuit voltage, V_{oc} (V)	48.62	48.75	22.45	22.2	44
Short circuit current, I_{sc} (A)	9.99	10.6	8.9	6.07	3.03
Module efficiency (%)	19.54%	19.7%	14.91%	12.93%	13.42%
Size (mm × mm)	1960 × 992	2034 × 1000	1490 × 675	1035 × 675	1035 × 675
Height (mm)	43	30	35	35	35
Weight (kg)	22	27	13	10.15	10.15
NOCT/NMOT (°C)	46±2	44±2	46±2	46±2	46±2
Temperature coefficient of P_{max} (%/°C)	−0.375	−0.36	−0.38	−0.38	−0.38
Temperature coefficient of V_{oc} (%/°C)	−0.282	−0.28	−0.28	−0.28	−0.28
Temperature coefficient of I_{sc} (%/°C)	+0.043	+0.05	+0.041	+0.041	+0.041

The solar charge controller protects the batteries from being overcharged by solar PV array/modules during the day. During night, when solar panels don't generate and have zero voltage, there will be a current flow from the battery to the solar panels. The charge controller provides controls by incorporating a blocking diode or relay to prevent the reverse flow of the current from the battery to the array to avoid draining the batteries during low solar irradiance or night time. This means that the batteries are not drained during the day and that the electricity does not run overnight back to the solar panels and drain the batteries. Some solar charge controllers, such as the one used for streetlight applications, take care of lighting control as an additional function. Additionally, a load control feature is also available with some solar charge controllers.

Types of charge controllers: There are two basic types of charge controllers used for small PV systems: the series and the shunt type. These are single-stage controllers that disconnect the array when the battery voltage reaches a high voltage level during charging.

Shunt controller—shunts part of the solar array and redirects the charging current away from the battery. During shunting of the solar array, heat dissipation takes place and a large heat sink is required to dissipate the excess current.

Series controller—series switch is incorporated between the array and the battery. The series switch in the controller will be opened based on the state of charge of the battery to interrupt the charging current from the PV array. This series controller has a limitation regarding the capability of the components to handle the current during switching operations.

At different stages of battery charging, there is a requirement of different levels of charging current. So, multistage controllers are used to provide a more efficient method of charging the battery. As the battery nears full SOC, its internal resistance increases, and using a lower charging current wastes less energy.

Broadly, there are two different topologies of solar charge controllers available: the maximum power point tracker (MPPT) and pulse width modulation (PWM). The performance of each of these controllers is not the same and MPPT has better performance as compared to the PWM controller.

7.2.2.1 Solar charge controller—PWM type

PWM has less-expensive solar charge controller topology. Solar PV modules/arrays as well as batteries are connected directly to the solar charge controller for operation; it is mostly used for solar home light or home power systems.

The solar PV module/array is connected continuously to the battery and the higher solar PV array voltage is brought down to the level of battery terminal voltage. While charging the battery with the solar PV array, the battery voltage increases and the charge controller ensures that the solar array output voltage is higher than the battery voltage.

A solar PV module known as a 12 V nominal voltage is suitably designed to give an output of around 18 V (considering the solar PV module output voltage variation due to temperature rise, cable voltage drop), which can charge the battery to 14.4 V (max charge for a flooded lead-acid battery). As may be understood, if the solar PV module and battery are of the same voltage, then charging the battery by a solar PV module would not be possible. Hence, it is required to design the solar PV module at a higher voltage than the battery.

For charging a 24 V battery bank, we need to connect two solar PV modules (with V_{mp} of 18 V each) and similarly for charging of a 48 V battery bank, it would need four solar PV modules ($V_{mp} = 18$ V).

If one uses two solar PV modules ($V_{mp} = 18$ V) in series to charge a 12 V battery, one would be wasting half of the solar PV module capacity, which is not an optimal use of solar PV module capacity. In a similar manner, if it is intended to charge a 24 V battery bank by a single solar PV module ($V_{mp} = 18$ V), one would not be able to charge the battery and will end up in discharging the battery. Figs. 7.4 and 7.5 explain the loss of energy with the connection of a 12 V module with a 12 V battery and a 24 V module connected with a 12 V battery.

PWM solar charge controllers are constant voltage controllers with two-stage regulation. In the first stage, the controller will charge the battery with a higher voltage so that the battery can be 100% charged. In the second stage, after the battery is fully charged, it will lower the voltage from the solar array to trickle charge the battery, so that the battery is kept charged at a 100% charge level. This type of charging maintains a 100% charge level of the battery while minimizing water loss and overcharging of the battery.

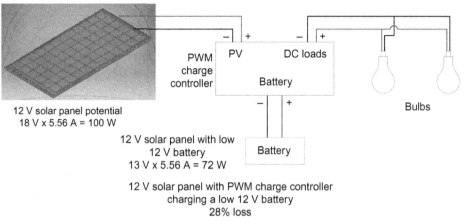

12 V solar panel potential
18 V x 5.56 A = 100 W

12 V solar panel with low
12 V battery
13 V x 5.56 A = 72 W

12 V solar panel with PWM charge controller
charging a low 12 V battery
28% loss

FIG. 7.4

Loss of energy with connection of a 12V module with a 12V battery with PWM charge controller.

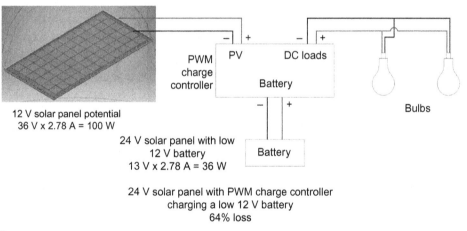

12 V solar panel potential
36 V x 2.78 A = 100 W

24 V solar panel with low
12 V battery
13 V x 2.78 A = 36 W

24 V solar panel with PWM charge controller
charging a low 12 V battery
64% loss

FIG. 7.5

Loss of energy with connection of a 24V module with 12V battery with PWM charge controller.

7.2.2.2 Solar charge controller—MPPT type

The MPPT controller is considered a better solar charge controller as compared to the PWM controller.

The maximum power point of the $I–V$ curve depends on the operating temperature of the module and the radiance falling on the module. So, the V_{mp} point of the solar array continuously changes due to changes in weather conditions. The controller tracks the V_{mp} (voltage at maximum power point) and steps down the solar array voltage to that of the battery voltage. In view of lowering the V_{mp}, as the power is the product of voltage and current, there is eventually a rise in current to keep the solar array power at the same level and the array power generated at that instant is available for use.

Ideally, a solar PV module with 36 solar cells with a V_{mp} of 18 V can be used to charge a 12 V battery bank. Also, one can make use of a typical solar PV module with more than 36 cells (used for a grid-tied system: say, a 60 cell solar PV module) to charge a 12 V battery. Similarly, two solar PV modules can charge a 24 V battery bank and three solar PV modules can charge a 48 V battery bank. Using solar PV modules with more than 36 cells with MPPT controller paves the way for using other solar PV modules for off-grid systems with flexibility.

7.2.2.3 Functions of charge controllers
Overcharge protection

A battery that reaches a 100% state of charge (SOC) cannot accept energy coming from the solar array. When the solar array supplies current at a full rate, it is likely that the battery voltage increases to a high level. This extra energy is not absorbed and splits the water in the electrolyte to hydrogen and oxygen gases. This splitting of water causes water loss in the battery and could cause a small explosion. Also, this overcharging will degrade the battery and is likely to overheat the battery. Also, the higher voltage can lead to shutting down the inverter or may cause problems to loads.

Hence, overcharge protection is required, which will help in reducing the amount of energy into the battery reaching the specified overvoltage limit. And when the voltage is reduced due to low solar irradiance or an increased load usage, the controller shall allow a maximum charge. The above phenomena is called voltage regulation or overcharge protection, which is the most important function of a solar charge controller.

Low voltage cut-off and overdischarge protection

The life of the battery depends on the depth of discharge. When loads are connected continuously, the battery can be deep discharged and drained out. By this deep discharge, it reduces the capacity and the life of the battery by a specified amount. If the battery sits in this overdischarged state for days or weeks at a time, it can be damaged quickly. Overdischarge protection in charge controllers is usually accomplished by open circuiting the connection between the battery and the electrical load when the battery reaches a preset or adjustable low voltage load disconnect (LVD) set point. Once the battery is recharged to a certain level, the loads are again reconnected to the battery.

Over load protection

When a higher current is maintained than the specified current limit, it leads to an overload situation, which is not safe for the circuitry. This overloaded or overcurrent situation will lead to overheating and create a fire hazard. Overloading can be caused by a short circuit in the wire or faulty loads. The solar charge controller has built-in overload protection with a reset push button. Built-in overload protection is very useful and is done by either a circuit breaker or a fuse.

Charging of lead-acid batteries

The deep cycle lead-acid battery has to be charged in a controlled manner with solar array-generated power. There are four stages of charging: constant-current charge (bulk stage), topping charge (absorption stage), and float charge, as shown in Fig 7.6. Equalization is the fourth stage followed to reverse stratification.

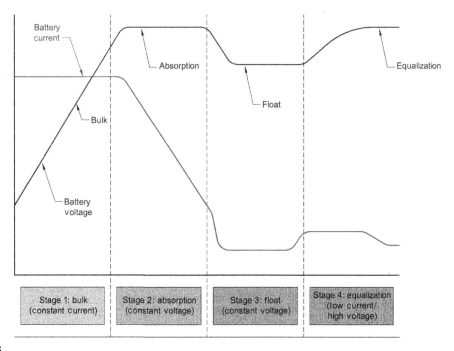

FIG. 7.6

Different stages of lead-acid battery charging.

Stage 1—Bulk charging: If the battery is at a low state of charge such as 20%, the charge controller charges the battery in constant current mode at high voltage. This process fills up the battery bank to the program-specified capacity, which is anywhere around 80%–90% of its full capacity. This is called constant current charging or bulk charging.

Stage 2—Absorption charging: Once the bulk charging stage is passed, the battery will be at around 80%–90% of its full capacity. The voltage is at its peak and the amperage is then tapered down as the charge controller's internal resistance is increased.

Stage 3—Float charging: When this stage is reached, the battery bank is just about full. However, if the charge trickles out of the battery bank by various means outside of normal battery usage, the charge controllers send in small tickles of charge into the battery as maintenance charging. The feeding voltage is usually just above the battery bank's voltage to prevent overcharging.

Stage 4—Equalization and desulfation: Equalization is the process of deliberately overcharging the battery bank in order to reverse negative chemical effects such as stratification and sulfation. This maintenance feature is done after the float stage when the battery capacity is full, so sometimes it is called the fourth stage of charging. Equalization is not always done after the completion of three charging stages. The charging for equalization is usually done once a month to a few times per year.

7.2.3 **Battery bank**

A battery cell is an electrochemical energy conversion device that converts chemical energy contained in the constituent components directly into electrical energy. A battery is a storage device that stores electrical energy. The basic building block of a battery is a cell. A battery has one or more of these cells connected in series and parallel depending on the required output voltage and capacity. The primary functional components of a cell are the anode, cathode, electrolyte, seperator, terminals, and casing. In a rechargeable battery, the composition of the anode and cathode materials can be restored by reversing the direction of current by charging using an external source.

Oxidation and reduction reactions are responsible for the generation of current, which is the transfer of electrons from one electrode to the other. At the site of oxidation, electrons are released and received at the site of reduction. The electrolyte facilitates the exchange of anions and cations between the positive and negative electrodes. The voltage of the battery cell is the difference between the electrochemical potentials of the two electrodes. The chemical reactions taking place at each electrode-electrolyte interface creates electrochemical potentials. During discharge, oxidation takes place at anode which has +ve voltage and there is a −ve voltage at cathode, where reduction takes place. The difference of the voltage between the electrodes gives the battery cell voltage and the electron flow takes place from anode to cathode.

Each electrochemical cell consists of two half-cells connected in series by a conductive electrolyte.

One half-cell includes the electrolyte and the electrode to which ions move, that is, the anode or negative electrode. The other half-cell includes the electrolyte and the electrode to which ions move, that is, the cathode or positive electrode. The electrochemical potential difference between the two half-cells corresponds to the voltage of the battery that drives the load, and the exchange of electrons between the two reactions corresponds to the current that passes through the load.

In a battery, two plates of similar or dissimilar materials are immersed in an electrolyte. Both plates, which are called positive and negative electrodes, along with the electrolyte are kept in a container. The electrodes are connected externally, as shown in Fig. 7.7.

Lead-acid and Li-ion batteries are used for electrical energy storage for off-grid applications in the PV industry.

7.2.3.1 *Characteristics of a battery*

Specific energy (Wh/kg)—The nominal battery energy per unit mass.

Specific power (W/kg)—It is the maximum available power per unit mass. It determines the battery weight.

Energy density (Wh/L)—It is the nominal battery energy per unit volume and depends on the battery size.

Power density (W/L)—Specific power is the maximum available power per unit volume and it determines the battery size.

All the above are the key performance parameters of a battery and depend on the chemistry of the battery and its packaging.

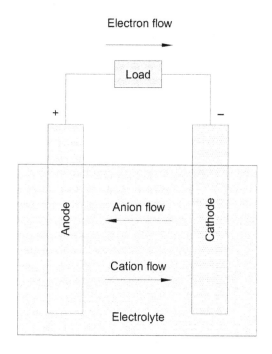

FIG 7.7

Operation principle of a battery.

Battery capacity

It is a measure of a battery's ability to store or deliver electrical energy and it is expressed in units of ampere hours (Ah). An ampere hour is equal to a discharge of 1 A over 1 h. For example, a battery that discharges 15 A to a load in 10 h is described as having delivered 150 Ah. The capacity of a battery depends on several factors such as the quantity of active material, the number and physical dimensions of plates, and the specific gravity of the electrolyte. The battery capacity also depends on the operational conditions such as the load, discharge rate, depth of discharge, cut-off voltage, temperature, and cycle history of the battery. Usually, the battery capacity will be specified for a given discharge/charge rating or C rating. The storage capacity of the battery is also expressed in watt hours or Wh. If V is the battery voltage, then the energy storage capacity of the battery can be $Ah \times V = $ watt hour. For example, a nominal 12 V, 150 Ah battery has an energy storage capacity of $(12*150)/1000 = 1.8\,kWh$.

Cold temperature reduces the battery capacity, because the chemical reactions go slower and fewer active materials can be accessed and converted.

Fast discharging also reduces capacity. During fast discharging, the reaction products such as water get in the way of the electrolyte, so the capacity is limited. In contrast, if the battery is discharged slowly, the fresh electrolyte can be penetrated more efficiently into the plates and more capacity is made available for discharging.

Battery voltage

The voltage between the battery +ve and −ve terminals during operating conditions is known as the nominal voltage. It varies with the discharge/charge current and the state of charge of the battery. It may be 2/6/12/24 V, etc.

Battery lifecycle

A discharge followed by a recharge is considered one cycle. It is the number of complete charge-discharge cycles a battery can experience before the nominal capacity decreases less than 80% of its initial rated capacity. The discharge can be very small or shallow, or it can be very severe or deep. The lifecycle, that is, the operating life of the battery, is controlled by the rate of charge/discharge, the depth of discharge, the number of charge/discharge cycles, and other ambient conditions such as temperature and humidity. The lifecycle is dependent upon the daily depth of discharge. The batteries used in solar photovoltaic applications will definitely be subjected to cycling on a daily basis, and are also cycled.

Discharge/charge rate or C—Rate

Different batteries will have different capacities and can be charged/discharged with different currents. The rate of charge or discharge of a battery is expressed as a ratio of the nominal battery capacity to the number of hours it takes for a full charge or discharge in order to normalize against battery capacity. A C1 rate means that the entire battery capacity shall be discharged in 1 h with a specified end voltage. For example, a 75 Ah battery is discharged at the rate of 3 A; the time to completely discharge a fully charged battery at this rate would be 75 Ah/3 A = 25 h. So, it can be said that the battery is discharging at 25 h rate or at C/25. Batteries used in typical solar PV systems experience very low rates (based on number of days of storage 3/5/7 days) of charge and discharge compared to industrial applications.

Self-discharge

It is the electrical capacity lost when a battery is not being used due to the internal electrochemical process within the battery. The self-discharge increases with an increase of temperature. The batteries can be stored at lower temperatures to reduce self-discharge.

Cut-off voltage

The cut-off voltage is the lowest voltage at which a battery system is allowed to reach in operation. The battery manufacturers often rate capacity to a specific cut-off or end of discharge voltage at a defined discharge rate.

Round trip efficiency

During the conversion and storage of energy in a battery, there is a loss of efficiency. The round-trip efficiency can be expressed in the % of energy that the battery discharges as compared to the energy it received for charging.

7.2.3.2 Types of batteries
A. Lead-acid batteries

Lead-acid batteries are commonly used as energy storage devices in solar PV system applications. The cathode is PbO_2 and the anode is Pb and both the positive and negative electrodes are immersed in dilute sulfuric acid, which serves as an electrolyte. These plates are housed inside the hard containers. Lead-acid batteries can be either 2/6/12 V type in a hard plastic container. There are different types of lead-acid batteries, as shown in Fig. 7.8. The flooded cell type or sealed/gel and AGM type batteries are popularly used for off-grid solar PV systems.

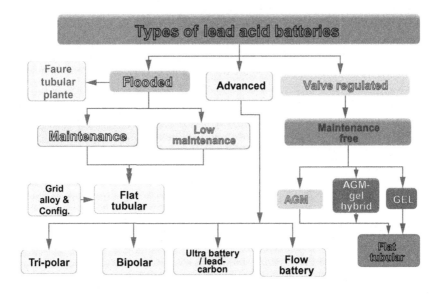

FIG. 7.8

Different types of lead-acid batteries.

Courtesy: Prof. S. Venugopalan.

A1. Flooded lead-acid battery. This is the most commonly used battery for solar PV systems today due to its reliable performance and cheaper price. There are two versions of flooded lead-acid batteries: flat and tubular plate.

The positive and negative electrodes are immersed in dilute sulfuric acid, which serves as an electrolyte. These plates are housed inside a hard container that has removable caps to allow filling of lost water from the electrolyte.

While charging the flooded batteries to a full state of charge, hydrogen and oxygen gases are generated from the water by the chemical reaction at the negative and positive plates. There is a need for proper ventilation for the escape of gases so as to avoid fire hazards. The gases are allowed to go out through the vents of the battery. This requirement demands the periodic addition of water to the battery.

These lead-acid batteries in view of long operation are considered to be rugged and low-cost. With proper maintenance, they provide reliable performance. These batteries are open vented, which calls for proper positioning so as to avoid spilling of the electrolyte. With regular maintenance by means of adding distilled water and charging, they provide long years of reliable service in field conditions for a variety of solar PV power system applications.

A2. Sealed/gel battery. These batteries contain an immobilized form of the electrolyte. The sealed maintenance-free lead-acid batteries are also called valve-regulated lead-acid (VRLA) batteries. These are of two types: gelled electrolyte and absorbed glass mat.

VRLA batteries will have less electrolyte-freeing problems compared to FLA batteries. During the charging process, hydrogen and oxygen gases are produced from water due to the chemical reactions at the negative and positive plates. These gases recombine to form water, thus there is no need for the addition of water. These are easily transportable, have no need for the addition of water, and are suitable for remote applications because of fewer maintenance requirements.

The internal design of a gel battery is of similar design to that of the flooded lead-acid battery. The silicon dioxide is added to the electrolyte and it becomes a warm liquid. When the electrolyte with silicon dioxide is added to the battery, it become a gel after cooling. So, the electrolyte has been immobilized by the addition of a thickening agent, which looks like petroleum jelly. The hydrogen and oxygen produced during the charging process are transported between the positive and negative plates through the cracks and voids in the gelled electrolyte. A gel battery can be mounted horizontally as well as vertically and doesn't release hydrogen gas during normal operation. Gel batteries are costlier and have better deep-cycling capability as compared to flooded lead-acid batteries. Gel batteries can be shipped by air, as they are sealed and are leak- and spillproof.

A3. Absorbed glass mat (AGM) batteries. In AGM batteries, glass mats are used. The glass mats are placed in between the battery plates, and these glass mats hold the electrolyte. The oxygen molecules from the positive plate move through the electrolyte in the glass mats and recombine with hydrogen at the negative plate to form water. Hence, oxygen and hydrogen are combined within the battery while not being vented out.

These batteries are sealed and are designated as valve-regulated lead-acid (VRLA) batteries. As compared to open vented plugs in flooded batteries, these batteries have built-in valves, which helps in the release of excess gases in the event of the overcharge of batteries.

Glass fiber mats are compressed between each plate and hold the electrolyte like an absorptant while supporting the plates. The AGM/gel batteries are sealed and hence allow mounting either horizontally or vertically. Because of sealing, AGM batteries need fewer ventilation requirements as compared to flooded lead-acid batteries.

These batteries can be shipped by air due to sealing. They can accept higher charge and discharge rates and provide higher performance at low temperatures. Overcharging of AGM sealed batteries will lead to a severe loss of water and can lead to drying out.

The self-discharge for gel and AGM batteries is 1%–3% per month, which is lower than FLA batteries. The round-trip efficiency for gel and AGM batteries is 80%–90%. Both gel and AGM batteries require controlled charging. In these batteries, generally lead calcium electrodes are used to minimize gassing and water loss. Voltage and current must be controlled below C/20 rate.

Lead-acid batteries will have a high specific energy of 30–50 Wh/kg with a 250–750 lifecycle. These batteries are environmentally friendly and able to go for deep discharging. They have high efficiency, simple operation, and low cost.

A4.　*Advanced lead-acid (lead-acid carbon) battery.* The battery with lead-acid has low cost and high reliability and performance (70%–90%). It is a popular choice for power quality storage, UPS, and some reserve spinning applications. Nevertheless, its energy management use was very limited due to its short lifecycle (500–1000 cycles) and low energy density (30–50 Wh/kg), due to the inherent high lead content. Lead-acid batteries also have poor performance at low temperatures, and thus require a thermal management system. Lead-acid batteries have always been used in applications for industrial and large-scale energy management. The depth of discharge also effects the life of a battery; discharges beyond about 50% will shorten the battery life. Colder operating temperatures will yield a little extra life, but they will also lower the capacity of lead-acid cells. High temperatures yield higher capacity, but they have a detrimental effect on life.

Advanced lead-carbon batteries use carbon combined with lead as the cathode. This helps them to show a high-rate characteristic in both charge and discharge characteristics. It doesn't show detrimental effects, as is usually observed in valve-regulated lead-acid (VRLA) lead acid batteries. This feature allows the carbon batteries with lead-acid to produce and accept high current levels that are only possible with higher-cost Li-ion batteries.

By adding carbon to the negative plate, the advanced lead-carbon (ALC) battery avoids the accumulation of sulfate experienced in a standard lead-acid battery.

The ultrabattery incorporates lead and carbon in a twin negative electrode. Though the ALC is low-cost, larger, and heavier than Li-ion, it works at subfreezing temperatures and does not require active cooling.

Lead carbon can operate at a charge state of between 30% and 70% without sulfating. In ALC, there is a sudden drop in the voltage during discharge that resembles a supercapacitor.

Lead carbon battery benefits
- Very high lifecycle—2200–4500 cycles as against 200–1800 cycles for lead-acid.
- High charging efficiency (95%).
- Excellent charge acceptance.
- Reduced sulfation.
- Hugely improved partial state of charge (PSoC) performance.
- Low maintenance and no watering.
- Sealed VRLA construction—almost zero gassing.
- Lead-acid batteries are >96% recyclable.
- High reliability and predictable performance.
- Wide temperature tolerance (−30 to 60°C).

B. Nickel-cadmium (NiCd) batteries

In an NiCd battery, the positive electrode is made up of cadmium and the negative electrode is nickel hydroxide separated by nylon separators immersed in potassium hydroxide electrolyte placed in a stainless steel casing. It has a longer deep lifecycle and is temperature tolerant compared to the lead-acid battery. Metal hydrides replace cadmium due to regulations. The memory effect degrades the capacity of the battery when kept idle for a long time.

The memory effect is the process of remembering the depth of discharge in the past. If the battery is discharged to 25% repeatedly, it will remember it, and if the discharge is greater than 25%, the cell voltage will drop. To recover the full capacity of the battery, it should be reconditioned by fully discharging and then fully charging once in few months.

C. Nickel-metal hydride (NiMH) batteries

It is an extension of NiCd batteries with high energy density. The anode is made up of metal hydride instead of NiCd. It has less memory effect and delivers high peak power. It is more expensive than NiCd batteries and overcharging damages the battery easily. These batteries will have a high specific energy of 65–75 Wh/kg with a 700 lifecycle. These batteries are environmentally friendly and able to go for deep discharging. But these have a high self-discharge, low efficiency, and are costlier.

The following are performance metrics/figures of merit used to compare various battery technologies:

- Energy density and specific energy.
- Power density and specific power.
- Cycle and calendar life.
- Safety, toxicity (environmentally benign), power rating, form factor.
- Cost.
- Ease of operation (orientation free)—plug and play, no start up or shut down.
- Wide operating temperature range.
- Efficiency η_{Wh}—no waste of heat generation.
- Maintenance free.
- Self discharge.
- Resistance to vibration and shock.
- Reliability.
- Commercial availability—technology readiness.
- Hermeticity.

Table 7.2 gives a comparison of different performance parameters with different battery technologies.

D. Lithium-ion batteries

The energy density of Li-ion batteries is three times that of Pb-acid batteries. The cell voltage will be 3.5 V, and a few cells in series will give the required battery voltage. The lithium electrode reacts with the electrolyte and creates a passivation film during every discharge and charge operation. This is compensated by the use of thick electrodes. Because of this fact, the cost of a Li-ion battery is higher than NiCd batteries. Fig. 7.9 shows the operation principle of a Li-ion battery. Generally, two electrodes and an organic electrolyte compose a lithium battery. The cathode, the positive electrode is made up of lithium metal oxide, with layered or tunneled structures with aluminum collector.

Table 7.2 Performance parameters of different battery technologies [3].

Parameter	Lead acid	NiCd	NiMH	Li FeO$_4$
Specific energy (Wh/kg)	30–50	45–80	60–120	90–120
Internal resistance (mΩ)	<100 for 12 V pack	100–200, 6 V pack	200–300, 6 V pack	25–50 per cell
Lifecycle (80% discharge)	200–300	1000	300–500	1000–2000
Fast charge time	8–16 h	1 h typical	2–4 h	1 h or less
Overcharge tolerance	High	Moderate	Low	Low
Self discharge/ month (room temp.)	5%	20%	30%	<10%
Cell voltage	2 V	1.2 V	1.2 V	3.3 V
Charge cutoff voltage (V/cell)	2.4, float 2.25	Full charge detection by voltage	Full charge detection by voltage	3.6
Discharge cutoff voltage (V/cell, 1C)	1.75	1	1	2.8
Peak load current	5C	20C	5C	>30C
Peak load current (best result)	0.2C	1C	0.5C	<10C
Charge temperature (°C)	−20 to 50	0 to 45	0 to 45	−0 to 45
Discharge temperature (°C)	−20 to 50	−20 to 65	−20 to 65	−20 to 60
Maintenance	3–6 months (topping charge)	30–60 days (discharge)	60–90 days (discharge)	Not required
Safety	Thermally stable	Thermally stable, fuse protection common	Thermally stable, fuse protection common	Protection circuit is required

The anode, the negative electrode is made of graphite carbon cell on copper collector. The electrolyte can be nonaqueous solution made of an organic solvent and a dissolved lithium salt or a solid polymer or a solid-state conductor. Lithium ions move back and forth between two electrodes during charging and discharging processes. While charging, lithium ions flow from the positive metal oxide electrode to negative graphite electrode and during discharge, it flows in reverse. The ions are inserted or extracted from the interspatial space between the layers of electrode materials.

The Li-ion technology is considered the best frontier of energy storage due to high energy density (160–200 Wh/kg), fast response time (milliseconds), low self-discharge rate (2%–8% per month) and high efficiency (up to 97%) of batteries. The batteries can operate in a wide temperature range of 0–45°C for charging and 40–65°C during the discharge process. They have a higher round trip efficiency of

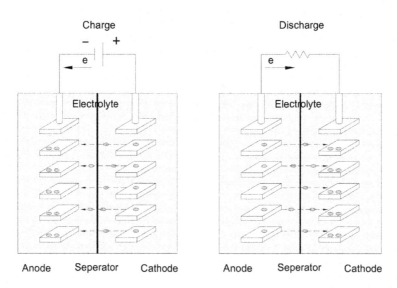

FIG. 7.9

Principle of operation of a Li-ion battery.

78% with 3500 cycle life. These batteries can be fabricated in different sizes with different capacities from 0.1 to 160 Ah.

Li-ion batteries are classified based on different types of chemistries. They have the highest energy density and are considered safe. To prolong the battery life, there is no requirement of scheduled cycling and there is no memory effect. Li-ion batteries are used in electronic devices such as cameras, calculators, laptop computers, and mobile phones, and are increasingly being used for electric vehicles.

Advantages: High charge and discharge capability, higher specific energy, long lifecycle, and extended shelf-life; maintenance-free; high capacity, low internal resistance, good coulombic efficiency; simple charge algorithm, and reasonably short charge times.

Disadvantages: Thermal runaway, degradation at high temperature and high voltage. Impossibility of rapid charge at freezing temperatures ($<0°C$, $<32°F$). They exhibit potential degradation when discharged below 2 V at 65°C. There will be a risk of danger, if batteries are overcharged. To take care of all the issues for the reliable operation of the battery, a battery management system (BMS) is required. Transportation regulations are required to be followed when shipping in larger quantities.

D1. Types of lithium-ion batteries. *Lithium cobalt oxide (LiCoO$_2$)*: In this battery, the cathode is a lithium compound of cobalt oxide and the anode is a graphite/carbon material. During charge, ions move from the cathode to the anode and vice versa on charge. These batteries exhibit relatively low thermal stability. They also have a short lifespan and limited load capabilities.

Lithium manganese oxide (LiMn$_2$O$_4$): Lithium manganese oxide construction forms a three-dimensional spinel structure. This spinel structure improves the ion flow on the electrode, which results in lower internal resistance and improved current handling capability. An additional advantage of the spinel structure is that it provides high thermal stability and enhanced safety, but the lifecycles are limited.

Lithium nickel manganese cobalt oxide (LiNiMnCoO$_2$, or NMC): It is a cathode combination of nickel-manganese-cobalt. While the exact material ratios differ by manufacturer, typically 60% nickel, 20% manganese, and 20% cobalt are the typical combinations. On the other hand, manganese offers low internal resistance but has the drawback of low specific strength. Furthermore, the combination of the two increases the strengths of each other, making them ideal for EV powertrains and cordless power tools.

Lithium nickel cobalt aluminum oxide (LiNiCoAlO$_2$) (NCA): NCA battery has come into existence since 1999 for various applications. It has long service life and offers high specific energy around good specific power along the lines of NMC. Safety and costs are less flattering. NCA gives greater stability to the chemistry by adding aluminum as it is further developed version of lithium nickel oxide. NCA batteries are not conventional in the consumer industry but have promise for EV manufacturers. They require special safety monitoring measures to be employed for use in EVs. They are also costlier to manufacture, limiting their viability for use in other applications.

Lithium iron phosphate (LiFePO$_4$): Li-phosphate as the cathode material offers good electrochemical performance with low resistance. This is made with nanoscale phosphate cathode material. These batteries provide high current rating and a long lifecycle in addition to good thermal stability, enhanced safety, and tolerance. The low-resistance properties, improving the protection and thermal efficiency, making these batteries ideal for electric vehicles. The only downside is their low-voltage power which offers less energy than other Li-ion battery types.

Lithium titanate (Li$_4$Ti$_5$O$_{12}$): Li-titanate replaces the anode material graphite in a typical lithium-ion battery. The lithium titanate nanocrystals, forming as anode provides a larger surface area over carbon, enabling the electrons to easily enter and exit the anode. This, in effect, makes it one of the Li-ion category's faster-charging batteries. The cathode material can be lithium manganese oxide or NMC. The nominal cell voltage is 2.4 V. It provides fast charging and delivers a high discharge current. The lifecycle is higher than that of a regular Li-ion. Li-titanate has excellent low-temperature discharge characteristics, and has a good capacity and performance at $-30°C$. They have lower intrinsic voltage and lower ratings of actual energy than traditional lithium technologies. These are one of the safer thermal tolerance platforms, making them extremely safe for use in EVs and e-bikes with potential in the military and aerospace industries. The electrical parameters of different Li-ion batteries are compared in Table 7.3.

7.2.4 Inverter

The inverter, which is also known as the power conditioning unit, is the key element in any standalone PV system with AC loads. DC operating voltage of the system, the AC load, the output voltage, the variation in output wave form, and the frequency are to be considered in specifying the inverter. On the input side, the DC voltage, surge capacity, and acceptable voltage variation must be specified. With an off-grid system, the inverter must provide high-quality stable power and operate around the clock for decades without failure. It should be capable of providing surge power well beyond its rated capacity to support reactive loads and sensitive loads simultaneously.

Table 7.3 Comparison of Li-ion batteries of different chemistries [4].

Parameter	Lithium cobalt oxide	Lithium manganese oxide	Lithium nickel manganese cobalt oxide	Lithium iron phosphate	Lithium nickel cobalt aluminum oxide	Lithium titanate
Specific power	L	M	M	H	M	M
Specific energy (Wh/kg)	150–200 up to 240	100–150	150–220	90–120	200–260, 300	50–80
Safety	L	M	M	H	L	H
Lifespan lifecycle	500–1000	300–700	1000–2000	1000–2000	500	3000–7000
Cost	L	L	L	L	M	H
Performance	M	L	M	M	M	H
Cell voltage	3.6 V nominal	3.7 (3.8 V) nominal	3.6 (3.7 V) nominal	3.2 (3.3 V) nominal	3.6 V nominal	3.6 V nominal, 1.8–2.85 V
Operating voltage	3–4.2 V	3–4.2	3–4.2	2.5–3.65 V	3–4.2	
Charge rate	0.7–1C	0.7–1C, 3C max	0.7–1C, 2C possible	1C, 3.65 V	0.7C, 4.2 V	1C, 5C max, 2.85 V
Discharge rate	1C, 2.5 V cut off	1C, 2.5 V cut off; 10C possible, 30C (pulse 5 s)	1C, 2.5 V cut off; 2C possible	1C, 2.5 V cut off; 25C possible, 40 A (pulse 2 s)	1C, 3 V	10C, 1.8 V cut off; 30C (pulse 5 s)
Thermal runaway	150°C	250°C	210°C	270°C	150°C	High safety

L, *low;* M, *medium;* H, *high.*

Characteristics

Standalone inverters typically operate at 12, 24, 48, or 120 VDC input and create 120 or 240 VAC at 50/60 Hz. The selection of the inverter input voltage is an important decision because it often dictates the system DC voltage. The shape of the output waveform is an important parameter. Inverters are often categorized according to the type of waveform produced such as square wave, modified sine wave, and sine wave. The pure sine wave inverter has to be used for off-grid PV system applications.

Inverter specifications and features

Power conversion efficiency—The ratio of output power to input power of the inverter is defined as the power conversion efficiency. The power conversion efficiency of standalone inverters will depend upon the connected load, the ambient temperature, and other factors.

Rated power—The inverter rating is designated by VA or KVA. This is the rated power of the inverter and the selection depends on the size of the maximum AC load it has to support. Rated power represents how much power the inverter can produce continuously.

Duty rating—This rating is defined as the delivery of rated power by the inverter over a specified time. Exceeding this time may cause failure of the inverter or shut down the inverter.

Input voltage—This is defined as the rated DC input voltage range of the inverter and is dependent on the total connected load in watts. Generally, the larger the load, the higher the inverter input voltage to limit the input current.

Surge capacity—This is important for standalone and bidirectional inverters. Loads such as a motor requires a starting current 3–7 times their operating current for several seconds. The inverter must be capable of providing the surge power.

Standby current—The amount of current/power consumed by the inverter during standby conditions, without loads being connected. This results in a power loss from the battery bank.

Voltage regulation—This describes the regulation of the output voltage of the inverter.

Voltage protection—The inverter needs to be provided with protection for variations in DC input voltage. Built-in circuits provide shutdown or cut-off of the inverter in the event of DC voltage variation beyond its limit.

Maximum AC power and power factor—The AC output power consists of two parts: real power and reactive power. It describes the ratio of the real power to the apparent (total) power. For resistive loads, the power factor will be 1.0, but for inductive loads, the most common load in residential systems, the power factor will be as low as 0.5. The power factor is determined by the load, not the inverter.

Maximum input current—MPP tracker input

The maximum current, which is maintained through the DC side of the solar PV inverter, should always be lower than the limiting current of the inverter.

The inverter converts the direct current from the solar array and the battery to single-phase or three-phase alternating current compatible to the grid.

Ambient temperature range—Due to seasonal variations, there is a change in ambient temperature. It should work for a temperature a range of −20°C to 50°C without a change in the performance.

Bidirectional inverter

The heart of an off-grid PV system is a bidirectional inverter that can convert DC-based PV and battery power into AC power to supply the connected loads. DC-AC conversion can synchronize with an AC source such as a grid or a DG set. The same convertor acts as a grid charger by converting AC to DC to charge the batteries using grid supply whenever required.

The inverter section has built-in galvanic isolation using a power transformer of rated capacity, which ensures a rugged design under extremely fluctuating and impure grid conditions. It also provides isolation between the DC and AC sides.

Other than converting DC power to AC, features and other considerations in choosing an inverter to produce high quality power include the following:

Using a surge protection device on the inverter input to protect against nearby lightning strikes is recommended for most areas.

Should produce low noise emissions and should have minimum standby/night power consumption.

To track performance and troubleshooting of the system, monitoring features including a remote monitoring system are required.

For overcurrent protection of both the input and output circuits of the inverter, fuses or circuit breakers are introduced.

Grounding faults can arise with waterlogging, a broken sheath, or insulation of cables and can cause tripping of the inverter. So, ground fault interrupter circuits should be provided. The inverter trips if the fluctuations in the grid are beyond the operating levels of the inverter. However, if the protection system of the inverter doesn't work, it can also cause burning of circuits. So, it should have grid management features, including power output variation, frequency and voltage management, and dynamic grid support.

7.2.5 Module mounting structure

Solar modules are to be mounted outside to get exposed to sunlight for the generation of electric power. Solar PV modules have an aluminum frame on four sides with mounting holes to anchor with the racking arrangement. Solar module mounting or racking systems are used to fix solar panels on surfaces such as roofs or the ground. These mounting systems are generally made up of hot dip galvanized mild steel, with pregalvanized steel, or galvaluminum and aluminum alloy. In general, we come across ground-mounted systems and rooftop-mounted systems. The roofs are classified as reinforced cement concrete (RCC) and industrial roofs with a metallic sheet. The following are the functions and requirements to be considered in the design of module mounting structures (MMS).

The MMS and the associated equipment should hold the solar modules firmly and should withstand storms, hurricanes, swirling winds, etc. To meet this, the MMS has to be designed by selecting a suitable material to withstand the wind load of the site location with a safety factor of more than one. The MMS design depends on the tilt angle, which is decided based on the latitude of the plant site. The design of the MMS of PV module will have the proper circulation of air to provide a self-cooling effect to improve the performance. The weight of the structure and its associated components and its anchoring foundations should not exceed the load-bearing capacity of the building. The anchoring arrangement with the roof should not cause water seepage in the roof.

The MMS design is such that it should harvest the incidence of maximum irradiance on the solar modules to maximize the annual energy generation. The MMS should work for 25 years in the hostile environments of rain, high and cold temperatures, winds, high humidity, and a salty atmosphere. So, all structural materials and other supporting components used in the solar PV installation have to be of high quality and should be corrosion-resistant. The structural components made up of hot dip galvanized mild steel, pregalvanized mild steel, or galvaluminum are used for ground-mounted systems, whereas for sheet roofs, lightweight anodized aluminum alloy profiles are preferred. If stainless steel SS 304 material bolts, nuts, fasteners, clamps, and other hardware are used to mount the module and fix the structural elements, there will not be a corrosion problem. The MMS should be optimally designed to occupy minimum space while providing maximum output from solar PV modules. The structure is designed such that repair and maintenance activity could be done easily and shall be in line with the site requirements.

The alignment and tilt angle of the solar PV modules is fixed in such a manner that they provide the maximum energy output on an annual basis in accordance with the latitude of the place of installation. The clearance between the lowest part of the module and the ground level is maintained as per the local standards. It should be around 400–500 mm to provide ample space for the access of the rear side for maintenance and protect the modules from unforeseen floods for the ground-mounted systems. No waterlogging shall take place in any location due to the installation of a solar photovoltaic module. The design arrangement of the solar photovoltaic is done to prevent overturning and/or damage due to high wind speeds.

For ground-mounted systems, the soil report of the area has to be considered in designing the MMS and its foundation and the design guide lines of the country are to be followed. For solar car parking, the design shall be such that there won't be any water leakage inside the parking area/shed through the rooftop of the parking.

For mounting solar PV modules on a flat RCC rooftop, some customers prefer not to penetrate the surface of the RCC roof. In this case, the design with ballast structure is to be preferred, or a cement foundation of MMS to be bonded to the RCC surface using an adhesive.

The total load on roof shall be well distributed and shall not exceed the load-bearing capacity of the roof. The load-bearing capacity of the roof is followed as per the local standards of the country.

For solar photovoltaic mounting on a slanted metallic sheet rooftop, the better design prefers the anchoring of module frames with steel members of the roof structure. The load-bearing capacity of the steel members of the roof is checked as to whether they can bear the additional load due to solar photovoltaic modules considering dead load, live load, and winds at the specified speed of the site. Solar PV module frames are laid parallel to the roof surface. The hole drilling for the purpose of anchoring Al channels is done only at the ridge locations of the corrugated roof sheet. Customers prefer minimum puncturing of the roof. Water tightness in the existing roof with suitable sealants/washers will ensure waterproof roofing. A permanent maintenance walkway over the roof is designed to avoid damage to the roofs during cleaning and maintenance of the modules. Providing an adequate permanent staircase access/cage ladder to access the rooftop from the ground level will ensure the safety of personnel during work.

The load of the solar PV modules/catwalks/maintenance bays shall not come on the metallic sheet of the roof. For this, a grid structure with an Al channel is formed by anchoring the Al channels with the rafters and purlins of the roof structure.

There are different types of mounting configurations such as fixed tilt, seasonal tilt, single-axis tracker, dual-axis tracker, ballast type, east-west mounting, and super structure types that are being followed for RCC roof and ground-based systems.

To provide the slope for the module to face the sun, triangular mounting arrangements are being followed in the industry. For RCC and ground mounted systems, the MMS consists of the following structural elements, as shown in Fig. 7.10:

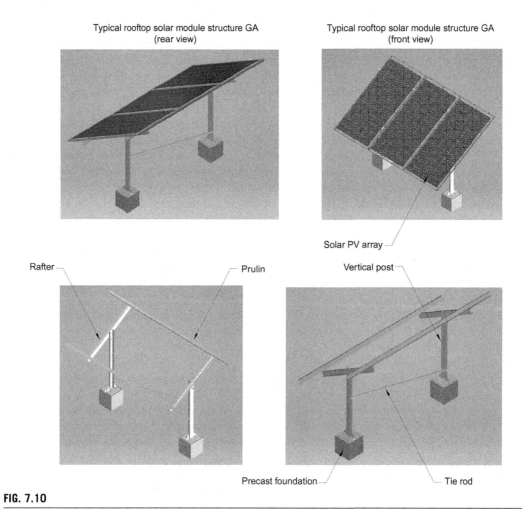

Typical rooftop solar module structure GA (rear view)

Typical rooftop solar module structure GA (front view)

Solar PV array

Rafter — Prulin

Vertical post

Precast foundation — Tie rod

FIG. 7.10

Module mounting structure fixed with modules.

Vertical column/leg: This is the load-bearing element that anchors the structure to the ground or soil via the foundation. It bears the weight of the module and provides support to the other components. The vertical column anchored with the foundation provides a counter-balancing force to the upward thrust caused due to wind loads.

Purlins: These are run in a horizontal direction to provide support to the modules for mounting.

Rafters: These are run in a tilted vertical direction, anchored with the vertical leg, and provide support to the modules for mounting.

Bracings: These are connected from the vertical leg to the rafters to provide extra support.

Hardware elements: Nuts and bolts for connecting the structural elements to form the MMS and to fix the solar PV modules to the MMS.

7.2.5.1 Design methodology of MMS

Different standards are followed in different countries for the design of MMS system. An example of wind load calculation considering Indian standard is described. For a wind speed of 234 kmph, considering a 2×10 matrix solar module configuration, 20 modules are arranged in portrait configuration. The module dimensions are $1960 \times 990 \times 40$ mm with a weight of 22.5 kg. There are four vertical legs anchored with four foundations and there are four purlins and four rafters, and two bracings for each rafter.

The wind load estimation is given in Table 7.4.

The wind load estimation is as follows:

Number of module rows per structure—2
Number of module in each row—10
Number of supporting purlins—4
Total number of panels—20
Total weight of the panels—450 kg
Load on each purlin—112.5 kg
Length of each purlin—10.08 m
Uniformly distributed load, UDL on each purlin—11.161 kg/m
Upward wind pressure estimation is taken up considering the pressure coefficients for mono slope roofs mentioned in the local standards.

For a tilt angle of 20 degrees, the pressure coefficient for the modules is 1.3, as per Table 7.8, of IS 875 (part 3) 2015, whereas the pressure coefficient for the structural members is 2.05 as per Table 29 of the standard.

Table 7.4 Wind load estimation.

Wind load on module	Wind load is calculated as per IS 875 part 3
Wind speed, V_z	234 kmph = 65.0 m/s
Category 2 and class B	
k_1	0.89 (25 years)
k_2	1 (Height < 10 m) category-2 considered
k_3	1
k_4	1.15 (industrial structure considered)
Design wind speed, V_b	$k_1 \times k_2 \times k_3 \times k_4 \times V = 66.53$ m/s
Wind pressure, P_z	$0.6 \times V_b^2$ (clause 5.4) = 2.656 kN/m²
Design wind pressure, P_d	$K_d \times K_a \times K_c \times P_z$
K_d	1 For cyclone area
K_a	1 (1.94 m² average area of module)
K_c	1
P_d	2.656 kN/m²

Whereas V_z is the design wind speed at height z in m/s; P_z is the pressure due to wind load; k_1 is the risk coefficient probability factor; k_2 is terrain roughness and height factor; k_3 is risk topography factor; k_4 is the importance factor for a cyclone region; $V_z = V_b k_1 k_2 k_3 k_4$; K_d is the wind directionality factor; K_a is the area averaging factor; K_c is the combination factor; and p_d is the pressure due to wind load = $K_d K_a K_c p_z$.

Wind pressure on panels is $3.452 \, \text{kN/m}^2$.

Wind pressure on structural members is $5.444 \, \text{kN/m}^2$.

The purlins, vertical legs, rafters, and bracings are suitably selected to meet the design requirements. Based on the properties of structural elements and the arrangement drawing, a stadd file is made. The stadd file is given to the Staddpro software and different load simulations are done and the responses at different nodes are obtained. Based on the simulation results, the MMS elements will be designed.

The simple type of MMS on an RCC roof is shown in Fig. 7.11.

The ballast structure with counterweights is shown in Fig. 7.12.

The foundation-based MMS for a ground-mounted system is shown in Fig. 7.13.

This type of MMS is widely used where soil has a poor cohesion properties. It requires an RCC foundation over which vertical columns are mounted and fastened with a J-bolt. This type of structure is composed of hot rolled close annealing (HRCA) duly hot dip galvanized as per the relevant standards and is capable of withstanding a high wind load. Fig. 7.14A and B shows the superstructure-based MMS. This type of superstructure is for rooftop design to tackle shadows and offer roof space utilization.

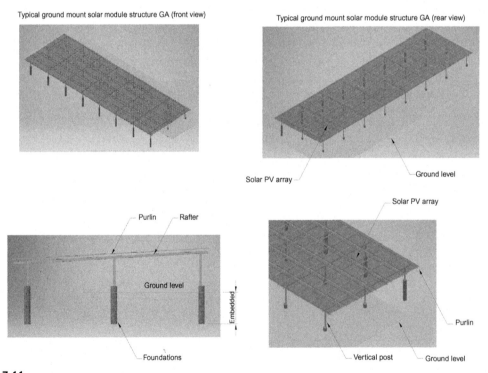

FIG. 7.11

Simple module mounting structure for an RCC roof.

FIG. 7.12

Ballast structure with counterweights.

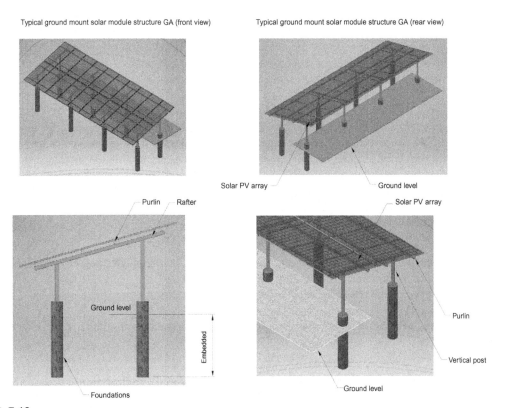

FIG. 7.13

Foundation based module mounting structure for Ground.

(A)   (B)

FIG. 7.14

MMS with a superstructure. (A) Superstructure with a higher column height and (B) top view of superstructure.

Courtesy: Grand solar pvt Ltd, Chennai.

7.2.6 Array junction box

Solar PV module strings are connected with cables and brought to the string combiner box or array junction box (AJB) to combine the strings. These cables have a single conductor connected with connectors that are wired onto the solar PV modules. The output of the PV string to the AJB is one larger size single-wire or two-wire conductor in the conduit. The AJB incorporates a security circuit breaker/fuse for each string as well as a surge protection device. It will have a DC circuit breaker as disconnect switch.

The AJB provides protection to the DC side. AJB is provided between the solar PV array and the solar charge controller/inverter. It has two busbars, one for positive and the other for negative. It includes a string protection device, overvoltage surge protection, and a DC output switch disconnector. Two in two out, three in three out, and four in four out AJBs are used based on the capacity of the system. A sample AJB is shown in Fig. 7.15. In an off-grid system two in one out, three in one out, and four in one out are used to get the required power outputs by combining the strings from the solar arrays. Each solar PV string is provided with fuses of adequate rating to protect the solar arrays from accidental short circuits.

AJBs shall have the following features:

They are free from entry of water and dust and are made of thermoplastic, polypropylene, etc.
The AJB complies to protection class II as per IEC 61439-1.
The enclosure material is UV-stabilized, chemically resistant, and has a temperature rating of $-10°C$ to $55°C$.
The enclosures will have suitable dielectric strength and withstand an insulation voltage of 1000 or 1500 VDC.
It is ideal to have breather glands in the AJB to prevent explosions and overheating.

Surge protection devices in the AJB protect the electrical and electronic equipment from power surges and voltage spikes. Surge protection devices divert the excess voltage and current from the transient or surge to the earthing system. So, surge protection devices are incorporated in the AJB to protect them from surge currents.

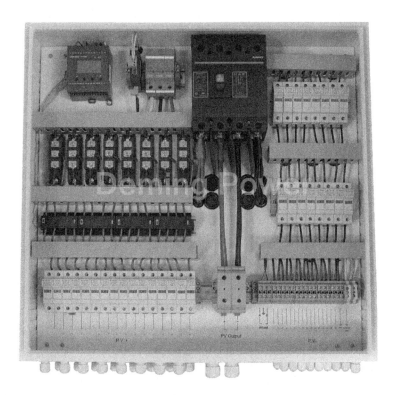

FIG. 7.15

Typical array junction box.

The AJB is equipped with suitable cable entry points with cable glands of appropriate sizes to allow incoming and outgoing cables.

7.2.7 AC distribution board

An AC distribution board (ACDB) is also known as a panel board or breaker panel. It interfaces the inverter with the grid. The ACDB controls the electrical power into the grid while providing a protective fuse or circuit breaker for each circuit.

The ACDB is an important part of a solar PV power system to provide electrical protection between the inverter output and the AC grid. The ACDB also acts as an isolator between the inverter and the AC grid during break down of the inverter or AC grid. Also, the ACDB has an energy meter and Fig. 7.16 shows the image of ACDB.

The ACDB shall have the facility to terminate single or multiple inverter AC outputs. The ACDB is usually designed as per the applicable local standards and shall have the following minimum features:

There shall be a provision in the system for the measurement and display of solar power being fed to the load through each inverter.

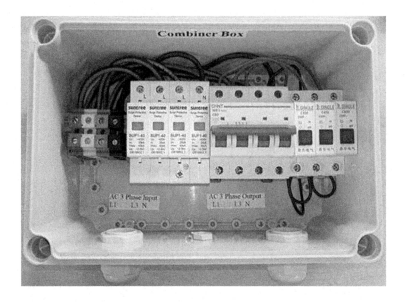

FIG. 7.16

Three-phase ACDB.

Surge protection on the AC side is generally incorporated. Class I + II surge protection devices (as per IEC 62305; IEC 61643, and IEC 60364-5-53) give safe and reliable operation.

A metal-encapsulated spark gap solution provides better fireproof operation at the site. It withstands a total discharge capacity/lightning impulse current at 10/350 µs. The nominal discharge current at 8/20 µs shall be a minimum 100 kA for a three-phase power supply system and 50 kA for a single-phase power supply system.

The discharge capability of line-neutral connected devices shall be 25 kA at 10/350 and 8/20 µs. It will have sufficiently rated MCBs or MCCBs.

7.2.8 Cables DC and AC

For DC cabling, XLPE or XLPO insulated and sheathed, UV-stabilized single-core flexible unarmored copper cables are used. The minimum cable size used for this type of system is 4 sq. mm copper.

The total voltage drop on the cable segments from the solar PV modules to the solar grid inverter shall not exceed 2.0% for a better harvest of energy.

The total voltage drop on the cable segments from the solar grid inverter to the nearest building distribution board shall not exceed 2.0%.

All cables and conduit pipes are clamped to the rooftop, walls, and ceilings with thermoplastic clamps at intervals not exceeding 50 cm. In three-phase systems, the size of the neutral wire size is equal to the size of the phase wires.

The size of the cable from the ACDB to the nearest LT panel is designed normally for two times of the designed current capacity and the voltage drop shall not exceed 2.0%. The standard derating factors of the cables with respect to the laying pattern in buried trenches/on cable trays are considered while sizing the cables.

All the cables of size 50 sq. mm. or higher are armored to fetch mechanical protection of the sheath, insulation, and conductor.

If the data transmission length is more than 100 m or as required, single-mode optical fiber cables throughout the entire system are used in the industry for communication networks.

The control cables are designed for outdoor application with a continuous ambient temperature of 50°C, 1.1 kV grade, heavy duty, stranded copper/aluminum conductor, XLPE insulated, rated for minimum 90°C, galvanized steel wire round armored, and flame-retardant low-smoke (FRLS) extruded PVC type outer sheathed insulation is used for cables.

For AC cabling, PVC or XLPE insulated and PVC sheathed single or multicore flexible aluminum cables are commonly used.

Connectors

MC4 connectors are used to connect solar PV modules. The MC in MC4 stands for the manufacturer, multicontact and the 4 for the 4 mm diameter contact pin. Fig. 7.17 shows male and female MC4 connectors. Straight connectors are suitable to mate with MC4 or equivalent solar module connectors. Connectors will have watertight sealing conforming to IP 68 and are available as "male (plug)" and "female (socket)" types and suitable for crimping $1C \times 4 \text{ mm}^2$ and $1C \times 6 \text{ mm}^2$ XLPE insulated copper cables. The connectors withstand a rated current, 15 A, and impulse withstand voltage of 16 kV. The proper connections have to be selected to have compatibility with the module connectors.

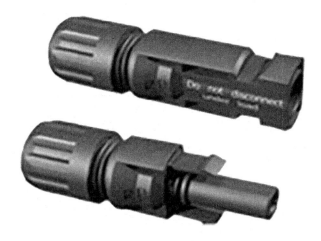

FIG. 7.17

MC4 cable connectors.

7.3 Design of off-grid or standalone solar PV system

A. Daily energy requirement

It is important to consider the daily energy requirement with the load diversity factor and estimated load operational hours. The daily energy requirement can be estimated using an Excel sizing sheet as input to calculate the solar PV array capacity (series/parallel solar PV module combination) to match the daily energy production as compared to the daily energy requirement, which has to be a surplus on an average monthly basis.

B. Choice of system DC voltage

The choice of system DC voltage plays an important role in deciding the various system components with cost optimization and maximum performance and is also dependent on solar module/array capacity. The system voltage is dependent upon the available solar charge controller, battery, inverter operational voltage range/input voltage, and the array capacity. Table 7.5 gives the system voltage for different solar array capacities.

C. Loss of load probability

The loss of load probability (LOLP) of a standalone or off-grid solar PV system is a function of normalized solar array output and energy availability at the battery with a specified load curve and period for optimal size. Various models used to calculate the LOLP [5] to get the optimal size of the off-grid solar PV system in terms of number of solar PV modules (solar array size in Wp), battery capacity and cost of the system as per the defined LOLP (from 0.00 to 0.10 in steps of 0.01).

D. System availability

It is defined as the percent of time the solar PV power system is able to meet energy supply to loads. For example, a system designed for 98% availability is expected to supply the energy requirement of the designed loads for 95% time with only 5% down time. This is linked to the battery capacity and recharging from a solar PV array. To maintain the best uptime of the plant, it is required to select reliable components and maintain the plant properly. For higher system availability, the battery capacity has to be higher and the cost of the system also increases. For critical systems, the system availability should be higher.

Table 7.5 System voltage for different loads.

System voltage (V)	Array capacity (Wp)	Load (W)
6.0	<10.0	2–3
12.0	11–100	10
24.0	101–1000	100
48.0	1001–5000	1000
96.0	5001–10,000	2500
120.0	10,001–15,000	5000
>120.0	>15,000	>5000

E. Highest achievable efficiency—component selection

The total system efficiency is a multiplier of the individual efficiencies of the components used in the system.

$$\text{Total combined efficiency} = \text{solar pv module efficiency} \times \text{battery efficiency} \times \text{charge controller efficiency} \times \text{inverter efficiency}$$

Hence, it is very important to choose the right components such as the solar PV module, the charge controller, the battery, and the inverter with high efficiency to have the best efficiency and reliability of the system.

F. Optimized cost

The optimized cost of an off-grid or standalone solar PV system is very important, considering customer expectations while keeping the integrity of the system performance. One has to strike a balance between the overall optimized cost while considering system availability and the optimized performance of the solar PV system.

7.3.1 Design methodology

(i) Site survey

It is important to capture all details of the existing load details and their operation hours per day and the local solar irradiance data with other weather data such as temperature, wind speed, etc.

(ii) Solar irradiance assessment

The local source of data or from other met data sources (NASA/Meteonorm/IMD data, Solar GIS, PVgis, NREL, etc.) needs to be collected to do the system sizing (broadly, the selection of the solar PV module/array, the battery capacity, the solar charge controller, the inverter, etc.). The average sun hours per day or monthwise global horizontal irradiation (GHI) for the specific location can be obtained taking the latitude and longitude of the place from the NASA meteorological data. Meteonorm or Solar GIS data can also be used to obtain the GHI. The monthwise solar irradiance in the plane of the array (for the tilt angle of the array) has to be obtained from the above-mentioned sources. For the winter season, the GHI will be low. As the off-grid system has to support the loads during the month of lowest radiation, the December value of GHI can be considered for the locations of northern hemisphere. By dividing the value with the number of days of the month, the GHI/day, the average sun hours per day can be obtained.

(iii) Sizing of system components

In order to determine the various components of a standalone/off-grid solar PV system, one could make use of Excel sizing sheets or standard system sizing software.

7.3.2 Design methodology (steps)—Excel sheet sizing

The step-by-step procedure to determine the components of a standalone/off-grid solar PV system is as follows:

Table 7.6 Estimating the daily load in terms of Ah/day.

Details of load	Operation (DC/AC)	Load wattage (W)	Load Qty	Total load wattage (W)	Daily duty cycle (h)	Load input voltage (V)	Load input current (each load) (A)	Total load input current	Total load Ah per day
Total									

1. Determine the daily load energy requirement using the Excel spread sheet by including load details with operational hours. Calculate the total watt-hours per day for each load. Identify each load, that is, the wattage of the load that is going to operate and the number of hours of operation in a day. List all the appliances and their wattage and hours of operation in a day. Enter the voltage and current of each appliance and multiplication gives wattage. List the AC loads and DC loads separately, as shown in Table 7.6.

Add the watt-hours needed for all appliances together to get the total watt-hours per day that must be delivered to the appliances. For the conversion of AC loads into DC, a correction factor of 1.15 has to be applied, considering the inverter efficiency of 0.85. By dividing the Wh/day with the system voltage, the load requirement in Ah/day will be obtained.

2. Need to enter the monthly solar irradiance and temperature data either from local met department or from Meteonorm or other reliable sources. The monthwise irradiance in the tilted angle of the solar modules is required.
3. Determine the tilt angle, either by choosing the fixed tilt angle (latitude ±5 degrees from horizontal) or using a standard off-grid/hybrid system sizing software and determining the best tilt angle considering the annual energy yield.
4. The choice of system voltage is important and is guided by various parameters while keeping in view the guidelines above for the selection of system voltage as per Table 7.5.
5. Choice of other loss parameters such as array power loss due to temperature, misorientation, dust, charge controller loss, and battery loss is important before performing the system design.
6. Then you need to input the number of parallel strings so that the excess energy in each month is positive.

For sizing the array, the total daily energy in watt-hours is required and the DC voltage of the system (VDC) must be finalized. By dividing the watt-hours per day with the voltage, the demand of load in Ah/day is obtained. To avoid undersizing, losses must be considered by dividing the total power demand in Ah/day by the product of efficiencies of all the components in the system to get the required energy.

The system efficiencies, including the efficiency of the charge controller, battery charge/discharge efficiency, and the loss due to self-discharge of the battery, are to be incorporated. The soiling losses, module voltage and current mismatch losses, cable losses, inaccuracy in array orientation, and the temperature affect the performance of the solar array. So, all these factors are to be incorporated in sizing the solar array to meet the requirement of the load. The actual load met by the solar array is obtained by dividing the daily energy consumption/requirement with the efficiency of system components.

$$\text{Total load to be met by solar array} = \frac{\text{Demand load in Ah/day}}{\text{Charge controller efficiency} \times \text{Battery charge} - \text{discharge efficiency}}$$

This gives the total current required in a day from the solar array.

The solar modules are available with different wattages and with nominal voltages of 12, 30, and 24 V with 36, 60, and 72 solar cells, respectively.

The selected solar modules have parameters of I_{sc}, V_{oc}, V_{mp}, and I_{mp}.

Calculate the V_{mp} of the module for the highest temperature of the location in summer using the temperature coefficient of V_{mp} and applying the degradation factor of 15% for 25 years. This gives the end of life (EoL) V_{mp} of the module.

If the battery voltage is divided by the EoL V_{mp} of the module, the required number of modules to be connected in series can be obtained. For direct and rough estimation, the battery voltage can be divided by the nominal voltage of the module to get the number of modules in a string.

The I_{mp} of the solar module has to be derated for its temperature and all other loss factors such as soiling (2%), cabling (2%), mismatch (2%), array misorientation (2%), and degradation (8%) for 20 years are to be applied to get the EoL I_{mp}.

The I_{mp} value slightly increases with temperature. By applying the temperature coefficient, the I_{mp} value of the module can be obtained for the operating temperature of the module.

The monthwise solar irradiance in the inclined plane of the array for the site of the plant can be obtained from NASA, Meteonorm, and other sources. This value gives the global inclined irradiance (GII), which is one sun equivalent hours of operation in a day. The EoL I_{mp} can be multiplied with GII (the number of hours) and the number of modules in parallel in the system to get the generated energy in all the months in Ah/day. This generation has to be matched with the required load by estimating the number of modules in parallel (N_p). By iterating the N_p value, the array will be sized such that the energy balance in a day is positive.

Modules must be connected in series and parallel according to the need to meet the desired voltage and current in accordance with: First, the number of parallel modules, which equals the total load demand current divided by the rated current of one module at EoL condition. Second, the number of series modules, which equals the DC voltage of the system divided by the EoL V_{mp} of the module. The result of the calculation is the minimum number of PV panels. If more PV modules are installed, the system will perform better and battery life will be improved. If fewer PV modules are used, the system may not work at all during cloudy periods and battery life will be shortened.

7. Similarly for deciding the battery capacity, we need to give input on the Ah capacity of the selected battery bank with the number of parallel battery strings (not more than four parallel) and use the following thumb rule for battery sizing.

$$\text{Battery capacity (Ah)} = \frac{\text{Daily Wh} \times \text{No. of days of storage}}{\text{VDC} \times \text{DoD} \times \text{CAF} \times \text{CDF}}$$

where VDC is DC system voltage; DoD is depth of discharge; CAF is capacity appreciation factor; and CDF is capacity degradation factor.

The battery recommended for a solar PV system is a deep cycle battery. First, the amount of backup energy to be stored has to be calculated for the given application. This is usually expressed as the number of sunny days, in other words, for how many cloudy days the system should operate using the energy stored in batteries. This depends on the application, the type of battery, and the system availability desired.

To specify the battery, first establish the efficiencies of the inverter and battery, battery voltage, and the allowed depth of discharge of the battery. The days of autonomy, that is, the number of days that the system needs to operate when there is no power produced by PV panels, has to be specified. The battery voltage can be considered based on a lower voltage drop due to cables, the operating voltage of the DC loads, the total array capacity, etc.

The sizing of the battery can be done as follows:

Calculate total watt-hours per day used by appliances\loads.
Divide the total watt-hours per day used by inverter efficiency (0.85) to get the energy required at the input of the inverter.
Divide the energy required at the input of the inverter by the nominal battery voltage to get the Ah of the battery.
Divide the Ah of the battery by the battery charge-discharge efficiency (0.93).
Divide the answer obtained above by the depth of discharge of the battery (0.8).
Multiply the answer obtained above with the days of autonomy to get the required ampere-hour capacity of the deep-cycle battery.

Battery capacity is calculated with the given formula:

$$\text{Battery capacity} = \frac{\text{Total Watt} - \text{hours per day} \times \text{Days of autonomy}}{\text{Inverter efficiency} \times \text{Battery efficiency} \times \text{DoD} \times \text{Nominal battery voltage}}$$

In cold temperatures, the capacities of batteries are greatly reduced. Most battery manufacturers will publish a temperature versus capacity chart. The temperature correction factor has to be applied in sizing the battery.

8. Similarly, to decide the solar charge controller rating, we need to calculate the maximum solar array current that each of the solar charge controllers will handle (it will be different for a series/shunt controller).

The solar charge controller is typically rated against amperage and voltage capacities. Select the solar charge controller to match the voltage of the PV array and batteries and then identify which type of solar charge controller is right for your application. Make sure that the solar charge controller has enough capacity to handle the current from the PV array.

The sizing of the controller depends on the total PV input current that is delivered to the controller and the PV panel configuration (series or parallel configuration). According to standard practice, the sizing of the solar charge controller is to take the short circuit current (I_{sc}) of the PV array and multiply it by 1.3. A safety factor of 30% has been considered.

$$\text{Solar charge controller rating} =$$
$$\text{(total short circuit current of PV array (number of parallel strings} \times \text{module } I_{sc}) \times 1.3)/\text{temp.derating factor}$$

A positive or negative grounding controller is to be selected based on system grounding selection.

9. For deciding the inverter capacity, the load table needs to be considered (with the surge rating of each load) as well as the derating due to ambient temperature and other conditions.

An inverter is used in the system where AC power output is needed. The input rating of the inverter should never be lower than the total wattage of all appliances. The inverter must have the same nominal voltage as the battery. The system operating voltage has to be defined.

For standalone systems, the inverter must be large enough to handle the total amount of watts that it will be using at one time. The inverter should have 25%–30% extra over-load capacity to the total wattage requirement. In case the appliance type is a motor or compressor, the inverter size should be a minimum 3–4 times the capacity of those high surge loads and must be added to the inverter capacity to handle surge current during starting.

The total connected AC load watts or those that will be used simultaneously as computed above are divided by the direct current system voltage. This will provide the maximum continuous direct current amps that the system will experience. The inverter specification should be made on the basis of the AC watts because essentially it should supply this load while considering the surge power requirement of the inductive or capacitive loads connected to the system.

For safety, the inverter should be considered at a 25%–30% higher capacity of all total connected loads:

$$\text{The maximum input current to the inverter} = \frac{\text{AC load (W)}}{\text{DC voltage (lowest inverter cut off voltage)}}$$

$$\text{Inverter capacity (VA)} = \frac{\text{Total load in Watts}}{\text{Temperature derating factor} \times \text{power factor}}$$

10. Cable sizing and cable voltage drop calculations.

The cable selection is of paramount importance and will determine the efficiency of the system with a minimum loss of power as well as the voltage drop (both DC and AC side). For deciding on interconnecting cables (for solar PV module interconnections, array to AJB, AJB to MJB (main junction box), MJB to input of the charge controller, charge controller to the inverter and battery bank, etc.), one needs to make the cable voltage drop calculations (based on cable lengths/cable resistivity, cable laid conditions, etc.).

7.3.3 Examples of system design

A. Example 1: Standalone system for home power requirements
Example: A house located in Delhi, India, has the following electrical load/appliance usage

- Three 9 W LED lamps used 6 h per day.
- There are two fans of 70 W used for 4 h per day.
- One TV with 200 W that runs 6 h per day
- One 150 W refrigerator that runs 24 h per day, with a compressor that run 12 h on and 12 h off.
- A computer of 250 W that runs 4 h per day.

The system voltage selected is 48 V, considering a load of 767 W. As per Table 7.7, the average load in terms of energy per day is 6522 Wh and it is 98.38 Ah/day.

A 325 Wp PV module is considered in this example. It has electrical parameters of I_{sc} 9.55 A, I_{mp} 8.82 A, V_{mp} 36.85 V, and V_{oc} 45.35 V.

Table 7.7 Load estimation of a solar home power system.

Load description	AC/ DC	Load (W)	No. of loads	Total power (W)	Daily duty cycle (h/ day)	Load duty factor	Daily load (Ah)	Daily load (Wh)
LED lamp	AC	9	3	27.0	6	1	3.38	162
Fan	AC	70	2	140	4	1	11.67	560
TV	AC	200	1	200	6	1	25.00	1200
Refrigerator	AC	150	1	150	24	0.5	37.50	3600
Computer	AC	250	1	250	4	1	20.83	1000
				767			98.38	6522

Total energy required for all the appliances $= 6522$ Wh/day.

$$\text{Daily load in terms of Ah is} = \frac{6522}{48} = 98.38 \, \text{Ah}$$

Total energy required from solar array =

$$\frac{\text{Average energy demand per day (Ah/day)}}{\text{Charge controller efficiency} \times \text{Battery efficiency} \times \text{Loss due to self discharge of battery}}$$

$$= \frac{98.38}{0.95 \times 0.93 \times 0.995} = 112 \, \text{Ah/day}$$

To meet 112 Ah/day, the array has to be designed.

I_{mp} of the 325 W module at 25°C is 8.82 A.

I_{mp} has to be derated considering 2% for each soiling factor, cable loss, module mismatch, and misorientation. For the obtained value, an array degradation factor of 8% has to be considered. The power of the solar module degrades 0.7% per year and overall there will be 20% degradation in Power in the end of 25 years.

The contribution of the current in the 20% degradation of power will be 15% for 25 years. For 12 years life of the system, 8% has been considered in the degradation of I_{mp}.

Derated I_{mp} is $8.82 \times 0.98 \times 0.98 \times 0.98 \times 0.98 \times 0.92 = 7.48$ A.

Cell operating temperature considered is 60°C.

Temperature coefficient of current is 0.0069 A/°C.

For a 35°C rise in temperature, the increase in module current is 0.24 A.

I_{mp} used for system design is $7.48 + 0.24 = 7.72$ A.

For January, the GII is 4.39 as per Table 7.8.

For 4.39 h of operation, one solar module generates $7.72 \, \text{A} \times 4.39 \, \text{h}$, which is 33.89 Ah.

To meet 112 Ah/day, the number of solar modules required is 112/33.89, which is 3.3.

So, 3–4 modules can be considered in parallel based on energy balance.

System voltage selected is 48 V.

V_{mp} of the solar module at STC is 36.85 V.

Table 7.8 Monthwise energy generation versus requirement considering three modules in parallel.

Month	Ambient temp. (°C)	Design temp. (°C)	Horizontal radiation (kWh/m²/day)	Inclined plane radiation (kWh/m²/day)	Number of parallel modules	Daily load (Ah)	Solar output (Ah)	Excess capacity (Ah)
Jan.	15.51	40	3.4	4.39	3	110.63	100.67	−9.97
Feb.	17.75	40	4.2	5.01	3	110.63	115.06	4.42
Mar.	23.85	40	5.3	5.78	3	110.63	132.72	22.08
Apr.	29.63	40	6.1	5.98	3	110.63	137.22	26.58
May	33.79	40	6.3	5.87	3	110.63	134.79	24.16
Jun.	33.8	40	6.4	5.77	3	110.63	132.40	21.77
Jul.	31.85	40	5.4	4.96	3	110.63	113.84	3.21
Aug.	30.65	40	5.0	4.81	3	110.63	110.36	−0.27
Sep.	29.89	40	5.6	5.84	3	110.63	134.01	23.37
Oct.	27.16	40	4.6	5.28	3	110.63	121.17	10.54
Nov.	21.58	40	3.6	4.60	3	110.63	105.63	−5.01
Dec.	16.57	40	3.1	4.10	3	110.63	94.08	−16.55
Average excess capacity Ah/month								8.69

Number of modules in series $48/36.85 = 1.3$.

The nominal voltage of the module is 24 V.

Hence, the number of solar modules in series is 48/24, which is 2.

If three modules are considered in parallel, there will be an energy deficiency in January, August, and December and the total deficit in energy generation is 8.69 Ah. If four modules in parallel are considered, there will be better system availability.

The total number of modules considered is $2 * 3$, which is six modules, and the solar array capacity is 1.95 kW. The monthwise energy balance is shown in Table 7.8.

Inverter sizing

Total wattage of all Loads/appliances $= 767$ W.

For safety and to consider surge power, the inverter should have 25%–30% overload capacity. In addition, the temperature derating factor of the inverter needs to be considered.

$$\text{Inverter capacity(VA)} = \frac{\text{Total wattage} \times \text{Overload capacity}}{\text{Temp derating} \times \text{Power factor}} = \frac{767 \times 1.3}{0.8 \times 0.8} = 1558\,\text{VA}$$

Considering available off-grid inverters, 2000 VA, 48 V has been selected.

Battery sizing

Load (total Wh consumption of all loads) = 6522 Wh, that is, 98.38 Ah @48 VDC.
Nominal battery voltage = 48 V.
Days of autonomy = 2 days; supports extra 2 days of rainy other than the normal days+1 normal day.

$$\text{Battery capacity} = \frac{\text{Total load in Ah} \times \text{No. of days of autonomy}}{\text{Inverter efficiency} \times \text{Battery efficiency} \times \text{DoD of battery}}$$
$$= \frac{98.38 \times 3}{0.95 \times 0.93 \times 0.8} = 417.55 \text{ Ah} \cong 450 \text{ Ah can be considered}$$

Total ampere-hours required are 450 Ah.

So the battery should be rated 48 V, 450 Ah for 3 days of autonomy.

Solar charge controller sizing

Solar PV module specifications are
$P_{max} = 325$ Wp; $V_{mp} = 36.85$ VDC; $I_{mp} = 8.82$ A; $V_{oc} = 45.35$ V; $I_{sc} = 9.55$ A.
Solar charge controller rating = (three strings $\times I_{sc}$ 8.82 A)$\times 1.3 = 37.25$, that is, 40 A and add the temperature derating to determine the final current rating of the solar charge controller.

So, the solar charge controller should be rated 40 A at 48 V or greater.

B. Example 2—Standalone system for telecom tower

One base transceiver station, BTS load of 1200 W operates for 24 h in a day. One air conditioner of 880 W operates for 12 h in a day. One lighting set and battery cooler operate 15 h in a day. All are AC loads.

The telecom tower load analysis is given in Table 7.9.

Table 7.9 Load analysis of a telecom system.

Details of load	Operation (DC/AC)	Load wattage (W)	Load Qty	Total load wattage (W)	Daily duty cycle in hours	Load input voltage (V)	Total load input current (A)	Total load Ah per day
BTS load	AC	1200	1	1200	24	48	25.00	600
Air conditioning	AC	880	1	880	12	48	18.33	220
Lighting and battery cooler	AC	400	1	400	15	48	8.33	125
Total				2480.00			51.67	945

Total load in wattage is 2480 W.

Total load input current is 51.67 Ah/day.

Total load is 945 Ah/day.

The system voltage considered is 48 V. So, the battery voltage is 48 V.

Battery charge-discharge efficiency is 93%; losses due to charge controller are 2%.

Actual daily load in Ah to be met by SPV System is 1032 Ah as per Table 7.10. Tables 7.11 and 7.12 show derating factors for I_{mp} and I_{mp} estimation at the end of life of the PV system.

To meet the actual daily load requirement of 1032 Ah by the solar array, monthwise radiation data in the inclined plane, along with the load and the energy generation, is compared as shown in Table 7.13.

The EoL I_{mp} of a 325 W solar module is 7.48 A.

Table 7.10 Net daily load Ah requirement.

Total load Ah	945.00	Ah per day
Losses due to controller (2%)	18.9	Ah per day
Losses due to battery charge derating (7%)	67.47	Ah per day
Losses due to self-discharge battery (0.1%)	1.03	Ah per day
Actual daily load Ah to be met by SPV system	1032.4	Ah per day

Table 7.11 Derating to be considered for I_{mp}.

Degradation factor in I_{mp} over lifespan considered	8%
Losses due to dust deposit or 1%, whichever is higher	2%
Losses due to module/branch mismatch or 2%, whichever is higher	2%
Losses due to line resistance of cables, losses in field wiring, and array wiring	2%
Losses due to panel orientation other than specified one as per insolation data	2%

Table 7.12 Estimation of I_{mp} at end of life for solar array sizing.

System nominal voltage	48	V
Number of solar modules in series	2.0	No
Array working voltage	48	V
I_{mp} of module at 25°C	8.82	A
Maximum battery voltage (24 cells with 2.4 V charge voltage)	57.6	V
Soiling loss factor 0.98, mismatch factor 0.98, misorientation 0.98, cable loss 0.98, array degradation 0.93 due to temperature		
I_{mp} after derating (applying above factors)	7.48	A
Cell temperature (ambient+20°C)	60	°C
Increase in temperature w.r.t STC	35	°C
Increase in module current due to temperature coefficient	0.24	A
I_{mp} used for system design (considering temperature factor also)	7.72	A

Table 7.13 Energy balance considering 24 modules in parallel.

Month	Ambient temp. (°C)	Design temp. (°C)	Horizontal radiation (kWh/m²/day)	Incline plane radiation (kWh/m²/day)	Number of parallel modules	Daily load (Ah)	Solar output (Ah)	Excess capacity (Ah)
Jan.	23.4	40	4.7	5.59	24	1032.40	1036.37	3.97
Feb.	25.66	40	5.4	6.05	24	1032.40	1121.59	89.19
Mar.	29.42	40	6.2	6.43	24	1032.40	1191.86	159.45
Apr.	32.09	40	6.5	6.31	24	1032.40	1169.79	137.39
May	34.03	40	6.4	5.99	24	1032.40	1110.53	78.12
Jun	30	40	5.2	4.80	24	1032.40	889.86	−142.55
Jul.	28.01	40	4.2	3.97	24	1032.40	735.57	−296.84
Aug.	26.93	40	4.3	4.12	24	1032.40	763.08	−269.33
Sep.	27.05	40	5.0	5.05	24	1032.40	935.59	−96.82
Oct.	26.62	40	5.1	5.51	24	1032.40	1021.42	−10.98
Nov.	24.14	40	4.7	5.40	24	1032.40	1001.71	−30.70
Dec.	22.64	40	4.4	5.32	24	1032.40	986.14	−46.27
Average excess capacity Ah/month								**−35.45**

As per Table 7.13, the radiation in the inclined plane for January is 5.59 kWh/m²/day. It means per day, the equivalent sun intensity is available for 5.59 h.

For a day in January, one module generates $7.48 \times 5.59 = 41.81$ Ah. $7.72 \times 5.59 = 43.15$ Ah

To meet the daily energy demand of 1032.4 Ah, the number of modules in parallel required is $1032.4/43.15 = 23.92 \approx 24$ modules.

By considering 24 modules parallel in the Excel sheet, the monthwise energy generation is compared with the requirement of 1032.4 Ah. There is a shortage of energy in the months of June, July, August, September, October, November, and December. So, the number of modules in parallel is increased to 25 and the energy balance looks as shown in Table 7.14.

Though there is a shortage in some months, there is excess energy in a year with the consideration of 25 modules in parallel.

The number of modules in series is two. The number of modules in parallel is 25. The total number of modules is 50 and the solar array capacity required is 16.25 kW.

Battery sizing calculations are shown in Table 7.15.

7.4 Cost of off-grid/standalone solar PV systems

The cost of a standalone PV system varies from country to country due to variations in the cost of the equipment and installation charges. As an example, a 9.75 kWp off-grid solar PV system with battery backup has been considered for Indian conditions for costing. The bill of materials required for the system is given in Table 7.16.

The cost of a 9.75 kW system is given in Table 7.17 and the cost breakdown is shown in Fig. 7.18.

Table 7.14 Energy balance considering 25 modules in parallel.

Month	Ambient temp. (°C)	Design temp. (°C)	Horizontal radiation (kWh/m²/ day)	Incline plane radiation (kWh/m²/ day)	Number of parallel modules	Daily load (Ah)	Solar output (Ah)	Excess capacity (Ah)
Jan.	23.4	40	4.7	5.59	25	1032.40	1079.55	47.15
Feb.	25.66	40	5.4	6.05	25	1032.40	1168.32	135.92
Mar.	29.42	40	6.2	6.43	25	1032.40	1241.52	209.12
Apr.	32.09	40	6.5	6.31	25	1032.40	1218.53	186.13
May	34.03	40	6.4	5.99	25	1032.40	1156.80	124.40
Jun.	30	40	5.2	4.80	25	1032.40	926.93	−105.47
Jul.	28.01	40	4.2	3.97	25	1032.40	766.22	−266.18
Aug.	26.93	40	4.3	4.12	25	1032.40	794.87	−237.53
Sep.	27.05	40	5.0	5.05	25	1032.40	974.57	−57.83
Oct.	26.62	40	5.1	5.51	25	1032.40	1063.98	31.58
Nov.	24.14	40	4.7	5.40	25	1032.40	1043.45	11.05
Dec.	22.64	40	4.4	5.32	25	1032.40	1027.23	−5.17
Average excess capacity Ah/month								6.10

Table 7.15 Battery sizing calculations.

Battery charge discharge eff.	93.00%	Per day
Self-discharge of battery	0.10%	Per day
Charge voltage	2.4	V
Nominal cell voltage	2	V
Number of cells in series	24	No
Minimum battery state of charge (SOC)	20%	
Number of days of autonomy	3	Days
Excess capacity	5%	
Daily load Ah requirement	945.00	
Battery capacity for meeting autonomy	2835	Ah
Battery capacity due to minimum SOC	3543.75	Ah
Battery capacity due to excess capacity	4075.31	Ah
C10 capacity @ 20°C (provided)	450	Ah
C100 capacity @ 20°C	518	Ah
Number of cells in series	24	No
Number of batteries in parallel	8	No
Total batteries (2 V cell) 500 Ah	189	No
Battery bank voltage (nominal)	**48**	**V**
Battery bank capacity @ C10	**3544**	**Ah**
Battery bank capacity @ 100	4075	Ah
Excess/spare capacity @ C100	**−3558**	**Ah**

Table 7.16 Bill of materials required for a 9.75 kW solar PV system.

Component rating	Quantity
Solar photovoltaic modules: polycrystalline Si 325 W (30 modules)	9750 W
Off-grid solar inverter (PCU): 10 KVA/120 V	1 no.
Solar battery (C10) LMLA 12 V: 150 Ah/12 VDC	20 no.
Array junction box 6 in 1 out	1 no.
AC distribution box: rating—9.75 kW/1 ph	1 no.
DC cables: 1C × 4 sq.mm 1.1 kV, PVC insulated Cu cable, for array connection	55 m
DC cables: 2C × 16 sq.mm 1.1 kV, PVC insulated, UV protected unarmored Cu Cable (from AJB to inverter connection)	15 m
AC cables: 2C × 10 sq.mm 1.1 kV, PVC insulated, UV protected Cu cable (from inverter to ACDB connection)	15 m
DC cables: 1C × 16 sq.mm 1.1 kV, PVC insulated, UV protected Cu cable (for battery connection)	28 m
Module mounting structure: hot dip galvanized/pregalvanized	1 set
MC4 connectors: male and female connectors for cable connections—30 A	7 pairs
Lightning arrestor: 2000 mm long 16 mm dia. with five spikes and base plate made in high grade aluminum	1 set
AC cables: 2C × 35 sq.mm 1.1 kV, PVC insulated, UV protected unarmored Cu cable (from ACDB to main distribution connection)	20 m
Earthing cable: 1C × 4 sq.mm 1.1 kV, PVC insulated, UV protected unarmored Cu cable	30 m
Earthing Cable: 1C × 16 sq.mm 1.1 kV, PVC insulated, UV protected unarmored Cu cable	30 m
Auxiliary items like fastener, cable tray, lugs, conduit pipe	1 set
Reinforced concrete foundation	1 set
Earthing kit: copper bonded rod earthing (dia. 17.2 mm, length 3000 mm, 250 micron copper coating) with 25 kg back filling chenical and clamp	3 sets
Earthing kit: 25 × 3 mm GI strip	30 m

Table 7.17 Costing of a 9.75-kW solar PV system.

Name of the item	Cost in $
Solar photovoltaic modules (30 nos. of 325 W modules)	3120
Inverter (off-grid)	1039
Battery	2532
Balance of system	2796
Total cost of system	9487
Installation cost (10%)	948.7
System cost including permissions + administration (15%)	12,001
	12,001

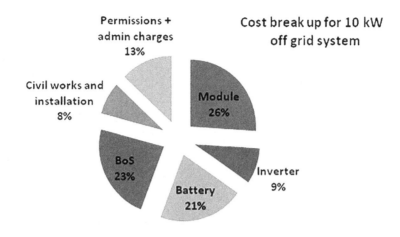

FIG. 7.18

Cost breakdown for a 10 kW off-grid PV system.

7.5 **Reliability of off-grid solar PV systems**

Generally, the main solar PV system is categorized into five subsystems according to their functions: PV modules, charge controllers, inverters, BOS, and battery subsystems. Furthermore, each of these subsystems is divided into subassemblies, as demonstrated in Fig. 7.19.

The reliability and performance of the solar modules depends on the quality of the materials used for solar module fabrication. So, a better quality solar module has to be used to get better reliability for the PV system. The battery has to be properly maintained to get reliable operation of the system and the specified life for the battery.

The solar module problems in the field are discussed in Chapter 8, "Rooftop and BIPV solar PV systems."

During operation, the lead-acid batteries show the different issues that are given below. So the batteries have to be maintained properly.

Acid stratification: In lead-acid batteries, there is a slight difference in density between the water and the acid. If the battery is left idle for a long time, the mixture of water and acid can separate into layers with water rising up and acid sinking down because of gravimetric effects. This can lead to corrosion of the plates at the bottom side. The stratification can be removed by stirring the electrolyte with air pumps or natural gassing of the battery at high voltages.

Sulfation: Sulfation forms during normal operation of the battery. During the discharging process, a thin layer of sulfate forms on the battery plates. The layer dissolves into the battery acid during charging. When a hard crystalline layer is formed, it cannot dissolve during charging. When the sulfate crystals cover the surface area of the plates, it will reduce the battery efficiency by holding less charge. Sulfation occurs if the battery is kept idle for a long time or if the charging is not enough to dissolve the sulfate formed during the charging cycle. Incomplete charging for a long time and a high temperature will also lead to sulfation.

Desulfation can be carried out by equalization, which is the process of overcharging the battery. Desulfation can also be done by pulse conditioning by simply controlling the pulses or frequency components of frequencies ranging from 2 to 6 MHz.

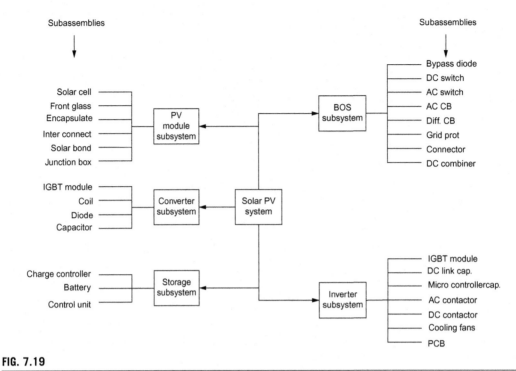

FIG. 7.19

Different components in subassemblies of off-grid solar PV systems.

Corrosion: The application of high positive potential at the positive electrode causes the corrosion of the lead grid. It is an irreversible process and results in the decrease of a cross-section of the grid, which leads to the increase in grid resistance. Further formation of layers of lead oxide and sulfates between the grid and the active material increases the contact resistance, which results in an increased voltage drop during the charging and discharging processes. The factors on which the corrosion depends are the electrode potential, temperature, grid alloy, and quality of the grid. For PV applications, batteries with thicker grids are suitable to minimize the corrosion effect and to increase the lifetime.

Erosion: The electrodes are subjected to strong mechanical loads during cycling due to the conversion of active materials into lead sulfate during discharge. Lead sulfate is 1.94 times larger than lead dioxide in volume per mole. Because of the change in volume, the active material loosens and gets separated from the electrode, forming sludge at the base of the battery. If the sludge volume becomes large, then it may cause a short circuit between the electrodes.

Short circuit: The plate connectors from the positive electrodes can also be subjected to corrosion and cause detachment of small layers of the connectors, which, when falling on the electrodes, will result in a short circuit. To avoid these problems, separators should extend upward over the electrodes. In case of lead-acid batteries, dendrites grow from the positive to the negative electrode through the separators. This growth can be accelerated by a low state of charge for long periods, which leads to low acid concentrations. This dendrite growth causes a microscopic short circuit, which will lead to the sudden and complete breakdown of the battery.

Low temperature: Low temperature will not accelerate any irreversible aging effect. Formation of ice must be prevented. Once ice is formed, then it is difficult to operate the battery and the cell housing may burst due to the increased volume. The surroundings will be affected by the scattering of sulfuric acid.

High temperature: Increase of battery temperature by high ambient temperature or by high current rate charging/discharging increases corrosion, sulfation, gassing, self-discharge, etc. For every 10°K rise of temperature, the lifetime of the battery is reduced by 50%. The normal operating temperature for batteries is 10–20°C.

7.6 Lifecycle cost analysis of off-grid solar PV systems

Lifecycle cost is the cost incurred in owning and operating a solar PV system during its lifetime. The overall cost of the project includes start-up costs, equipment and installation costs, operating and maintenance costs, repair and replacement costs, finance charges, and other nonfinancial benefits. This cost is influenced by several parameters such as capital cost, operating cost, maintenance and repair, replacement cost, other costs, and residual value. The residual value refers to the resale value of the plant at the end of its life.

It is a powerful economic assessment method whereby all costs incurred from owning, operating, maintaining, and finally disposing of a project are considered potentially important for the decision. It can be used to assess whether a project is economically viable and cost-effective. The lifecycle cost analysis (LCCA) usually contains several economic analyses, which are the lifecycle cost (LCC), the levelized cost of energy (LCOE), the net savings (NS), the savings to investment ratio (SIR), the net present value (NPV), the internal rate of return (IRR), and the payback period (PB).

Incentives for project construction and energy generation can also be incorporated.

In fact, it also requires analysts who understand the time value of money when comparing future return flows with the initial investment cost of a project.

Therefore, the values of the discount and the inflation rate play a significant role in determining the time value of money. The time value of money must be taken into consideration because the value of money in the present time will not be the same value in the future. For example, the value of $100 this year will not be the same as the value of $100 in the years to come and in previous years. Discount and inflation rate are the factors that cause the value of the money to vary:

$$\text{Lifecycle cost} = \text{capital cost} + \text{operation and maintenance cost} + \text{replacement cost of equipment}$$

Capital cost refers to the initial investment of power plants such as land, photovoltaic modules, transmission, system design, and installation costs. Operational and maintenance and repair costs include energy, water, operator's pay, inspection, insurance, property taxes, spares, and their replacement costs during the life of the system. The replacement cost is the total cost for the replacement of equipment required during the life of the system. The residual value of the system and other incentives and tax benefits for owning a PV system are considered in the equation of LCC.

For calculating the lifecycle cost, the present worth of future costs, the inflation rate (i) and the discount rate (d) has been considered. The present worth of the investment (X) that has to be made n years later would be lower by a factor called present worth factor for future investments ($Fpw\text{-}one$) [6]

$$Fpw - one = \frac{Futurecost}{Futurevalue} = \frac{X(1+i)^n}{X(1+d)^n} = \left[\frac{1+i}{1+d}\right]^n$$

If the present cost of a product is C_o, then the present worth of a one-time investment down the line of n years is

$$PW\ one = C_o \left[\frac{1+i}{1+d}\right]^n$$

The equation shows how to calculate the lifecycle cost of the system.

The lifecycle cost (LCC) method is used to estimate of the proposed standalone PV system cost. The following table gives the cost of the components of a 9.75 kW off-grid PV system and the lifecycle cost has been calculated in Table 7.18. The off-grid inverter has to be replaced 10 years after the

Table 7.18 Estimation of lifecycle cost for a 9.75 kW off-grid PV system.

Name of the item	Cost in $
Solar photovoltaic modules	3120
Inverter (off-grid)	1039
Battery	2532
Balance of system	2796
Total cost of system	9487
Installation cost (10%)	948.7
System cost including permissions + administration (15%)	12,001
	12,001
O and M cost 3%	360
Inflation rate	4%
Discount rate	8%
	0.963
Cost of inverter (replacement once in 10 years)	712.38
Cost of battery	4876.44
O and M charges for 20 years	4960.29
Capital cost	12,001
Inverter replacement cost after 10 years	712.38
Battery replacement cost once in 7 years	4876.44
O and M charges for 20 years	4960.29
Lifecycle cost	22,550

installation of the project. The replacement of the inverter is considered once in the system life of 20 years. The battery bank has to be replaced twice in the life of 20 years.

The lifecycle cost of a 10 kW off-grid residential solar PV system is $22,550.

References

[1] Solar Module Data sheets of reputed solar module manufacturing company.
[2] Risen Energy 405W data sheet.
[3] PV Education, https://www.pveducation.org/pvcdrom/battery-characteristics/summary-and-comparison-of-battery-characteristics.
[4] Battery University, https://batteryuniversity.com/.
[5] O.C. Otumdi, C. Kalu, I. Markson, Determination of loss of load probability for standalone photovoltaic power system, Eng. Phys. 2 (1) (2017) 7–12.
[6] A. Abd El-Shafy, Nafeh design and economic analysis of a standalone PV system to electrify a remote area household in Egypt, Open Renew. Energ. J. 2 (2009) 33–37.

Rooftop and BIPV solar PV systems

8.1 Introduction

There has been an increase in the demand for renewable energy-based systems worldwide to combat global warming issues arising out of fossil fuel-based energy generation systems. An increasing population and the depletion of fossil fuel reserves are paving the way for solar PV-based energy generation systems. Different governments across the world are proposing subsidies and incentives as well as buying power with certain tariff rates from the people to encourage setting up of solar PV plants. There is a shortage of land to implement the proposed gigawatt-scale solar PV plants by the governments of different countries. The electricity tariff rates are steeply increasing for commercial and industrial establishments. Hence, there is an increased interest in rooftop solar PV plants.

Grid-connected solar PV systems in the world account for about 99% of the installed capacity compared to off-grid solar systems, which use batteries [1]. Solar rooftop PV systems are cost-effective and easy to maintain. The generated power is directly supplied to loads in real time and the excess power is exported to the utility. It enables system developers to reduce their electricity bills.

8.2 Rooftop solar PV systems

The solar PV systems that can be installed on the rooftops of buildings are called solar rooftop PV systems. The roof areas can be used to mount solar PV modules and other components. The roofs may be of reinforced cement concrete (RCC) or industrial sheds with metallic sheets or tiled roofs. The PV systems can be installed on the rooftops of residential, commercial, and industrial buildings. These rooftop solar PV systems are of two types: (i) grid-connected solar rooftop PV system without a battery, and (ii) grid-connected solar rooftop PV system with storage using a battery.

In grid-connected solar PV systems, the solar modules mounted on the rooftops of buildings/sheds generate DC power. The DC power is given as an input to the inverter. The inverter converts the DC power to single/three-phase AC power of 220 or 440 V. The inverter output is connected to the grid and will be in synchronization with the grid voltage and frequency. The AC power is directly used to run the loads and the excess power is injected into the grid. For industrial applications, if the power requirement is 11 or 33 kV, then the inverter output will be stepped up to 11 or 33 kV using a transformer and injected into the grid. In case there is a grid outage, the solar PV system will not supply power to the loads, as there will be no reference voltage to the inverter from the grid.

Solar PV Power. https://doi.org/10.1016/B978-0-12-817626-9.00008-3

Solar PV systems can be installed on buildings to support power for residential and industrial applications. To set up a distributed energy generation source, the rooftops of the buildings can be used to install solar PV systems. On the rooftops of buildings, grid-tied systems as well as grid-connected systems with battery backup can be installed. For any country, the availability of roof areas of residential, commercial, and industrial buildings will help to reduce the land requirement for gigawatt-level addition of solar capacities.

For consumers, rooftop-based grid-connected solar PV systems can be coupled with batteries for energy storage to reduce the dependency on grid power and mitigate diesel generator dependency. The grid-tied system with battery backup acts as a long-term reliable power source with a lower cost without consuming any fuel. As the cost of solar-generated power is getting lower to the cost of commercial power, the rooftop system is most suitable for commercial establishments to get maximum generation during sun shine hours.

Solar PV systems can be set up on pitched roofs, RCC-based rooftops, industrial roofs with metallic sheets, and on car parking structures. Grid-connected power packs in the range of 1–50 plus kWp can be used for residential applications. Grid-connected power systems in the range of 10–100 kWp or more can be used for commercial applications. The buildings with RCC roofs or metallic sheets can be utilized to set up solar PV systems in the range of 100 kWp–5 MWp for industrial applications. The grid-connected rooftop system can supply excess power to electricity distribution agencies and thus can work on a net metering basis, wherein the beneficiary pays the utility on the net meter reading basis only. As an alternate, a bi-directional meter can also be installed to measure the export and import of power. The mechanism based on gross metering at a mutually agreed tariff can also be adopted.

8.3 Components of grid-tied rooftop solar PV systems

Fig. 8.1 illustrates a typical schematic of a grid-tied rooftop solar PV system without a battery.

Solar modules are connected in series to form a string. By connecting the solar PV modules in series, the voltages are added up to give a higher DC voltage. The module strings are combined in an array junction box (AJB) for some inverters. The AJB will have a fuse for protection as well as a surge protection device to suppress any transient voltages arising out of lightning. The output of the AJB is connected to the MPPTs of the inverter. The generated solar array power is passed through the AJB and is converted to AC power by the inverter. If more than one inverter is used in the system, the output of the inverters will be combined in an AC distribution board (ACDB) and the output will be fed to the LT panel of the building through a bidirectional net metering system. If there is a grid outage, the inverter shuts down and there will not be any supply of power to the load. The inverter will have all protections. Nowadays, it has an integrated string combiner facility in its input.

Solar PV modules, string combiner boxes, inverters, inverter combiner boxes, ACDB/LT panels, DC and AC cables, a net metering system, module mounting structures with foundations, an earthing system, and lightning arresters are the components used in a grid-tied solar PV system without battery backup. Other than solar PV modules and inverters, the components such as cables, switchboards, junction boxes, meters, module mounting structures, earthing systems, circuit breakers, fuses, etc., are called the balance of system (BoS).

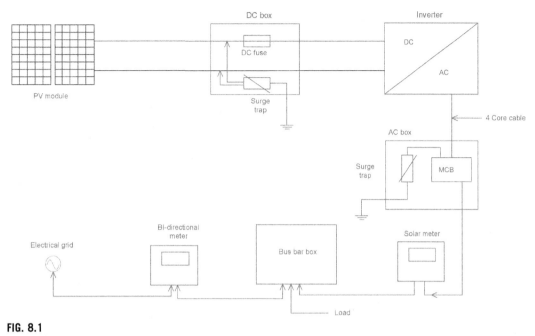

FIG. 8.1

Schematic of a grid-tied solar PV system.

Some aspects of inverters are discussed below, as the other components are discussed in Chapter 7.

8.3.1 Inverters

An inverter is the heart of the solar PV system. It converts solar array-generated DC power into AC power in synchronization with the grid. The inverter is also known as a power conditioning unit (PCU), as it has additional functions such as maximum power point tracking, grid monitoring, and antiisland-ing protection in addition to DC to AC inversion. Solid-state switching circuits are responsible for DC to AC conversion. The inverter contains magnetic circuits, capacitors, AC and DC disconnects, electromagnetic and radio frequency interference filtering equipment, a cooling system, a ground fault detection interruption circuit, and command and control and communication components to reliably feed high-quality AC power into the grid in a safe manner.

The method by which DC power from the PV array is converted to AC power is known as inversion. The physical principle involved in the conversion of DC current into AC current is electromagnetic induction. As per electromagnetic induction, if a conductor is exposed to a varying magnetic field, an electromotive force is induced in the conductor. The example is a cycle dynamo in which a rotating magnet induces electric current and voltage in a coil.

If there are two coils, one is considered as primary coil and the other as secondary coil. If the direction of current is continuously changing in the primary coil, the alternating magnetic field in the coil will induce AC current in the secondary coil. The DC current is made to flow through switches that are on/off with certain frequencies and times using a pulse width modulation technique creates

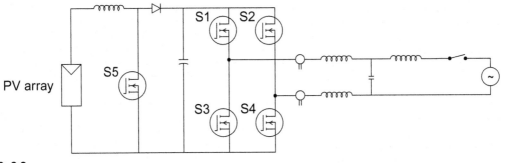

FIG. 8.2

Schematic of an inverter with an H-bridge.

Courtesy: Anil Kumar, Suryalogix.

square wave. The square wave is modified to a pure sine wave using digital signal processors. In an inverter, DC power from the PV array is inverted to AC power via a set of solid-state switches—MOSFETs (metal oxide field effect transistor) or IGBTs (insulated gate bipolar transistor)—that essentially flip the DC power back and forth, creating AC power [2]. Fig 8.2 shows a basic H-bridge operation in a single-phase inverter [3].

The solar array generates variable output such as current and voltage depending on the irradiance and temperature of the site. The input circuit of inverter consisting of the inductance, IGBT switch and diode tracks the MPPT of solar array and boosts to high voltage. When the IGBT switch S5 is closed, current flows through the inductor in the direction of the bottom side in Fig. 8.2. When the switch opens, the flow of current stops through the switch and the voltage is boosted to higher value. Thus under PWM control, by turning IGBT switch on and off, the voltage is boosted. There are four power semiconductor switches. In the first half cycle, the switches S1 and S4 are closed and the DC voltage is inverted to positive. In the second half cycle, the switches S2 and S3 are closed and the DC voltage is inverted to negative, creating a rectangular AC waveform. Thus, using PWM, the H-bridge converts the high voltage DC from the boost stage to a high-frequency AC signal. In the final stage, the inductors filter out the high-frequency switching cycles to create a smooth, grid quality sine wave with a voltage of 240 V AC using digital signal processing techniques. Both the input and output circuits of the inverter are protected with fuses or circuit breakers and are accessible. Using a surge protection device on the inverter input to protect against nearby lightning strikes or any other transient voltage spikes is recommended for most areas. The inverter manufacturers provide the features in the inverter such as DC and AC overvoltage protection, DC and AC short circuit protection, out of voltage range protection, and DC inverse polarity protection. Also, the inverter will have DC fuses at the input, an AC output breaker, a ground fault monitoring device, insulation monitoring, HVRT (high voltage ride through), LVRT (low voltage ride through), and antiislanding features for reliable and safe performance.

8.3.1.1 Single phase/three-phase string inverters

Solar modules are connected in series to form a string, which is connected to the inverter. As the solar module strings are connected to the inverter, they are called string inverters. Solar inverters convert solar array-generated DC power into AC power, synchronizing with the utility grid. Grid-connected inverters are classified into three main categories: module-level inverters, string inverters, and central inverters.

Module-level inverters (matches with the rating of the solar PV module, say 240–425 Wp) that are integrated with the solar PV module are the easiest to define because one inverter is dedicated to a single solar PV module.

A central inverter (typically 100 kVA to several MVA) is a grid-direct inverter with multiple parallel strings connected into a single MPPT circuit. Central inverters are used for ground-mounted grid-connected solar PV power plants.

The term "string inverter" by definition is an inverter that is dedicated to a single string of modules. It is understood that the term applies to smaller capacity inverters (between 10 and 200 kVA) with a single-phase or three-phase AC output. While all have a high DC input voltage, they are dedicated exclusively to a single series string input. More strings are added in parallel to the MPPT of the inverter to get the required current. String inverters are used in both single-phase and three-phase applications. The most common use for grid-direct string inverters is in single-phase applications, which are primarily residential installations. String inverters not only convert DC to AC power, but also provide antiislanding protection, AC leakage current fault protection, DC reverse-polarity protection, AC surge protection, DC surge protection with a metal oxide varistor, DC overvoltage protection, ground fault protection, and IP-65 ingress protection. Inverters are available in the wattage range from 1 to 200 kW with peak efficiency around 97.2%, MPPT efficiency >99.5%, THD <3%, power factor 0.8 leading to 0.8 lagging, transformer-less type, with a night standby power consumption of <2 W. The inverters produce a noise level <40 db under full load operation and provide an RS 485/232/WIFI/GPS/Ethernet or an LAN-based communication interface for remote monitoring.

In recent times, the power capacity of inverters has grown larger, but for low voltage 400/415 V on-grid inverters, the maximum capacity could reach up to 200 kVA. Also, the inverter has to manage more and more devices such as solar PV modules, DC cables, and even AC distribution boards and grids. Further, the communication functions have become more powerful. An inverter needs multiple USB, RS485, and RS232 ports to connect with the computer, the flash disk, the data logger, the meter, and the current transformer. The power capacity of the inverter has been increasing from 10 to 200 kVA for residential systems. The increasing power not only brings a larger current, but also the complexity of the control algorithm. Consider for example an 185 kVA inverter with 12 MPPTs. The inverter needs to control 12 different circuits and the controlling difficulty increased by six times compared to single MPPT based inverter. The present day inverter has more functions such as string monitoring, I/V curve diagnosis, AFCI (arc fault current interrupter) detection, panel PID healing, grid error recording, power factor adjustment, and grid harmonics adjustment.

If there are roofs with multiple orientations for solar panel mounting, multiple inverters with more MPPTs are required. If the roof installation needs to use solar modules of different technologies, multiple inverters are required. Multiple inverters are to be planned if the system capacity is extended over time or instances where the array capacity exceeds the capacity for any one single-phase string inverter. In general, with the growth of PV markets, larger and more complex systems are becoming common, even in residential applications. A certain number of these complex projects will always require multiple inverters.

The schematic block diagram of a typical three-phase inverter is shown in Fig. 8.3.

The variable output voltage of the solar array is connected to the input of the inverter. The inverter will have a DC fuse as protection in the input side. The voltage is made to pass through an EMI filter to remove the ripples in the current. The output is fed into the MPPT circuit of the inverter. The MPPT tracks the maximum power and boosts it to higher voltage. The voltage source inverter with a boost converter in the DC link covers the whole voltage range of the solar array under different environmental

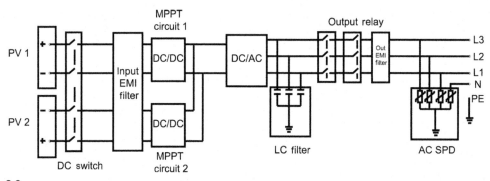

FIG. 8.3

Schematic block diagram of a three-phase inverter.

conditions. The inversion circuit using semiconductor switches in the H-bridge configuration provides a square wave. With the use of a digital signal processor, the square wave is modulated to a pure sine wave.

The following are the advantages with string inverters, as compared to central inverters.

There is a design flexibility to optimize the ratio between the inverter's DC input capacity and the size of the PV array with the use of string inverters with a distributed system approach. An optimized inverter-to-array ratio can be obtained with string inverters using long string lengths.

Central inverters use separate DC string combiner boxes to combine and connect the PV source circuits to the inverter. The use of string combiner/array junction boxes with a central inverter drives up the cable sizes and the associated costs. The use of string inverters with integrated junction boxes eliminates the need for separate DC source-circuit combiners [4]. Hence, there is a decreased DC cable cost. But, if multiple inverters are used, AC junction boxes are required to be used to combine the output of the inverters.

With the use of multiple string inverters in a distributed design approach, there is an increase in the number of inverter-direct monitoring points. The system with a central inverter design uses zone- or string-level monitoring. Hence, there is cost-effective granular monitoring of the system with the use of string inverters.

For central inverter design, a large control room or shelter is required to protect the inverter from harsh weather conditions. This consumes extra space and adds cost to the system. There are reduced space and infrastructure requirements with the use of string inverters, as they can be mounted on an existing wall, a module mounting structure, or with a localized shelter.

If a string inverter fails, the energy loss is equal to the capacity of the inverter, and it can be replaced immediately. The loss of energy generation is much less during the period of string inverter's failure, compared with power loss with the failure of the central inverter. In the event of any failure, with the easy replacement of the string inverter, system uptime improves, compared to central inverters as repairing or replacing a central inverter takes a longer time. There is an increased design flexibility with the use of string inverters in the distributed system design approach. As each string inverter employs one or more MPPT circuits, the string inverters can be deployed in adverse site conditions, such as localized shading, variable array orientations, and different pitches of module mounting structures.

String inverter designs are increasing in popularity, even for large-capacity ground-mounted projects. With central inverter design, a large number of strings are connected to a single MPPT, whereas string inverters have multiple MPPTs and a maximum of six strings can be connected to one MPPT of the inverter. So, there is less mismatch in the strings and there will be increased energy generation with the use of string inverters compared to central inverters. The increased design flexibility, more MPPT points, simpler operation and maintenance, and a better financial outcome such as lower installation and BoS costs have led to the system design with string inverters becoming popular.

String inverters have a limited power range and large solar power systems need several such inverters. The grid-connected string inverters are cost-effective and are widely used in residential, commercial, and industrial applications of solar PV systems. But, they have some drawbacks. When a part of a module in a string is shaded, the shaded panel output will decide the power output of the string and there is a power loss in the system. So, even lightly shaded areas cannot be used for installing solar modules and hence the building area cannot be effectively utilized for the PV system. For the efficient working of the inverter, all strings of an MPPT will have the same number of solar modules, all the solar modules will be mounted at the same tilt angle, and all the modules will have a similar technology. The modules in a string will have the same number, the same facing tilt angle, and the same technology. It is quite challenging to design a rooftop PV system for a small building with multiple roof orientations with a string inverter. If all roofs of different orientations are to be used, extra inverters are required for each roof; this may not be a cost-effective solution. If the area of a specific oriented roof is not sufficient to accommodate all the modules to form a string, the string inverter may not be useful.

There is a safety issue with traditional string inverter PV systems. During an emergency, the utility grid can be shut down, but not the operation of the solar modules that are connected to the inverter. The solar panels keep producing energy as long as sunshine is available. During a grid power cut, while AC power is not available on the grid side, the solar array will be producing power (during daylight hours) and the cables from the rooftop to the inverter will be energized. To sort out this problem, a rapid shutdown of the solar PV array will have to be incorporated into the system.

8.3.1.2 Microinverters

A microinverter is an alternative solution for the string inverter. There are two types of microinverters for solar PV system applications. One type directly converts DC power to AC power in the module level itself. The other type boosts the DC voltage in the module level using a DC optimizer and connects to the inverter. Microinverter-based options are called module level power electronics. Each module is connected in these systems to an inverter that is usually connected just below the panel. Each panel with one inverter prevents the dependence of the generation of one module on other modules. Each panel is going to be independent. This type of system is best suited to a place where shadow problems occur on one or more number of modules, roofs with different directions, etc. In these types of systems, module level monitoring is possible. It helps in the easy maintenance of the system. One module or one inverter failure wouldn't affect the generation of the rest of the system.

The electrical parameters of each module can be monitored and the data can be sent to a database center where the performance details can be monitored and analyzed. It is of great benefit for the operation and maintenance of solar plants. These microinverters are more efficient and produce more power compared to standard string inverters. Though they are expensive, they have better durability and a longer life.

Microinverter—AC module

A microinverter is a device that is used in a solar PV system to convert DC power generated by a solar module to AC using power converter topologies. The function of one big inverter is split into many inverters. For every solar PV module, one inverter is connected on its rear side and it converts the solar module-generated DC power into AC power directly; this is called the AC module. The schematic of the PV system with microinverters is shown in Fig. 8.4.

The output from each AC module is directly and independently connected to the AC combiner box. Thus, several microinverters can be combined together and fed into the electric grid. When one panel is affected by shading, soiling, orientation, or current mismatch in modules, its performance will not affect the performance of other modules [5]. For every module, the right part of the maximum power is obtained by the MPPT and converted to the high voltage of AC. The configuration of the system with a microinverter is flexible. Solar panels that are installed at different tilt angles can be connected to an inverter and different types of panels with different technologies can be connected to different microinverters. When there is a grid outage, the microinverter immediately shuts down and stop supplying power to the cables of the system. Though it is expensive compared to a standard string inverter, the microinverter is gaining popularity for residential PV systems. The Enphase microinverter is shown in Fig. 8.5. The holes can be used to mount the rear side of the module using the bolts.

In the event of grid failure, it is required that any independent power-producing inverters attached to the grid turn off in a short period of time. Thus, it has an islanding feature to prevent the inverters from continuing to feed power into small sections of the grid. It has rapid shut down features also. It has a Class II double-insulated polymeric enclosure. It complies with advanced grid support, and voltage and frequency ride-through requirements. The data can be monitored remotely and updated to respond to changing grid requirements. It has an EU efficiency of >96%. It has an MPPT range of 16–62 V

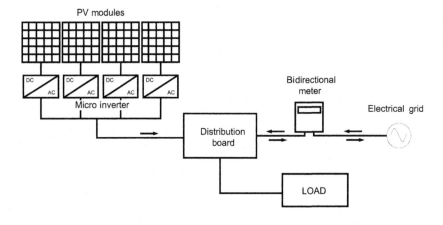

FIG. 8.4

Schematic of the solar PV system with microinverters.

FIG. 8.5

Enphase microinverter—rear view.

suitable for 60 or 72 cell-based solar PV modules. It has an adjustable power factor for 0.7 leading to 0.7 lagging. The output voltage is 230 V.

DC optimizers and inverters

In a series connected solar array string, the voltages are added up at constant current. If the current of each module in the string is same, the output of the string gives same current. However, under certain conditions, such as shadow some modules generate a smaller amount of current. The lower current generating module will pull down the current of other modules and affects the performance of the entire string. The *PV* system output power will be reduced as a consequence of mismatch effects and environmental factors. Mismatch in current of different modules occurs due to passing clouds over the plant, power measurement in accuracy of modules, partial shading of modules, uneven soiling, differences in the orientations and inclinations of solar surfaces, differences in temperature, or irradiance in the modules and variable degradation. Different voltage drops in the strings and accumulated wear and tear cause mismatch in voltage. There are different solutions available in the market to take care of current mismatch effects of solar modules in a string. Multiple MPPT-based string inverters, AC modules that are called microinverters and DC-DC optimizers are available with different system architectures and topologies to solve the mismatch effects of the string and PV system.

A DC optimizer with an inverter option is more affordable compared to an AC module-based microinverter. This option divides the standard inverter into two parts. One is a DC optimizer and the other is a power inverter. The power optimizer is connected to one or two panels and can be mounted on the rear side of the panels, turning them into an intelligent/smart solar PV module(s). These optimizers track the MPPT at lower voltages. This helps in power generation, even in low-light and intense shadow conditions. Each of these optimizers sends the DC output to the string inverter at the end, after optimizing the output of each module individually. The optimizer has built-in device to provide the module-level monitoring of electrical parameters. Due to this, the power optimizer maximizes energy by real-time monitoring and adjusting of the *I-V* curve of the module as per the requirements. It enables much larger strings and different types of solar panels can be added. During a power cut, the optimizer reduces the voltage of each panel to a minimum voltage, enabling safety.

Fig. 8.6 shows a schematic of a PV system with DC optimizers and the inverter. A DC optimizer is a DC/DC converter, which is a power electronics device connected to solar modules to mitigate the mismatch effects and harvest the maximum energy yield. The PV module output is connected to the input of the optimizer and many optimizers are connected in series as per the design requirement and the output is connected to the inverter. Modules affected by shading and modules mounted in different orientations can be connected in a string with the attachment of DC optimizer. Flexible design and longer strings enable to go for higher DC-to-AC ratio and reduced BoS cost such as in cable length and cable sizing achievable with the usage of optimizers.

Fig. 8.7 shows the Huawei DC optimizer and schematic of the connection of modules with optimizers to inverter.

Solar Edge, Tigo Energy, Huawei, and Maxim Energy manufacture DC optimizers. The DC optimizer, which is connected to one or two solar modules, allows each or a couple of modules to work more independently of the other modules in the string connected to the inverter. Each brand of DC optimizer does the optimization slightly differently.

The Solar Edge DC optimizer. The solar edge power optimizer is a DC-DC buck/boost converter with MPPT controller. It uses an input control loop, and maximizes the energy using a module-based MPPT. One or two solar modules are connected to DC optimizer depending on its wattage and other electrical parameters. All the optimizers fitted with modules are connected in series to form a string and multiple strings are combined and connected in parallel in a junction box. The output of junction box is connected to the solar edge inverter. The solar edge inverter is a single-stage current source and it doesn't have MPPT controller. The input voltage to the inverter is fixed. The output voltage of the string should be maintained constant as per the fixed input voltage of inverter, whatever the conditions of weather and shadowing on the modules [6].

Each power optimizer works independently and enables the inverter to automatically maintain a fixed string voltage at the optimal point for DC-AC conversion by the inverter. It enables performance monitoring of each module. The input voltage to the inverter is controlled by a separate feedback loop.

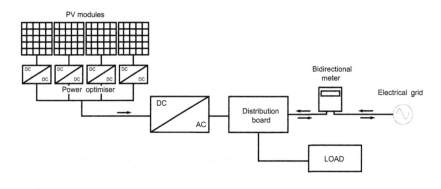

FIG. 8.6

Schematic of a PV system with DC optimizers and the inverter.

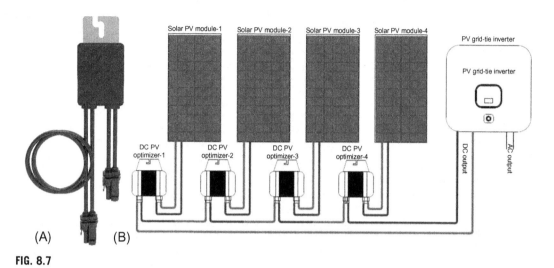

FIG. 8.7

(A) DC optimizer of Huawei. (B) Schematic connection of modules with DC optimizer and to inverter.

((A) Courtesy: Huawei.)

It boosts the voltage to a higher value if the input voltage is lower or lowers the voltage as per the inverter input voltage requirements. The optimizer continuously adjusts the voltage and current such that the sum of the voltages of all the optimizers should be constant in any condition. There is no MPPT circuit in the inverter and it continuously adapts the current it draws from the PV array in order to keep the input voltage constant. So, the optimizer has to modify the voltage and current according to the requirements of the inverter. The solar edge power optimizer is highly efficient, maintaining more than 98.8% weighted conversion efficiency over a wide range of conditions.

The power output of each solar module is maintained at the module's maximum power point by an input control loop within the corresponding power optimizer. The input voltage to the inverter is controlled by a separate feedback loop. Each module operates at its maximum power point, regardless of operating conditions. In optimizers, both up and down DC/DC conversion is automatically used, depending on environmental conditions. In solar edge optimizer, the voltage continuously decreases and increases to modify the current and voltage as per the requirements of the inverter voltage.

Solar Edge optimizers cannot be used with any other make of inverter. They have to be exclusively used with Solar Edge inverters. During normal operation conditions, the optimizers are forced to continuously modify the current voltage throughout the day to achieve the maximum power point as per the input requirement of the inverter.

Tigo DC optimizers. The Tigo Energy optimizers are also DC/DC converters. The Tigo Energy optimizer operates on the principle of impedance matching. It gives more independence to each solar module to work and perform. The power output of each module doesn't depend on the power of the other modules connected in a string to an inverter.

The optimizer can be mounted on the rear side of the solar panel using the bolts. Every solar panel has bypass diodes connected in reverse bias condition to protect the module from hotspots in the event of shading. Once a portion of the cell or cells in a module is shadowed, the bypass diode gets forward biased and pulls the power of the unshadowed panels in the string. The Tigo optimizer continuously tracks the current and voltage of the module. When the solar module gets shadowed, the optimizer senses the impedance caused by the shadow on the panel and opens a bypass tunnel to match the mismatched current, even before the activation of the bypass diode connected to that part of the cell in the panel [7]. This allows the optimizer to work at a higher current without affecting the solar panel bypass diode. The solar module with an optimizer prevents a shadowed panel from pulling down the power of the other panels, which are connected in series. To prevent the inverter from turning off from undervoltage, the optimizer/inverter reduces the current so the bypass diodes did not get activated. This means that each panel still produces higher voltage but with low current.

The shaded panels connected with optimizers in a string produce low current compared to unshadowed panels and perform poorly. The unshadowed solar panels connected in string produce higher current, but the shadowed panel will not allow it to pass through it. If the bypass diode is activated, it bypasses all the current by operating in forward bias condition. The Tigo optimizer connected to the shaded panel redirects the current generated by unshadowed panels to a bypass tunnel. Tigo optimizers are only required to work when there is a mismatch of current among the modules due to the shading of one or more modules or a crack in one of the modules. Even the partially shaded modules provide partial production with the usage of optimizers. With the use of Tigo optimizers, any brand of inverter with an MPPT circuit can be used. The inverter harvests the maximum power from all the optimizers using maximum power point tracking. The Tigo optimizers get off easily during normal conditions and are only required to work when their panel is impeding the other panels. If there are one or two panels that will ever be in the shade, those can be connected with optimizers selectively. Optimizers not only increase solar production, they also prevent bypass diodes from burning out. Tigo optimizers perform better in heavy shade, so they can be selectively deployed and are only required to work during shaded times. Tigo's optimizer has monitoring technology that is designed to reduce operational and maintenance costs by remotely detecting and diagnosing performance issues [8].

Huawei DC optimizer. Huawei DC optimizer works in buck mode or in bypass mode. It consists of a buck circuit that steps down the input voltage to lower values and steps up the current from input to output. It interacts with the inverter while tracking the MPPT of the module. In bypass mode, i.e., during normal operation of nonshadowing conditions, it delivers the same voltage and current of the module generated by MPPT tracking.

The DC optimizer tracks maximum power and captures maximum energy yield. It shuts down the module voltage to a safer one during times of fire. It enables more precise module-level operation via a physical-level view in the data monitoring system.

The Huawei solar optimizer only bucks the voltage of poorer performing panels in order to increase the current of the shadowed module equal to higher performing panels. The Huawei inverter then boosts the voltage as per the requirement. The function of an optimizer is to adjust the panel voltage and current depending on the sunshine and shading so that the current in each module is the same. This allows achieving maximum power from the system. The conceptual diagram of a DC optimizer is shown in Fig. 8.8 [9].

The input port of the optimizer connects to the PV module. Through a DC/DC circuit, the PV input voltage is converted into a required voltage and the maximum power point tracking (MPPT) function is implemented. The communication circuit is used for communication, and the escape circuit enables the

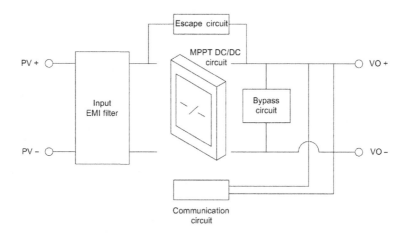

FIG. 8.8

Conceptual diagram of a Huewei DC optimizer operation.

Courtesy: Huewei.

optimizer to bypass itself due to a fault. The operating principles of the Huawei DC optimizer system are illustrated in the following example, and this has been taken from the website of Huewei [10].

Consider an example of a PV system that has a string with 10 number of 350 W modules and there is no current mismatch within the modules. Assume that the solar module generates 282 W power at certain temperature and irradiance conditions. The I_{mp} and V_{mp} values of the module are 8.7 A and 32.4 V, respectively. As 10 modules are connected in series the voltage will be added up and gives 324 V for the string. As there is no mismatch in current, all the modules are operating at 32.4 V giving 8.7 A. So the output power of the string is 10 × 32.4 × 8.7, which is 2820 W as shown in Fig. 8.9.

If one cell of one of the modules is shadowed, one of the bypass diodes gets activated and the current and voltages are reduced to 8.2 A and 15.3 V, respectively, providing 125 W. The shadowed low performing module pulls down the current and voltage of the un-shadowed modules to the values of 8.2 A and 33.1 V, so that the operating currents of all the modules are same. The combined characteristics of 1 shadowed module and 9 number of well-performing modules give a power of 271 × 9 + 125 is 2564 W. There is a loss of 9% power. The various system currents and voltages in this case of without usage of DC optimizers illustrated in Fig. 8.9.

Assume that each module has a power optimizer connected to it and the power optimizers are serially connected to form a string. Multiple of such strings can be connected in parallel to the same input of the inverter.

Assume that one of the modules of the string is shadowed and I_{mp} and V_{mp} are 7.9 A and 19.4 V, i.e., 153 W. The other well-performing modules generate 8.7 A. To match that current the power optimizer of shadowed module step down the voltage of the module to 153/8.7, i.e., 17.6 V. The optimizers of well-performing modules provide output current and voltage of 8.7 A and 32.4 V, respectively. The combined voltage, power and current of the string at the input of the inverter are (9 × 32.4 + 17.6 = 309.2 V), (9 × 282+153 = 2691 W), 8.7 A, respectively. There is a loss of 4.57% power. The various system currents and voltages with optimizers connected to modules illustrated in Fig. 8.10. The optimizer performs the functions of MPPT, module-level shutdown, module-level monitoring,

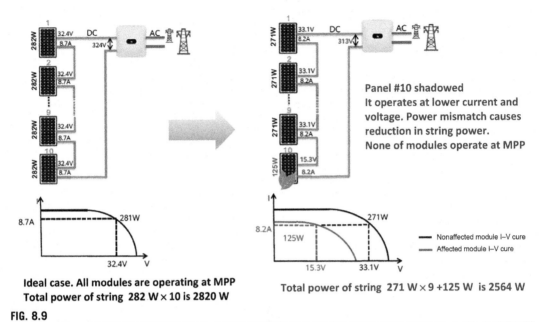

Panel #10 shadowed
It operates at lower current and
voltage. Power mismatch causes
reduction in string power.
None of modules operate at MPP

Ideal case. All modules are operating at MPP
Total power of string 282 W × 10 is 2820 W

Total power of string 271 W × 9 +125 W is 2564 W

FIG. 8.9

Illustration of various system currents without DC optimizers in ideal case condition.

(Courtesy: Huawei.)

automatic positioning, and *I-V* curve diagnosis. The optimizer and inverter communicate with each other over the smart PV safety box. It can be selectively used in the system for the panels wherever shadows are there. These optimizers can be connected with the inverter of any make having an MPPT circuit.

8.4 Building integrated PV (BIPV) systems

There are two basic ways of using photovoltaics (PV) in buildings. One is by adding PV on the envelope, which is called BAPV (building added photovoltaics), and the other is by substituting the parts of the building envelope using PV, which is called BIPV (building integrated photovoltaics). Solar PV modules are mounted on the elements of the building that act as part of the building structure; the system is called BIPV. The solar modules are integrated in building elements or material and function not only as energy generators but also as part of the building structure. They replace conventional building elements such as roof tiles, roof membranes, facade cladding, or facade or skylight glazing, rather than attaching to the roof or the façade.

In the case of BAPV, standard PV modules can be normally used and added on buildings by means of a suited mounting system. They are assembled when erection of a building is already completed. They perform the basic function of electric power generation by converting the incident solar radiation. They can be dismantled at any moment; the building integrity and reliability will not be affected. On the contrary, a photovoltaic module used in the substitution of traditional elements of the building envelope

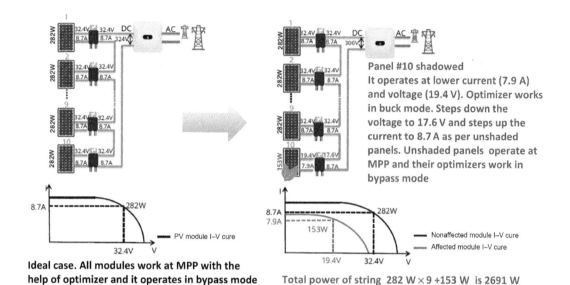

Panel #10 shadowed
It operates at lower current (7.9 A) and voltage (19.4 V). Optimizer works in buck mode. Steps down the voltage to 17.6 V and steps up the current to 8.7 A as per unshaded panels. Unshaded panels operate at MPP and their optimizers work in bypass mode

Ideal case. All modules work at MPP with the help of optimizer and it operates in bypass mode
Total power of string 282 W × 10 is 2820 W

Total power of string 282 W × 9 +153 W is 2691 W

FIG. 8.10

Illustration of various system currents and voltages with the usage of DC optimizers in the case of one module in shadowed condition.

(Courtesy: Huawei.)

(BIPV) has to ensure all the functions of the replaced element. It means that a BIPV module not only generates electricity, but also adds functionality to the building element. In Figs. 8.11 and 8.12, solar modules are integrated with the building and function as a building element.

8.4.1 Functions and requirements of BIPV systems

- Protection from intrusion, rain, wind, and noise to the insiders of the building.
- Regulation of users' comfort while reducing the use of nonrenewable energies for heating, cooling, and lighting to the minimum.
- Works as building envelope material and power generator. Can be interfaced to utility grid or standalone, off-grid systems.
- Provides weather protection such as insulation from winter cold and excessive summer, sun protection, thermal insulation, noise protection, and safety.
- Achieve sustainability and green goals.
- Improve thermal and acoustic insulation.
- Saves environment and reduce CO_2 emissions.
- Maintain the illumination level with BIPV facades.
  - In addition to the functional compatibility, it is important to ensure that the new multifunctional envelope system meets all building construction standards.

FIG. 8.11

Example for BIPV.

FIG. 8.12

Example of BIPV for an atrium.

○ The load of BIPV panels and their associated mounting structure should be correctly transferred to the load-bearing structure through appropriate fixing.

○ The BIPV system should withstand fire and weather wear and tear.

○ It should resist wind load and impact, and should be safe in case of damage.

○ The fixing of BIPV modules should avoid thermal bridges and the global thermal emittance value of the wall should not be negatively affected.

○ Vapor transfer through the wall should avoid condensation layers, and allow the wall to dry correctly.

○ The electric cabling of the PV should be studied to avoid shock hazards and short circuits, and measures should be taken to avoid fire.

○ Envelope materials in contact with the solar modules should withstand their high working temperature.

○ Fixing details and jointing should make BIPV panel material expansions compatible with those of the other envelope materials.

8.4.2 Advantages of BIPV and technology

Besides harvesting energy, well-integrated PV modules are suitable to contribute to the comfort of the building: they serve as weather protection, heat insulation, shading modulation, noise protection, thermal isolation, electromagnetic shielding, etc. Installation cost reductions in the building take place due to lower nonmodule costs by the elimination of racking hardware and the greater use of traditional roofing labor and installation methods. There are cost offsets for displacing traditional building materials due to lower supply chain costs. The other benefits of BIPV are improved aesthetics of the building, reduction in the thermal gain in facades, light and thermal management, effective shading, glare protection, innovative architecture, acoustic insulation, and replacement of claddings.

8.4.2.1 Where can BIPV be installed?

The solar module is the main component of the BIPV system. These can be mounted on building surfaces as an integrated material or elements of a building. Modules shaped like multiple roof tiles and shingles incorporating a flexible thin film cell can be mounted on pitched roofs. These can extend the normal life of the roof by protecting the insulation and membranes from ultraviolet rays and water degradation. BIPV can be integrated to the facade, skylights, railings, and windows of the building. Glass-based BIPV systems can be mounted on facades, window panes, cladding, curtain walls, skylights, atria, and the spandrels of the building.

Nonglass-based BIPV can be used as overlays and rigid and flexible materials as an alternative to roofing tiles. Nonglass-based walling can be used for wall attached, dedicated BIPV siding, and refurbished BIPV roofing. Solar tiles and shingles provide an aesthetic solution, mainly to residential pitched roofs. The flexible laminates cannot replace the function of the building element but work as BAPV. The flexible laminates have low efficiency and can be used for flat and curved roofs for commercial and industrial buildings. Fig. 8.13 gives the BIPV system with flexible modules. These modules cannot act as structural elements of the building.

FIG. 8.13

BIPV system with flexible modules.

The integration of solar panels with buildings can be done in common areas such as canopies in parks, shading at bus stands, and over parking lots, pedestrian pathways, skywalks, and footbridges.

8.4.2.2 BIPV market

The global BIPV market is estimated to reach revenues of around $7 billion by 2024, growing at a CAGR of approximately 15% during 2018–2024 [10]. The BIPV solutions are driven by the designing and planning of building structures with integrated and multifunctional facades and roofing PV panels, fulfilling technical, legal, and architectural demands. The growth of the BIPV market is fueled by a growing focus on self-sufficiency, energy efficiency, and thermal insulation. The rooftop PV market growth is in residential, commercial, and industrial buildings. The crystalline silicon segment occupied more than half of the market share in 2018, growing at a CAGR of approximately 17% during the forecast period. The BIPV market is categorized into roof, façade, and window. Facades and windows are the fastest-growing application segments in the worldwide market [10].

8.4.2.3 What technologies are involved with BIPV?

The module technology involved in BIPV systems is classified into crystalline silicon solar cells and thin-film silicon solar cells. Crystalline silicon solar cell glass/glass module technology is the most used. Thin film nonglass modules are flexible, can be easily integrated into the building envelope, and have a better response to indirect light. So, they are grabbing a lot of attention for integration to pitched roofs and other buildings, where there are no requirements of structural functioning. The following are the different types of technologies and modules used in BIPV applications: BIPV laminates for architectural glazing, frameless, glass/back sheet modules, frameless, glass/transparent back sheet modules, glass/back sheet module glazing with thicker glass, glass/glass modules with crystalline Si solar cells, glass/glass see-through modules, glass/glass amorphous Si-based thin film modules, and sun slates, sun tiles and shingles.

The types of solar cells used in BIPV are monocrystalline Si solar cells, multicrystalline Si solar cells, substrate and superstrate type single and triple junction amorphous solar cells, CdTe solar cells, CIGS solar cells with a glass/SS/polymer substrate, bifacial Si solar cells, and polymer solar cells.

The crystalline silicon solar modules are generally opaque in dark shades of gray or black with an optional antireflective coating over the protective glass; they are connected with solar modules. Frameless laminates fulfill the highest aesthetic demands of the building. There is no loss-inducing shading of the solar cells by the constructional elements. It avoids the deposition of power-reducing dirt on the solar laminates. These provide a uniform appearance of the photovoltaic glass facade. They offer electricity production combined with protection against the weather for the building. The frameless glass/back sheet modules and glass/glass laminates decrease the thermal load inside the buildings. The appropriate clearance between cells in glass/glass laminates provides suitable transparency and demands large glass dimensions. The back side glass of the double glass laminate prohibits moisture penetration, promising a longer lifetime, by keeping appropriate clearance from the module edge, which is also matched with the supporting metal parts of the frameless design. A double-glazing configuration is possible after laminating two sheets of glass for better heat insulation, condensation free and energy saving. The BIPV laminate with double glazing withstands high wind loads.

8.4.2.4 Specifications for BIPV laminates

Laminate size to meet the requirements of the shape and size of the building envelope.

Laminate structure and thickness to meet the wind load, thermal insulation, daytime light, and other requirements to protect the building as part of its envelope.

Mounting or jointing methodology of the laminate should withstand the wind load and not allow any water leakage.

The light-transmission ratio, shading coefficient, type of encapsulant used, junction type and its mounting methodology, cable routing, voltage of the module, solar heat gain coefficient, thermal conductivity, U-value, diode rating, type of connector used, type of cable, system voltage, gap between the cells and glass edge, type of cell used, and the snow load withstanding capacity are the specifications of laminates.

BIPV applications offer some unique challenges compared to standard PV installations, including:

Heat buildup: Limited airspace for convective cooling negatively affects the power output of the PV system and can also negatively affect the building performance by allowing heat transfer into the building.

Nonoptimal array orientation and shading: Architectural aesthetics are sometimes allowed to drive the location and orientation of BIPV modules, leading to nonoptimal performance. Additionally, landscaping or other building elements can create shading problems.

Limited access: Troubleshooting is often more complicated with BIPV systems. For example, the array wiring may be hidden under the BIPV roof or behind the BIPV facade.

As a result of these challenges, many in the solar industry believe that BIPV system performance is sacrificed in favor of aesthetics.

8.5 Design of rooftop and BIPV solar PV systems

This is the sizing and selection of the components required for a solar PV system. These systems are to be installed on the roofs, where grid connections are already available and the appliances are powered by the grid. The solar array generates DC power and converts it into AC power, supplying loads in real-time conditions. If there is an excess generation of power compared to the load requirement, it will be exported to the grid.

8.5.1 Sizing of the PV system

The following are the steps to be followed to size the system.

Step 1: Determine the monthly use of electrical energy and calculate the average daily requirement of energy

From the electricity bill, how much energy has been consumed in summer and winter and consumption trends can be determined. The daily average consumption of energy can be estimated by adding up all the monthly kWh of the year and dividing it by 365. The main idea of installing the rooftop solar PV system is to the reduce the electricity bill, as the grid power is available to the loads. So, the average daily consumption figure can be used to design the system.

Step 2: Assess the solar radiation—average peak—sun hours per day available at the site

In order to find the resource availability, one needs to define the location's longitude and latitude and then refer to solar radiation maps that are available online for each country or locally by the country and government. To assess the solar radiation for a particular location, there are several online sources such as NASA, Meteonorm, NREL, PVgis, and Solar GIS. The solar radiation data have to be estimated for the particular location of the site. Monthwise radiation data will be available. The average global horizontal irradiance (GHI) per day can be considered. A more accurate number is required for the accurate prediction of the energy yield of the system.

Step 3: Calculate solar array wattage and selection of the array

To support all the loads from the solar PV system, that is, 100% offset the electric bills, the solar PV capacity can be obtained by dividing the average daily consumption of energy by GHI per day and by system losses. Most grid-tied PV systems don't try to make all the power, but to cut down the electricity bill. Half the calculated capacity of the solar array can be installed.

Divide the daily average electricity used by the average sun hours per day. For example, if the daily average electricity use is 40 kWh at a particular site, the system size would be 40 kWh/5 h = 8 kW AC. Multiply kW by 1000 to get AC watts. The GHI of particular site assumed is 5 kWh/m²/day.

The solar PV system has losses in energy generation, including soiling losses of the array, mismatch losses of modules, quality loss, temperature loss, low irradiance loss, cable loss, and inverter conversion loss. All these losses amount to about 18%, depending on the location, and the system derating factor is 0.82.

To calculate the total required nameplate power of the PV array, divide the AC watts from the above step by the system derating factor. Use a derating factor of 0.82 for most systems (this is the standard derate used by PV watts) [11]. For example, if an array size of 8000 WAC is calculated in the above step, divide 8000 Wac by 0.82 to get 9756 Wdc based on the module's STC rating. This can be rounded to 10 kW.

Divide the system DC wattage in Step 3 by the nameplate rating of the chosen modules to calculate the number of PV modules needed to provide the desired AC output.

If 325 Wp solar module is considered, the number of modules required for 10 kW is 10,000/325 = 30.7, which can be rounded to 31 solar PV modules.

Step 4: Size and specify the inverter

The inverter has to be sized based on the size of the solar array. In most locations, the nameplate solar array size can be up to 1.25 times larger than the maximum inverter capacity. Solar radiation is rarely available in full 1000 W/m^2, which is the standard test condition. Most inverters now have two or more MPPT channels, some of which are limited to one string of modules, so it is best to use series strings at the highest voltage possible so long as the maximum voltage is never exceeded, even in the coldest conditions. A 10 kW inverter can be considered in this example.

Step 5: Estimate the string length

The solar modules are to be connected in series to form a string, which is connected to the inverter. The voltages of the modules are added up and give a high DC input voltage to the inverter. The number of modules required in a string is called the string length. The solar array output parameters such as voltage and current vary according to the solar irradiance and module temperature. The array gives the highest voltage during the lowest ambient temperature, that is, in the early morning hours of winter months. This high voltage, i.e., V_{oc} of array should not exceed the system voltage.

Step 6: Select the module mounting structure

8.5.2 Module mounting structure

The mounting structure is used to tightly clamp and secure the solar modules. The design of the module mounting structure should consider local site conditions such as wind speed, rainfall and temperature. As the solar plant is expected to have a 25-year lifetime, the module mounting structures are selected such that they should work for the lifetime of the project. Mounting structure life is highly effected by corrosion and it is important that the thickness of the structure galvanization generally should be around 80 μm to protect from corrosion. Hot dip galvanized cold rolled mild steel material is used for the vertical columns of the structure and the purlins and rafters. For some designs, pregalvanized mild steel sheets or galvalum sheets are used for rafters and purlins. Anodized Al profiles are used for industrial rooftop structures.

There are different types of module mounting methods, including rail free, rail in flush, rail-elevated, and non penetrating for sloped metallic sheet roofs.

For sloped roofs, the rafters and purlins of the roof have to be accessed and the railings have to be fixed. The solar modules have to be clamped with the railings. Sometimes, the rails will be clamped to a corrugated sheet in addition to the rafters of the roof. Fig. 8.14 shows the metallic roof fitted with solar modules using a railing system in flush with the modules. In this case, the temperature of the module operation will be higher, as there is no circulation of air to dissipate the heat. Figs. 8.15 and 8.16 show

FIG. 8.14

Solar roof with rail based in flush mounting of modules.

FIG. 8.15

Solar roof with inclined mounting of modules using rails.

the mounting of the modules on a sheet roof with elevated angles using a railing system fixed to the rafters and sheet.

In this case, fewer modules can be accommodated on the roof as space has to be left between the rows to take care of shadowing aspects. The generation in this case will be higher compared to a roof in a flush mounting system, as the module operating temperature will be lower and the inclination angle is elevated.

FIG. 8.16

Solar roof with inclined mounting of modules using rails, different configuration.

Without railings, the solar module frames can also be directly fixed to the rafters. In this condition, the modules are in flush with the roof, and there will be no or very less gap between the modules and the roof. The temperature of the module will be very high during its operation.

Some roof owners instructs not to puncture the metallic sheet of the roof to access the rafters for fixing the railings on the roof. For this type of nonpenetrating option, there are adhesives available that can be used to bond the railings on the corrugated/trapezoidal sheet. On the railings, the modules can be mounted. With this type of application, the adhesive bond strength and its withstandability for hostile weather conditions has to be checked and its reliability to be ensured based on testing.

For flat RCC roofs, the ballast-based, foundation-based or superstructure-based systems can be used.

Sometimes, an adhesive is used to bond the foundation blocks on the RCC roof. The bond strength of the adhesive has to be checked to ensure the reliable operation of the foundation of the MMS.

8.5.2.1 East-West mounting

The common methodology followed for module orientation is that they face southward for the locations in the northern hemisphere and northward for sites in the southern hemisphere. In this method for a fixed tilt angle to harvest the maximum radiation in a year, the tilt angle should be equivalent to the latitude of the location. It maximizes the energy yield in a year. This orientation requires more area to maintain the inter row distance to avoid shadows on the modules. There will be energy loss due to horizon-based shadows. To minimize the energy loss, the pitch between the MMS has to be increased, which will cause for the requirement of more area.

To optimally utilize the available space, the best method is to mount the solar PV modules in an east-west direction. In this method, more modules can be accommodated on a flat roof, as two rows of

FIG. 8.17

East-west mounting of solar modules on a flat RCC roof.

MMS are combined in an inverted V shape, and there is no need to leave more space for the interrow distance. There will be a reduction in the wind load on the structures as they are arranged in the inverted V shape. The east-west mounting methodology can be adopted for RCC roofs as shown in Fig. 8.17.

To get the advantage of the east-west directional mounting, near object shading has to be avoided. Every string should have the same number of solar PV modules. The inclination angle of all the modules in a string should be the same. The inverter should have more MPPT circuits. East-mounted modules should be connected to a separate inverter or to a separate MPPT circuit of the inverter. This should also be done for west-mounted modules. This is required to avoid current mismatch losses arising due to variations of irradiance falling on the modules mounted in both directions. In this method, the space required to mount a 1kW system is 7–8 m² area. The energy yield will be lower compared to south-oriented arrays. There will be greater energy generation during the peak demand hours of morning and evening and less during the midday peak when the electricity rate is cheap. This is useful for time of the day tariff schemes.

Step 7: Sizing of overcurrent protection devices, AC, DC cables

8.5.2.2 Selecting the circuit breaker

The standard determines that circuit breakers should handle 80% of their rated capacity for continuous loads (those being on for 3 or more hours) and 100% for intermittent loads. For safety reasons, assume that all loads are continuous. For example, for a 50kW inverter with a total current output of 77A, the calculation is as follows: 77A * 1.25 = 96.25A.

The designer should use a 100A MCCB because the immediately lower circuit breaker rating of 80A would not be enough for this load.

The resistance of the cable and different derating factors are to be considered for DC cable sizing to meet the fault current requirement of the source circuits. The DC cable loss should be less than 2%.

For AC cable design, the distance between the inverter and the point of power evacuation decides the cable size and number of runs to be done. The voltage drop should be minimized to 1.5%; otherwise, a voltage rise issue will arise and the inverter will shut down.

8.5.3 Some more aspects considered in system design
8.5.3.1 Site survey
There are a lot of aspects in doing a site survey for a designer and installer. The amount of detail a site surveyor collects depends on the scope of the project. The following points need to be covered during the site survey.

Orientation of the site/location and the total roof area available to be evaluated. The type of roof and its structure as they play a dominant role in the selection of the solar PV mounting system and other related issues. Possible routes for cables to be checked and location of the inverter, and the distances to be measured to estimate the cable sizing. Whether the rooftop would be a thatched, concrete, asbestos, or industrial roof with a metallic sheet and the load-bearing capacity of the roof has to be assessed. The profile of the metallic sheet should be known. Whether the roof will withstand the additional loads of the solar PV system to be checked. The drawing of the roof and the location of rafters is required. The location of the turbo ventilators or any other shadow-casting elements and their heights and distances from a reference point have to be measured. The most critical parameter is the identification of a shadow-free location. The required space needs to be identified appropriately, keeping in mind the capacity of the solar PV array that is going to be installed in a particular location.

The most vital and crucial parameters such as monthwise incidence of solar radiation, ambient temperature, and wind speed should be gathered. The shadowing aspect can be studied with a pathfinder or solemetric eye equipment. The shadow-casting objects can be modeled and the shadowing effect can be estimated using Google sketch software. This also can be done using PVSYST or Skelion software. The rooftop can be viewed through Google and the module layout can be made, considering the plot area.

8.5.3.2 Shading
Any shading on the solar module will drastically affect the performance of the solar array. If a module in any array is shadowed, it generates lower current and lower power and pulls down the other panels in the string. So, while designing the system, a shadow-free area has to be selected. All the shadow-casting elements such as water tanks, TV antennas, chimneys, trees, adjacent buildings, turbo ventilators, etc., on the roof are to be identified. Shadow analysis has to be done by feeding the coordinates of the shadow casting elements in the array layout. Google Sketch Up, PVSYST, and Skelion software are used to estimate the shadow-free zone. A well-designed PV system requires a shadow-free area from about 9 a.m. to 3 p.m. throughout the year. The worst-case shadow will occur in winter solstice conditions in the locations of northern hemisphere.

8.5.3.3 Orientation/tilt
The tilt angle of a photovoltaic array is the key to an optimum energy yield. Solar panels are most efficient when they are perpendicular to the sun's rays. In northern latitudes, by conventional wisdom, PV modules are ideally oriented toward true south. Generally, the optimum tilt angle of a PV array equals the geographic latitude to achieve yearly maximum output of power. An increased tilt favors

power output in the winter and a decreased tilt favors output in the summer. If the roofs are steep, the array can be made in flush with the roof.

8.5.3.4 Selection of solar array

Solar PV modules are available with different sizes, different wattages, different technologies, different brands, and different prices. The module should have a higher conversion efficiency, a lower temperature coefficient of power, cells of proven technology, have a lower annual degradation, and better performance abilities. The module should have been manufactured using high-quality materials, which decide the long-term performance of the module. The materials should have been tested and qualified by the manufacturer and they should have a history of usage. The module should have been manufactured using automatic or semiautomatic processes and technologies. The quality control plan should have been followed for the end-to-end process. The module should have undergone all the tests specified in IEC 61215 and IEC 61730 latest standards and, PID, salt spray, ammonia resistance, and sand and dust tests. Modules should have passed accelerated tests. The performance data of the modules if installed in the power plants can be sought from the supplier. The modules should have been thoroughly checked for microcracks and should have third-party power performance evaluation data.

The cost of the solar panel depends on different parameters such as the wattage, the dimensions, the brand, the material quality, the durability, the warranty period, and different certifications based on IEC or local country-based standards it has passed. The cost is one of the factors in the selection of the module. But, the cost should not be considered as the one and only factor for the long-term performance of the module. Choosing a good quality panel is more important than choosing the cheapest option for the reliable and long-term performance of the plant. The quality takes into account the quality of the materials used and the manufacturing processes followed in the fabrication of the solar panel. There are various types of solar panels available in the market that vary in performance, cost, and output.

8.5.3.5 Inverter selection

The solar inverter can be sized according to the size of the solar array. The inverter should meet the grid code requirements. The national grid code might require the inverters to be capable of reactive power control. In that case, slightly oversizing inverters could be required. The grid code also sees requirements on THD, which is the level of harmonic content allowed in the inverter's AC power output. It should have lower auxiliary power consumption. High inverter reliability ensures low downtime, maintenance and low repair costs. If available, the inverter mean time between failure values and track record should be assessed. The inverter should have the highest efficiency, a suitable MPPT window, and a lower thermal derating factor in power. It should have more number of MPPTs and should be equipped with all input/output protections and should have a better cooling design to avoid tripping while harvesting more energy.

8.5.3.6 String and DC voltage calculations—Example

Table 8.1 gives the solar module technical parameters and Table 8.2 gives the specifications of the 10 kW inverter considered in the example.

Table 8.1 Specifications of solar PV module at STC.

Peak power—P_{max} (W)	325
Open circuit voltage—V_{oc} (V)	45.9
Voltage at maximum power point—V_{mp} (V)	37.2
Current at maximum power point—I_{mp} (A)	8.76
Short circuit current—I_{sc} (A)	9.25
Module efficiency (%)	16.75
Module size (mm)	$1960 \times 990 \times 40$
Module weight (kg)	22.5
Temperature coefficient of P_{max} (%/°C)	−0.39
Temperature coefficient of I_{SC} (%/°C)	0.05
Temperature coefficient of V_{OC} (%/°C)	−0.32
Nominal operating cell temperature (NOCT) (°C)	46

Table 8.2 Specifications of the inverter.

Technical specifications of inverter	
Inverter capacity (kVA)	10
Max. input power (Wp)	12,000
Max. DC voltage (V)	1000
Min/start DC voltage (V)	180
MPPT voltage range (V DC)	160–850
Number of MPPTs/max. power per MPPT	2/6 kW
Strings per MPPT	2
Number of input connections/max. input current for each connection	2/11
Max. AC power output (W)	10,000
Nominal AC voltage (V)	230
AC voltage range (V)	−400 ± 20%
Phase/wire	3PH+N+E
Output frequency range (Hz)	50
Max. output current (A)	16.7
Power factor ($\cos\varphi$)	0.8 leading…0.8 lagging
Current THD (%)	<1.5%
Consumption (standby/night) (W)	<1
Max. efficiency	98.70%
Euro efficiency	98.10%
MPPT efficiency	99.50%
Cooling	Natural cooling

Location: Delhi
Lowest temperature of site: 7°C
Highest ambient temperature of site: 45°C; PV module wattage: 325 Wp
Inverter: 10 kVA three-phase

To determine the maximum number of modules required in series, first calculate the maximum voltage of the module at lowest temperature as follows:

$$V_{oc} \text{ at lowest temperature} = V_{oc} \text{ at } 25°C + ((T_{low} - T_{ref}) \times \alpha)$$

where T_{low} is the minimum temperature of the site; T_{ref} is the cell temperature at STC, that is, 25°C; and α is the temperature coefficient of V_{oc} ($-0.32\%/°C$), that is, $0.32/100 * 45.9 = 146.88 \text{ mV}/°C$ is the decrease in the voltage for a 1°C increase of temperature.

$$V_{oc} \text{ at lowest temperature} = 45.9 + [(7-25) \times (-0.1468)] = 45.9 + [-18 \times (-0.1468)] = 45.9 + 2.64 = 48.54 \text{ V}.$$

The maximum input voltage of the inverter is 1000 V, which is the system voltage. Divide the maximum inverter input voltage by the temperature-corrected open-circuit voltage and round down to the nearest whole number to determine the maximum number of modules in series:

$$\text{Maximum number of modules in series} = \frac{1000 \text{ V DC}}{48.54 \text{ V}} = 20.6 = 20 \text{ modules in series}.$$

Minimum number of modules in series: To determine the minimum number of modules in series, first calculate the per module minimum voltage as follows:

$$V_{mp} \text{ at highest temperature} = V_{mp} + [(T_{high} + T_{rise} - T_{ref}) \times \beta V_{mp})]$$

where T_{high} is the highest ambient temperature, T_{rise} is the rise in cell temperature expected considering the array mounting (typically 20–30°C), and βV_{mp} is the temperature coefficient of V_{mp} in volts, that is, $0.39\% + 0.02\%$ is -0.41%. [$-0.41/100 * 37.2$] is 152.52 mV. The temperature coefficient of V_{mp} of a module is not provided in the module data sheet. This can be considered 0.02% higher than the temperature coefficient of P_{mp} [12]. As the product of V_{mp} and I_{mp} provides P_{mp} and the I_{mp} has very low current variation with temperature, so the coefficient of V_{mp} will be nearest to the temperature coefficient of P_{mp}.

$$V_{min} = 37.2 \text{ V} + ((45°C + 25°C - 25°C) \times -0.15252 \text{ V}/°C) = 37.2 \text{ V} + (45°C \times -0.15252 \text{ V}/°C)$$
$$= 37.2 \text{ V} - 6.863 \text{ V} = 30.337 \text{ V}.$$

There will be variation in the V_{mp} in the 25 years lifetime of the solar module. Select and apply a multiplier factor to account for the degradation or change in V_{mp} due to combined effects of high AC grid voltage, array degradation for 25 years, and module voltage tolerance. With the fluctuations in grid voltage, the rise in inverter voltage causes increase in DC voltage of the solar array. For 3% rise in AC voltage, correspondingly, the DC voltage to be 3% higher during summer conditions. The annual degradation in P_{mp} which is the product of I_{mp} and V_{mp} of module is 0.5%. The degradation in V_{mp} per year can be considered as 0.25%. So, for 25 years the degradation factor for V_{mp} is $0.9975^{25} = 0.939$. So, 6% can be accounted for the V_{mp} degradation at the end of 25 years. There is a tolerance of -10% to $+10\%$ in the measured value of V_{mp}. Given the lack of information on module voltage tolerance available, it is best to err on the side of caution and assume an extra 5% DC voltage loss in the array [12].

By considering all the loss factors, a factor of 0.85 can be taken into account for estimation of minimum voltage of the module at the end of life (EOL) of the plant. By applying the degradation factor the minimum voltage of the module

$$V_{min} = 30.337\,V \times 0.85 = 25.79\,V$$

The minimum MPPT voltage of the inverter is 180 V. Divide 180 V with the minimum voltage of the module and round up to the nearest whole number to determine the minimum number of modules in series:

$$N_{min} = 180\,V/25.79\,V = 6.97 = 7\text{ solar modules in series.}$$

The maximum DC input current of the inverter is 22 A.

The I_{mp} of the module at the highest module temperature of 65°C is

$$I_{mp}\text{ at 65°C and at 1000\,W/m}^2\text{ is }I_{mp}\text{ (at 25°C)} + (65-25)*0.05/100*8.76 = 8.76 + 0.1971 = 8.957\,A.$$

The maximum number of strings connected in parallel to the inverter is 22 A/8.957 A = 2.46, so two strings can be connected.

The 10 kW solar plant capacity is with 32 number of modules and each module is of 325 W. So, two strings with each string of 16 modules in series can be connected and the plant capacity will become 10.4 kW.

The best array design for this case study calls for 20 modules per source circuit. However, the simplest string-sizing program specifies 16 modules per source circuit. More sophisticated string-sizing programs apply a margin of safety to the minimum expected DC voltage to account for high AC grid voltage, array degradation, and module-to-module voltage mismatch.

8.5.4 Design of BIPV PV systems

The following are the design requirements of the BIPV system, including laminates and modules.

The laminate should withstand all weather and environmental conditions.
Wind loads: double glazing; triple glazing is being used.
Rain, hail stones: Tempered glass on both sides will give protection.
Shadows: Bypass diodes across the group of cells protect the module from hotspots.
High and low temperature extremes: a high fatigue-resistant solar ribbon will sustain the thermal stresses that arise due to changes in temperature.
Humidity and high temperature environment: EVA sealant or PoE will take care of this.
Mechanical support that joins the modules and hold thems pointing toward the sun.
Should withstand corrosive environment. Anodized aluminum alloy-based frames are preferable for module mounting structures.
Protection to cells, interconnects, and diodes from weather is done with qualified glass/glass laminates.
To couple as much light as possible into the solar cell: Higher transmittance glass with antireflection coating is preferable.
To minimize the temperature increase of the solar cell, proper thermal design has to be followed.

The following design parameters are to be considered in designing the BIPV system:

Wind speed and direction, temperature, global horizontal and normal radiation falling on the building, shadowing effect of neighbor buildings, angle of incidence of irradiance, aspects of ventilation, thermal loads, and light management.

8.6 Cost of rooftop solar PV systems

As an example, a 52 kWp grid-tied solar PV system has been considered for costing and the bill of materials required is given in Table 8.3.

Array junction boxes are used to incorporate surge protection devices. For present inverters, there are multiple MPPTs and the strings can be connected directly to the input of the MPPTs.

To calculate the costing, the transportation charges for the equipments, duties, and taxes for the equipments, fees for the permits and approvals, installation cost of the plant, finance costs and

Table 8.3 Bill of materials required for a 52 kW solar PV system.		
Component rating	**Quantity**	**Unit**
Solar photovoltaic modules: polycrystalline Si—325 Wp	160	No.
Grid-tied inverter: 50 kVA/3 phase	1	No.
Array junction box/DCDB: 6 in, 6 out; polycarbonate IP65 enclosure with the following items: (a) DC fuse, (b) surge protection device (SPD), (c) DP MCB/DC disconnector at output, and (d) suitable cable glands and MC4 connectors at input and output	2	No.
AC distribution box: 50 kVA/3 ph, polycarbonate box IP 65 with the following items: (a) MCB, (b) surge protection device (SPD), and (c) changeover switch from solar to grid and vice versa, (d) energy meter	1	No.
DC cable: $1C \times 4 mm^2$ 1.1 kV, PVC insulated, UV protected Cu cable (for array connection)	190	m
AC cable: $4C \times 35 mm^2$ 1.1 kV, PVC insulated, UV protected Cu. cable (for inverter, ACDB connection)	20	m
Module mounting structure: Pregalvanized Mild Steel capable of withstanding a minimum wind load of 150 kmph (with hardware for structure interconnection and module mounting) module tilt angle: As per site location	40	Set
Male and female MC4 connectors for cable connections	30	Pairs
Lightning arrestor: 2000 mm long 16 mm dia with five spike and base plates made with high-grade aluminum	1	Set
AC cables: $4C \times 35 mm^2$ 1.1 kV, PVC insulated, UV protected unarmored Cu cable (for ACDB to main distribution board connection)	20	m
Earthing cable: $1C \times 4 mm^2$ 1.1 kV, PVC insulated, UV protected unarmored Cu cable (for earthing green color)	30	m
Earthing cable: $1C \times 16 mm^2$ 1.1 kV, PVC insulated, UV protected unarmored Cu cable (for earthing green color)	30	m
Earthing kit: Copper bonded rod earthing (dia, 17.2 mm, length, 3000 mm, 250 μm copper coating) with 25 kg chemical and clamp	3	Set
Earthing strip: $25 \times 3 mm$ GI strip	40	m
Aux. item such as fasteners, cable trays, lugs, conduit pipes with acc.	1	Lot
Reinforced concrete foundation	1	Lot

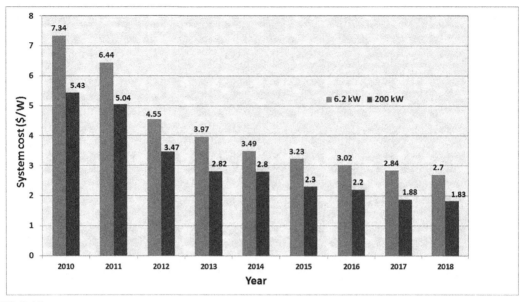

FIG. 8.18

Yearwise variation of rooftop system cost in United States for residential rooftop systems.

administrative expenses, and margin of profit can be considered. NREL has done benchmarking costing for residential and commercial solar PV systems [13]. Fig 8.18 shows the benchmark costs of a 6.2 kW residential system and a 200 kW commercial PV system from 2010 to 2018.

There is a reduction in the price due to increased efficiency and a price decrease of solar PV modules, reduction in the labor cost and finance costs. IRENA has compiled the data for the installed costs and system costs for residential and commercial rooftops for different countries. Fig. 8.19 shows the installed costs of a residential system and a 500 kW commercial PV system in 2019 in different countries [14]. The system costs are low in India and China. This is due to increased efficiency and power from the module and reduction in the installed costs and the financial costs.

8.7 Reliability of rooftop solar PV system

The reliability of a grid-tied PV system depends on the design, installation, and the quality of the components/equipment used as part of the system. The main components of the system are solar modules, inverters, module mounting structures, string combiner boxes, AC and DC cables and their connectors, cable conduits, LT panels, and earthing systems. The solar module is the heart of the solar PV system. The reliability and long-term performance of the modules depends on the quality of the materials used in the manufacturing of solar PV modules, the manufacturing process, and the followed quality system. There are defects observed in the modules installed in different solar PV plants, which are shown in Figs. 8.20–8.27.

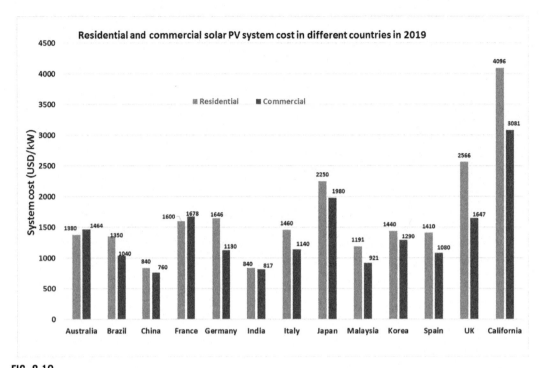

FIG. 8.19

PV System cost in different countries for residential and commercial rooftops in 2019 [14].

Glass breakage: It occurs due to bad handling of the module during installation and commissioning. Mechanical stress caused by an object cause breakage of glass.

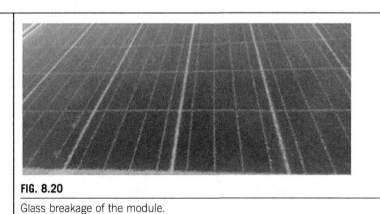

FIG. 8.20

Glass breakage of the module.

Burning of back sheet: This type of burning of solar PV module along with its back sheet occurs due to DC arc fault or due to lightning strike or due to earth faults.

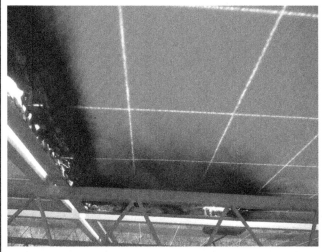

FIG. 8.21

Burning of back sheet and module.

Burning of junction box: It is also due to DC arc fault arising out of loose connections of the contacts inside the junction box. It also happens due to poor quality of junction box, the moisture penetrates and causes corrosion of contacts, which may become intermittent loose connections, creating DC arc faults leading to fire.

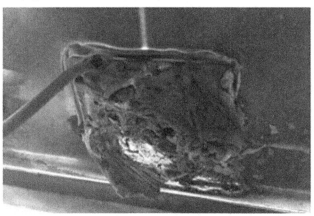

FIG. 8.22

Burning of junction box.

Continued

Solder joint failures: The burning of the back sheet at the locations of interconnecting solder joints might be due to series arc fault. The interconnects are connecting the solar cells might be in offset with the cell bus bars, and during high-temperature conditions, the interconnect might have detached from the cell bus bar causing intermittent loose contact creating series arc fault causing localized burning. Poor soldering and poor solder joint strength are also caused for DC series arc faults.

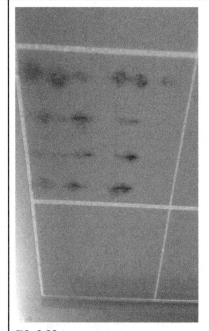

FIG. 8.23

Solder joint failures.

Hot spot: The burning spot on the backside of the module might be due to bypass diode failure or DC arc fault. If one of the bypass diodes of a module is failed due to open condition, more number of cells will come across the other bypass diode. If the module is shadowed, the voltage of un-shadowed cells reverse biases the shaded cell causing power dissipation and increase of temperature, which might cause burning spots in the module. The poor quality of the cell further aggravates the problem.

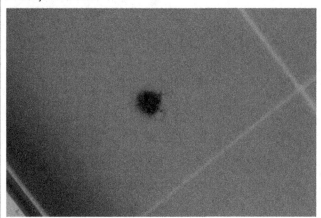

FIG. 8.24

Hotspot/burning spot.

Detachment of junction box from module: The fixed junction box (JB) has got detached from the module. This is due to the poor adhesion of the adhesive used to bond the JB on the module. If shelf life expired adhesive is used or improper process is followed during adhesive bonding and curing process, this type of junction box detachment occurs even with small strain on JB.

FIG. 8.25

Detachment of junction box from module.

Back sheet delamination

See Fig. 8.28

Bypass diode failure

See Fig. 8.28

Snail trails: The spreading of lines as shown in the figure in the solar cells of the module is called snail trails. These will not create power reductions in the module immediately. It is felt that the chemical reactions of EVA cause the formation of different chemicals react with silver metallic grid of solar cells and show these types of trails as snail paths. The other argument is that the microcracks are responsible for the snail trails.

FIG. 8.26

Snail trails.

Continued

Burning of solar modules—fire on the roof: The burning of the roof takes place due to fire in the solar modules. This type of fire is due to blind spots or DC arc faults.

FIG. 8.27

Burning of solar modules—fire on the roof.

Fig. 8.28 shows the commonly found defects in solar PV modules. The microcracks in solar cells, discoloration and browning of EVA causing delamination of glass and back sheet occurs in the modules. Bypass diodes failure in short condition reflects in I–V characteristics and thermography studies. The impact of hailstones completely damaged the solar PV modules as shown in Fig. 8.28I. For hail stone prone areas, the modules are to be tested for hailstone test with specific requirements. Improper design of the module affected with PID shows chessboard like pattern in thermography studies and I–V curves also reveals low fill factor and low power. The sealing of the laminate and improper fixing with frame causes the penetration of moisture and leakage of water inside the module. During the rainy season or at the time of cleaning of the modules, the insulation resistance of the module becomes zero and the inverter trips showing an earth fault of the string. For some modules, dry insulation resistance also lower, so the inverter trips with the connection of low insulation resistance module. All these defects will affect the performance of the solar modules and in turn the performance of the solar PV system.

The solar module should use high-quality solar cells. The solar cell should have a low reverse leakage current and a high shunt resistance while also being PID-resistant.

The solar module should pass all the tests such as the PID test, the salt spray test, the ammonia resistance test, and the sand dust test, in addition to all the tests mentioned in the IEC 61215 and IEC 61730 standards.

In the PV system, the inverter frequently trips due to string faults. The inverter should have passed all the tests. The inverter should not trip due to high ambient temperature. Failure of fans is a common problem causing malfunctioning of inverters. Malfunctioning of the IGBT in the inverter can lead to a distortion of the output waveform or a complete shutdown of the system. Fire is the most hazardous risk caused by DC arcs, short circuiting, or overheating. The inverter trips if the operating temperature of the inverter reaches more than 50°C. Grounding faults or module string faults arise during the rainy season with waterlogging or a broken sheath and insulation of cables,

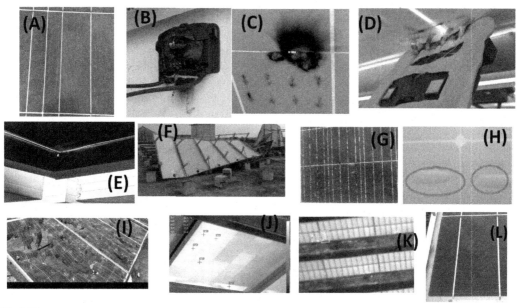

FIG. 8.28

Different observed defects in modules deployed in the field. (A) Coloration/browning of EVA, (B) junction box burning, (C) burning of back sheet and at solder joint locations due to DC arc fault, (D) detachment of JB from the module, (E) leakage in the frame fixed to laminate, (F) modules along with module mounting structures overturned due to high winds, (G) shattered glass due to burning caused by DC arc fault, (H) bubbles in the back sheet, (I) damage of the solar modules due to hailstones, (J) thermal image of PID affected module, (K) thermal image of panels with bypass diode issue, and (L) snail trails in the module.

causing inverter tripping. Selection of the cable and proper handling of the cable during installation and its routing through conduits to avoid rodents and soiling will minimize the string earth faults.

The inverter trips if fluctuations in the grid are beyond the operating levels of the inverter voltage rise issue. All these aspects should be taken into consideration in the design and installation of the system.

If the MMS is not properly designed, the modules will be flown during high wind conditions.

The counterweights of ballast design will be displaced from their locations during high wind conditions if the design is improper.

Grid fluctuations will cause a voltage rise in the inverter output and the inverter gets tripped. To avoid this problem, the voltage drop due to the AC cable should be 1.5%.

Shadowing of the modules will cause hotspot issues. To avoid energy generation loss, the design should avoid mounting the modules in continuous shadowed zones.

Arcing of connectors due to loose connections will cause a fire in the PV system. The ground faults, the lack of protection, and DC arc faults will even cause burning of the modules and the roof. If the systems are not properly grounded and there is no lightning protection, the systems will get damaged.

The following are some of the issues with PV systems.

8.7.1 **Voltage rise issue**

As per the grid standards, the nominal mains voltage is 230+10%–6%. This gives a range of 253–216.2 V. The grid-tied inverter will have minimum and maximum voltage set points for disconnection from the grid. The inverter disconnects from the grid if the inverter output voltage >255 V for 10 min. If the inverter output voltage exceeds 260 V any time, it immediately disconnects from the grid.

A solar array generates maximum power at noon and the inverter will be exporting the excess energy of home needs to the grid. The grids are made 10 or 15 years back and they will have grid impedance and grid voltage fluctuations. Due to the increased penetration of renewable energy systems, the grid voltage fluctuates. The inverter voltage should be a couple of volts higher compared to the grid voltage for energy export to the grid.

The inverter has to export the energy to the grid even if the grid voltage increases due to fluctuations. As per ohm's law, $V = IR$.

If the resistance R between the inverter output and the point of connection for the export is fixed and the generated current I is also nearly fixed, the output voltage V has to increase to support the system for the export of the power. The inverter output voltage rises with the increase of grid voltage, a phenomenon called the voltage rise of the inverter. With the grid voltage fluctuations, if the grid voltage is higher, the inverter tries to increase its output voltage to meet the requirement of energy export to the grid.

The inverter has lower and upper voltage set points, and beyond these voltages, the inverter gets tripped. If the grid voltage increases to its higher point, the inverter voltage automatically increases. If it exceeds the highest set point voltage, the inverter trips. After some time, if the grid voltage comes down, the inverter starts and supplies power to the grid. Once again, the grid voltage increases, the voltage rise takes place in the inverter, and it trips. The inverter trips on and off. Due to this tripping, there is a loss of energy and in turn a loss of money.

The following reasons cause a voltage rise, and remedial action can be taken [15].

Grid fluctuations and grid impedance contribute to a voltage rise. This is not under the control of the PV developer. The grid management agency should be asked to change the voltage tapping points in the transformer.

The cable resistance from the inverter output to the grid connection point causes a higher voltage drop. To avoid a voltage rise, the system/cable has to be designed/sized such that the AC cable voltage drop should be less than 1.5%.

The cable size should be increased and the inverter should be installed near the location of the point of connection to the grid as a remedial action to the voltage rise issue.

8.7.2 **Potential induced degradation**

Solar modules are connected in series to form a string and the string is connected to the inverter. The string will generate a voltage from 0 to 1000/1500 V depending on the irradiance and module temperature. In a solar module, the glass/back sheet laminate is fixed with the frame and the solar cell circuit is insulated with the frame. The module frame is connected with the structure. Using the frame grounding hole, the module frame along with the module mounting structure is earthed. The voltage between the negative pole and the positive pole of the circuit is the electric potential. This is not the absolute electric potential of the circuit. The potential measured between the positive pole and the ground and the negative pole and the ground will be different. The absolute potential depends on the topology of the inverter and the way the array circuit is connected. The inverter topology and different potentials with different connection configurations are given in Figs. 8.29 and 8.30.

The solar module consists of a p-type solar cell sandwiched between glass and a back sheet using EVA as an encapsulant. If the array is with a floating ground, that is, not grounded, there is a voltage potential between the solar cell and frame. The top sun-facing material of the module is glass, which is an insulator, touching the frame. Due to the biasing of DC voltage to glass, it induces ionic charges. Sodium ions are induced in the glass, as it is of borosilicate glass. The sodium ions are migrated to the solar cell via the top EVA and ARC layer of the cell. The sodium ions reach the junction and disturb the electrical transport phenomenon in the solar cell. The sodium ions cause shunting to the solar cell and the junction will not be able to produce the voltage for which it was intended. So, the V_{oc} of the module and the fill factor will be reduced. This occurs due to the voltage between the cell and the frame. The modules that are nearest to the negative pole of the string are affected more. The degradation in the module occurs due to biasing of the module. So, the phenomenon is called potential induced degradation.

There are PID solutions to be taken care of at the cell level, the module level, and the system level. PID can be avoided if the solar array circuit is negatively grounded (for p-type solar cell modules), so that the voltage between the cell and the frame will be maintained as zero volts during the biasing condition of the module. To avoid the PID effect, the solar cell should be PID-free. To make the cell PID-free, the silicon nitride, the antireflection coating of the solar cell, should be homogeneous with high density and treated with ozone. In the module level, the top-side EVA with high volume resistivity to avoid the movement of sodium ions can be used to mitigate the PID effect.

This PID effect has been observed with sun power solar modules that use n-type Si solar cells. In this case, if the array circuit is not grounded, the module will be positively biased, and the polarization of the glass/transparent conductive oxide creates ions. For this type of module, the solar array has to be positively grounded.

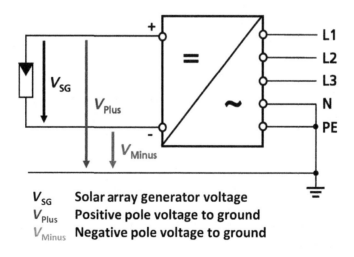

V_{SG} Solar array generator voltage
V_{Plus} Positive pole voltage to ground
V_{Minus} Negative pole voltage to ground

FIG. 8.29

Input and output configuration of an inverter.

Courtesy: Fraunhoffer Institute.

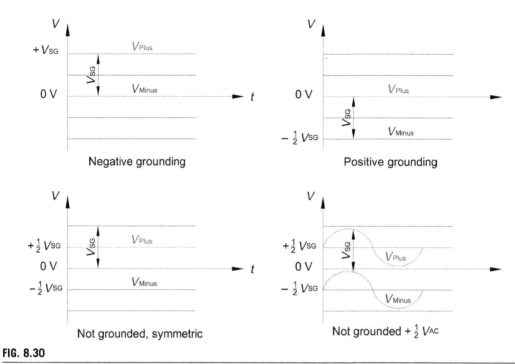

FIG. 8.30

Different types of wave forms and potentials with different array grounding configurations.

Courtesy: Fraunhoffer Institute.

The PID effect is accelerated in the presence of high humidity and high temperature. The PID effect is reversible, provided the solar cell junction is not shunted. By applying the current in the night, the PID effect can be mitigated. So, the inverters are equipped with anti-PID kits to reverse the PID effect of the modules. A PID-affected module shows a lower Voc, fill factor, and power. This can be detected using thermal imaging as well as electroluminescence and current-voltage measurements; the images are shown in Figs. 8.31 and 8.32. The chessboard like pattern of EL image in Fig. 8.31 confirms PID effect in the module. The I–V characteristics of the PID affected module is shown in Fig. 8.32. The wattage of the module at STC was 230 W. The outdoor I–V testing was done at 942 W/m^2 intensity and module temperature of 57.5°C. Due to PID effect, the fill factor of the module reduced drastically and gives 83.1 W at measuring conditions.

8.7.3 Electrical arcing and fire of roofs and systems

The following are the cause of faults in DC side of a solar PV system. Shaded solar cells in a module cause hotspot issue. Improper ingress protection of module JB and deterioration of module back sheet cause the issues in module. Rodent bites, pollution deposit cause failure of the insulation cable. Earth fault which is an unintentional path to ground with zero impedance occurs between the single line to earth.

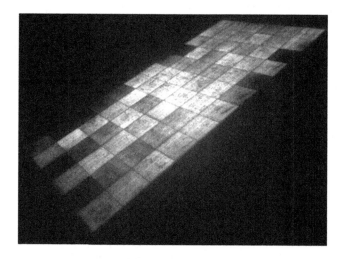

FIG. 8.31

Electroluminescence image of a PID-affected module.

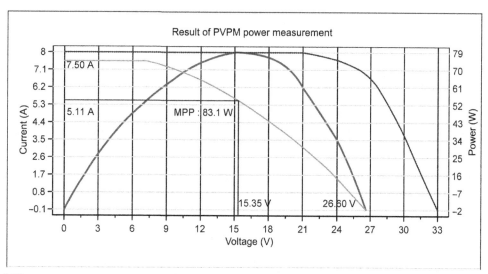

FIG. 8.32

Current-voltage curves of a PID-affected module taken with a mobile array tester.

Line-to-line faults: An unintentional path due to an accidental short circuit between two points in a string with different potentials. This is very rare to occur as the cables have double insulation. Most of the line-line faults originate as single line ground faults. The line-to-line faults create parallel arc faults. Arc fault can occur in the DC section at module level, at JB, at DC cable, string level, combiner box level, or at input section of inverter.

Series arc fault: An arc fault due to discontinuity in any of the current-carrying conductors resulting from solder disjoint, cell damage, corrosion of connectors, rodent damage, abrasion from different sources, broken interconnect, or a small cut in bus bar. It produces high resistance path and the voltage is dropped at that location and generates heat, plasma, and arc and leads to fire. It melts the metals and burns the insulation of the cable, back sheet, or JB and shatters the glass. A parallel arc fault occurs due to insulation breakdown in current-carrying conductors.

Bypass diode faults: Short circuit in case of incorrect connection. Bypass diode failure in short or open condition during operation in the field.

Degradation faults: Yellowing and browning, delamination, bubbles in the solar module, cracks in cells, defects in antireflective coating, and delamination over cells and interconnections lead to degradation and increasing of the internal series resistance.

Bridging fault: Low-resistance connection between two points of different potential in string of module or cabling.

Open-circuit fault: Physical breakdown of panel-panel cables or joints, objects falling on PV panels, and loose termination of cables, plugging and unplugging connectors at JBs.

The impacts of burning modules can be huge, particularly for house and commercial roof solar PV installations. The fire not only burns the modules but can also set fire to the structures on which solar modules are mounted, as shown in Fig. 8.27. If some strings are connected wrongly with opposite polarity, this type of burning takes place. DC arcing is the prime reason for fires in PV plant. The DC arc might occur due to the loose intermittent connections in the string connector to the JB, the JB to the cable, the cable to the cable joining with MC4 connectors, and the cables running through conduits. The following aspects such as loose wiring between the JB and cable, improper cable support, rodents chewing of wires, improper grounding (leading to a ground fault, line to ground) and line-to-line fault also contribute to creating arc faults and in turn causing the burning of DC cable and PV modules. Historically, the ground-fault detection blind spot has caused many latent ground faults and ultimately resulted in several PV fires in North America [16].

Latent ground faults can either be grounded conductor-to-ground faults or high-impedance ground faults on unearthed conductors. The initial ground fault is generally not a fire hazard, but will remain latent because the fault current is too low to trip the inverter's GFDI fuse.

In the event that a second ground fault occurs in the array, the fault current, which may be very large, will bypass the GFDI device, and the inverter's ground-fault protection system will not work as intended to prevent a fire.

Because latent ground faults are known to cause fires in PV systems, the solar ABCs stakeholder group identified ground-fault protection alternatives to fuse-based GFDIs. The group largely based these alternatives on European fault-detection techniques that vendors or integrators could implement using readily available products. The alternatives identified included current sense monitors, which measure current flow through the ground bond of DC-grounded systems; isolation monitors, which measure array resistance to earth in temporarily or permanently ungrounded PV systems; and residual current detectors, which measure differential current between the positive and negative conductors.

The following are some key design and installation tips—largely intended to minimize opportunities for conductor damage—that will reduce the occurrence of PV system ground faults.

Use shorter circuit lengths. Both copper and aluminum conductors expand and contract at different rates than steel raceways. Long-distance circuit runs magnify this difference and can cause significant problems at turns and terminations.

Avoid conduit bodies for 90-degree turns. When coupled with dissimilar expansion and contraction rates, the tight turns associated with these fittings are a common cause of conductor damage.

Avoid the need for expansion fittings. Whenever circuit runs exceed a distance of 100 ft, consider using a cable tray rather than a conduit to eliminate the need for expansion fittings.

Use aluminum for large circuits. To eliminate dissimilar expansion rates, use aluminum cable trays or raceways with aluminum conductors, which are also much lighter and cheaper than copper.

Terminate aluminum with care. Extra attention is warranted when terminating aluminum to ensure quality long-lasting circuits. Verify that the terminals are rated for use with aluminum conductors and use antioxidant on all terminations.

Engage both clips on connectors. Engaging only a single safety-locking clip on an MC4 or MC4-style connector dramatically increases the likelihood of the connection coming apart.

Follow the connector manufacturer's assembly instructions when preparing and assembling connectors in the field. Use the manufacturer-specified crimp tool and ensure that it is set appropriately for the conductor gauge and type.

Check the connector assembly. After assembling connectors, perform a pull test of about 30 pounds. The connectors should never yield to this amount of pull tension. This test can verify the mechanical integrity of both field- and factory-assembled crimp connections. Conductors slipping out of connector fittings are a common cause of arc faults.

Use connectors from the same manufacturer. There is no connector standard for interoperability. Though manufacturers of MC4-style connectors often claim that their connectors are compatible with Multi-Contact's MC4 connectors, it is unclear whether these companies—even those that are quite large—will back up the contractor in the event of an arc-fault fire. The simple rule, therefore, is to never mate connectors from different manufacturers, as this eliminates connector mixing as a potential contributing cause to a fire.

Do not strain junction box conductors. Fires have resulted from strain at module terminations compromising the electrical connection to the module. Conductor thermal cycling can exacerbate the tension on the terminations and increase the stress at these connections. This is particularly problematic for home runs to combiner boxes, as longer conductor lengths increase the probability of cable tensioning strain.

Torque, retorque, and mark terminations. Terminations can loosen over time due to thermal cycling, vibrations, and other strand movements. To minimize loose terminations in conductors smaller than 250 kcmil, torque the connection to the specified value and then move the conductor 2–3 in. side to side a few times. Moving the conductor in this manner while it is under stress can reposition the strands so that the conductor better fills the terminal cavity. Afterward, retorque the connection to the proper value and mark the terminal with a permanent marker.

Avoid using fine-stranded conductors. Whenever possible, use standard Class B rather than fine-stranded conductors and cables. Where fine-stranded conductors are required, never assume that a terminal is rated for these conductors without referring to supporting documentation. Few larger-diameter pressure terminals are listed and rated for use with fine-stranded conductors. The weight of a large, heavy conductor alone can cause an improper connection to work loose over time. Do not use flexible, fine-stranded cables with setscrew-type terminals or lugs.

8.8 Lifecycle cost analysis of rooftop solar PV systems

It is a powerful economic assessment method whereby all costs incurred from owning, operating, maintaining, and finally disposing of a project are considered potentially important for the decision. It can be determined through LCCA whether a project is economically viable and cost-effective. The lifecycle cost analysis (LCCA) usually contains several economic analyses, which are the lifecycle cost (LCC), the levelized cost of energy (LCOE), the net savings (NS), the savings to investment ratio (SIR), the net present value (NPV), the internal rate of return (IRR), and the payback period (PB). Here, the LCOE, LCC, and PB are discussed.

What is LCOE?. LCOE is a metric that describes the unit cost of energy generated by a project in $/kWh. Investors in the PV solar industry have to understand the price of their solar projects to estimate the return on their investments. It is required to know how to compare alternatives on an equal basis to select the right technology for an economic point of view. LCOE only considers the cost of the lifecycle and the amount of energy generated during the period. The LCOE has become a very practical and valuable comparative method to evaluate and analyze the cost of energy delivered by projects utilizing different energy-generating technologies. So, the PV solar industry has moved from a cost per watt estimation to a cost per kilowatt hour. The LCOE represents a comparative calculation on a cost basis and not a calculation of feed-in tariffs. It is one of the metrics most commonly used by the residential solar industry.

Uses of LCOE. It allows the cost comparison of alternative technologies to make a decision whether to accept or reject the investment. For example, the LCOE could be used to compare the cost of energy generated by a PV power plant with that of a fossil fuel-generating unit or another renewable technology. It is used to evaluate different designs of the system to know the cost of energy with variations in technology. Examples include the: (i) evaluation of a fixed tilt versus a single axis tracker for the PV system, (ii) PV system with solar modules using crystalline silicon versus thin film-based technology, and (iii) the effect of inverter loading more than 40% versus 10%. Thus, it is used to rank the options and determine the most cost-effective energy source and technology. The discom or utility can use this to decide the feed-in tariff rate from the results of the LCOE. The LCOE may also be used to compare the cost of energy from new sources to the cost of energy from existing sources.

The method of LCOE calculation is internationally recognized as a benchmark for assessing the economic viability of different generation technologies. It also can be used to study the viability of individual projects and enable the comparison of different energy technologies with respect to their cost [17].

Method of calculation. The LCOE is determined by dividing the total LCC of the project and its operation by the energy generated. It captures capital costs, ongoing system-related costs, and operation and maintenance costs. It may also incorporate any salvage or residual value at the end of the project's lifetime along with the amount of electricity produced, then convert that into a common metric: $/kWh. The basic formula to determine the LCOE for a project starts with equating the costs and revenues of the project. This can be represented in the simple formula below.

$$\text{LCOE} = \text{total lifecycle cost}/\text{total lifetime energy production}$$

This equation is an evaluation of the lifecycle energy cost and the lifecycle energy production. A low LCOE is better because it shows that less money is needed to produce one unit of energy. This basic definition of LCOE can be expressed mathematically in more complex ways to account for all the variables that impact the LCC and the total energy production for a PV system. Incentives for project construction and energy generation can also be incorporated.

The numerator of the equation is the LCC of the project.

In fact, it also requires analysts who understand the time value of money when comparing future return flows with the initial investment cost of a project.

Therefore, the value of the discount and the inflation rate play significant roles in determining the time value of money. The time value of money must be taken into consideration because the value of money in the present time will not be the same value in the future. For example, the value of $100 this year will not be the same as the value of $100 in the years to come or in previous years. The discount and inflation rate are the factors that cause the value of the money to vary.

Lifecycle cost (LCC). The LCC is the cost incurred in owning and operating a solar PV system during its lifetime. The overall cost of the project includes capital costs, fee for permits and approvals, operating and maintenance costs, repair and replacement costs, waste values, finance charges, and other nonfinancial benefits. This cost is influenced by several parameters such as module cost, plant installation cost, interest for finance and its repayment term and other costs like operating and maintenance cost and residual value. The residual value refers to the resale value of the plant at the end of its life.

$$\text{LCC} = \text{capital cost} + \text{operation and maintenance cost} + \text{replacement cost of equipment}$$

The capital cost refers to the initial investment of power plants such as land, photovoltaic modules, infrastructure for transmission line and evacuation system, system design, BoS of the plant and installation costs. Operational and maintenance and repair costs include the recurring expenses incurred throughout the life of the project such as labor costs pertaining to the upkeep of the equipment, insurance, energy, water, operator pay, inspection, property taxes, and repair costs during the life of the system. The replacement cost is the total cost for the replacement of equipment required during the life of the system. The residual value of the system and other incentives as well as tax benefits for owning a PV system are considered in the equation.

For calculating the LCC, the present worth of future costs of a product of present cost X can be calculated considering the inflation rate as i and the discount rate as d. The present worth of the investment that has to be made in later years would be lower by a factor Fpw-one. Fpw-one is the present worth factor for future investments and is calculated using the formula given below:

$$F_{pw} - \text{one} = \frac{\text{Future cost}}{\text{Future value}} = \frac{X(1+i)^n}{X(1+d)^n} = \left[\frac{1+i}{1+d}\right]^n$$

Table 8.4 Estimation of the LCC and LCOE for a 52 kW PV system.

Name of the item	Cost in $
Solar photovoltaic modules	16,640
Inverter (grid-tied)	2000
Balance of system	13,565
Total cost of system	32,205
Installation cost	2930
Total cost including permissions and administration expenses	40,405.25
	40,500
O&M cost	1215
Inflation rate	4%
Discount rate	8%
	0.963
Cost of inverter	1371.28
O&M charges for 20 years	16,739.49
Capital cost	40,500
Inverter replacement cost after 10 years	1372
O&M charges for 20 years	16,740
Lifecycle cost	58,612
Average energy generation for 1 kW per day in the first year (kWh)	3.7
Energy generation of 52 kW for 25 years	1,123,616
LCOE ($/kWh)	0.052

If the present cost of a product is C_0, then the present worth of one time investment down the line of n years is

$$\text{PW one} = C_0 \left[\frac{1+i}{1+d} \right]^n$$

The LCC method is used to estimate the proposed stand-alone PV system cost. Table 8.3 gives the list of the components of a 52 kW grid-tied PV system, and the life cycle cost of the system and LCOE has been calculated in Table 8.4. The LCOE is 5.2 US cents per kWh.

Fig. 8.33 shows the LCOE values for year 2019 in different countries for residential and commercial based rooftop systems compiled by IRENA [14]. The variation of LCOE in different countries is due to differences in involved soft cost, labor cost for system installation and operation and maintenance, and finance costs. The LCOE values for rooftop systems in India and China are lower. The reduction in solar PV costs has been driven by cost reductions in solar PV modules. The continued improvement in module efficiency and improvements in manufacturing, reduced labor costs through improved

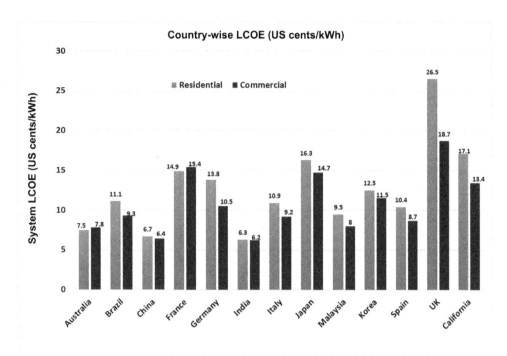

FIG. 8.33

LCOE variation for residential solar PV systems in different countries [14].

productivity and increased factory automation, economies of scale in manufacturing, along with vertical integration of the manufacturing process increased competition among suppliers caused the reduction in installed costs of solar PV plants.

References

[1] J. Sreedevi, N. Ashwin, M. Naini Raju, A study on grid connected PV system, in: IEEE, Power Systems Conference (NPSC), December, 2016, pp. 1–6.

[2] J. Worden, M. Zuercher-Martinson, How inverters work, in: Solapro 2.3, April–May 2009, pp. 68–85.

[3] R. Teodorescu, M. Liserre, P. Rodríguez, Grid Converters for Photovoltaic and Wind Power Systems, John Wiley & Sons, Ltd., 2011

[4] R. Erlichman, Distribute inverter design, in: Solarpro 6.5, August–September 2013.

[5] Installation and Operation Manual of Enphase IQ 7, IQ 7+, and IQ 7X Micros, May 15, 2018.

[6] Technical Note—SolarEdge Fixed String Voltage, Concept of Operation—9/2012.

[7] M. Cavanagh, In Blog, Inverters, Micro Inverters & Optimisers, Solar Panels, https://mcelectrical.com.au/blog/tigo-energy-solar-panel-optimisers/, September 27, 2018.

[8] Detecting Earth Faults With Module-Level Monitoring—Case Study by Tigo. https://www.tigoenergy.com/library/view/Detecting+Earth+Faults+with+Tigo%27s+M.

[9] SUN2000P-375W Smart PV Optimizer, Huawei Technologies Co., Ltd. Issue 04, 2019-07-05.

[10] Global Building Integrated Photovoltaics (BIPV) Market Outlook & Forecast (2019-2024), February 2019.

[11] A.P. Dobos, PVWatts Version 5 Manual, NREL, September 4, 2014.

[12] P.E. Bill Brooks, Solar Array Voltage Considerations, Solarpro Issue No. 3.6, October/November 2010, pp. 68–75.

[13] F. Ran, D. Feldman and R. Margolis, U.S. Solar Photovoltaic System Cost Benchmark: Q1 2018, 2018, National Renewable Energy Laboratory.

[14] IRENA, Renewable Power Generation Costs in 2019, International Renewable Energy Agency, Abu Dhabi, 2020, pp. 61–73.

[15] J. Berdner, Voltage rise considerations for utility—interactive PV systems, In: Solarpro 5.4, June–July 2012, pp. 14–20.

[16] B. Brookes, The Ground-Fault Protection BLIND SPOT: A Safety Concern for Larger Photovoltaic Systems in the United States, January 2012, A Solar ABCs White Paper.

[17] C. Kost, S. Shammugam, V. Jülch, H.-T. Nguyen, T. Schlegl, Levelized Cost of Electricity Renewable Energy Technologies, 2018, Fraunhofer Institute for Solar Energy Systems, ISE Publication.

Grid-connected solar PV power systems

9.1 Introduction

A grid-connected solar photovoltaic (PV) system is a power system that generates electricity using solar PV modules as the generator while the inverter is the converter and controller of the DC power generated by the solar array. The inverter, associated with other control electronics and devices, delivers power to the grid. The basic components of a grid-connected solar PV power system are the solar PV modules, the junction box (DC/AC), the power conditioning unit (PCU), the AC distribution board (ACDB), and the transformer, switchgear, switchyard, and other control and isolating devices. These systems could be small residential, commercial, or industrial to large-scale, land-based, megawatt-level grid-connected solar PV power systems.

In a grid-connected system, the solar power generated during the daylight hours is fed to the grid without any storage of energy. Of late, storage solutions have been integrated with large MWp-level ground-mounted or rooftop grid-connected solar PV power systems.

As of 2019, a total of 580 GWp of solar PV power systems had been installed around the world [1], out of which the maximum share is from ground-mounted grid-connected solar PV power plants. As per a PV magazine report, the global solar demand is expected to reach 144 GW in 2020 while in the following three years, new PV additions are forecast to total 158, 169, and 180 GW, respectively [2].

The grid-connected solar PV power plants are broadly divided into three segments:

1. Ground-mounted, grid-connected solar PV power plants.
2. Roof mounted, grid-connected solar PV power plants.
3. Building-integrated, grid-connected solar PV power plants.

This chapter deals with ground-mounted, grid-connected solar PV power plants while Chapter 8 dealt in depth with rooftop and building-integrated grid-connected solar PV power systems.

9.2 Components of grid-connected solar PV power systems

The schematic of a grid-connected system is given in Fig. 9.1. Solar modules are connected in series to form a string to get the required voltage. The number of strings is combined and paralleled in a string combiner box. The outputs of such string combiner boxes are connected to the input of the inverter. The MPPT of inverter tracks the maximum power of solar array and the inverter converts the DC power into

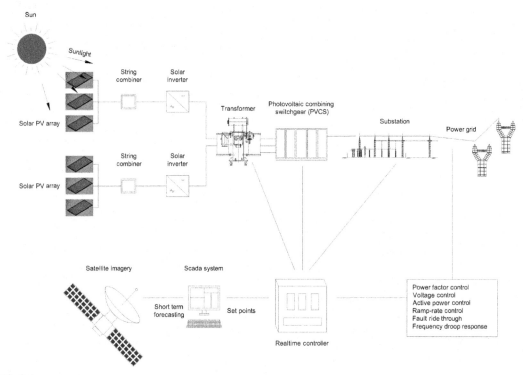

FIG. 9.1

Schematic of a grid-connected solar PV system.

three-phase AC power. The inverter output voltage is stepped up to the required voltage of 11 or 22 or 33 kV, depending on the requirement, using an inverter duty transformer. The outputs of the transformers are combined and paralleled in a switchgear. The switchgear with protection systems transmits the power to the switchyard and there, the power is evacuated to the transmission line. There will be a SCADA system to monitor the electrical parameters of the plant and weather details and provide the remote monitoring facility. Broad system components of a ground-based grid-connected solar PV power system are as follows:

1. Solar PV Modules
2. DC Array Junction Box (String Combiner Box)
3. Inverter/Power Conditioning Unit
4. AC Distribution Board
5. DC and AC Cable, HT cable
6. Module Mounting Structure
7. Single Axis Tracker (optional)
8. Transformer
9. Isolators
10. LT Panel
11. Switchgear/HT Panel

12. HT Switchyard/Substation
13. Metering Unit

9.2.1 Solar PV modules

Solar PV modules are available with different technologies, different wattages, different sizes, different brands, and with different costs. This discussion is confined to crystalline silicon solar cell technology-based modules.

There are solar cells available with mono- and multicrystalline types. Under the multicrystalline solar cell category, the standard BSF type and poly-PERC type are available. Under the monocrystalline solar cell category, standard Al-BSF-based p-type, p-type PERC, n-type IBC, HJT/HIT, PERT, and PERL cell-based modules are available. In addition to the above, half-cut cell technology-based and bifacial cell-based solar modules of different wattages are available.

Mono- versus multicrystalline Si solar PV modules

A monocrystalline Si solar cell has the potential to improve its efficiency because of its high-purity wafer. A monocrystalline Si solar cell has more conversion efficiency compared to a multicrystalline Si solar cell of the same size of wafer and a similar cell structure. Due to the presence of boron as a dopant in the p-type monocrystalline Si solar cell, it exhibits about 2% of light-induced degradation (LID), whereas a multicrystalline Si solar cell shows about 1% LID. Earlier multicrystalline Si solar cell-based Al-BSF technology was popular due to the cheaper price of polycrystalline wafers. In 2018, the market share of multicrystalline Si solar cell-based modules was 60%. With the adoption of diamond wire-based wafer cutting, the monocrystalline wafer cost has come down and monocrystalline wafers are dominating the industry.

Recently, poly-PERC and mono-PERC modules have been developed. These PERC technology-based modules exhibit a degradation in power when exposed to light and high-temperature conditions and this phenomenon is called light-enhanced temperature-induced degradation (LeTID). Solar cell manufacturers have solved the problem. Before going for PERC-based solar modules, it is required to know the LeTID mitigation technology and the LeTID test results while the performance of the modules installed in the field has to be checked.

N-type mono-based modules such as IBC and HJT have a higher conversion efficiency, a lower temperature coefficient of power, and better field performance. These modules are expensive, but based on generation, if LCOE is considered, they outperform normal standard technology-based modules. These modules are used in areas where there are space constraints to accommodate a greater capacity of solar arrays.

If the module efficiency is 19%, this means that in a one meter area, the module generates 190 W of power at standard test conditions. With the use of higher-efficiency solar PV modules, less space is required to accommodate more array capacity. The number of module mounting structures, cable lengths, and string combiner boxes are reduced and in turn the cost of the BOS items will be reduced. It is better to go for higher-efficiency solar PV modules.

Solar modules should have low temperature coefficients in power and voltage to get better generation during high-temperature conditions. The temperature coefficient of power of a conventional mono- or multicrystalline module system a decade ago was $-0.5\%/°C$. The value of the temperature coefficient of voltage depends on the band gap of the material and the dark current (I_o) of the solar cell. A decade ago, a $3\,\Omega$-cm resistivity conventional solar cell with a V_{oc} of $600\,mV$ showed a reduction of $2.3\,mV/°C$ in voltage with a rise of $1°C$ in temperature. With the improvements of technology such as passivation and selective contact technology, the dark current of the cell has been considerably reduced and the V_{oc} value has been increased. A mono-PERC cell gives $666\,mV$, an Al-BSF standard mono- or multicell gives $640\,mV$, an IBC cell gives $690\,mV$, and an HJT cell gives $703\,mV$ at STC conditions. The V_{oc} of the cell has been improved with a reduction of I_o due to the incorporation of advanced technologies in the crystalline Si solar cell. The reduction in dark current of solar cell is contributing for the lower temperature coefficient of V_{oc} of the cell. The temperature coefficients of solar modules with different brands of similar technology and construction should have the same values. The variations in the voltage temperature coefficients of solar modules with different brands is arising due to measurement accuracies in the testing.

60 Cell module versus 72 cell module

For 60 and 72 cell solar modules, the bypass diodes are connected across 20 and 24 cells, respectively. If one cell is shadowed in a 60 cell module, the voltage of unshadowed 19 cells will come across the shadowed cell in the reverse bias condition, whereas for a 72 cell module, 23 cells' voltage will be coming across the shadowed cell. In one cell shadowed condition, 19 cells' voltage will bias the shadowed cell in a 60 cell module, compared to 23 cells' voltage biasing the shadowed cell in a 72 cell module. In a 60 cell module, the shadowed cell will have less reverse bias stress compared to the 72 solar cell module in a frequent continuous shadowed condition. With the use of a 72 cell module, the number of strings will come down. The cable lengths and the quantity of string combiner boxes will be reduced. The space requirement will also be reduced compared to a 60 cell module. As the number of modules is reduced with the use of 72 cell modules, the module installation cost also will be cheaper compared to 60 cell modules.

The quality of modules can vary significantly, so due diligence should be exercised while selecting and procuring modules. This may include verifying the antecedents of the manufacturer, and independent checks on the quality of the module. Although most modules available in the market carry IEC certification, it should be noted that the IEC certification is really a certification of the module design and does not guarantee that the module will perform adequately throughout its intended life. The test data of accelerated testing has to be verified.

An independent audit of modules may include detailed testing of randomly selected modules during the manufacturing process and auditing the manufacturing process.

The evaluation of solar modules involves the following:

(a) Evaluation of manufacturing facilities—equipment and test facilities.
(b) Verifications of all certificates and test reports given by the testing agency and the accelerated testing data. The certificates include:
  (i) IEC 61215 crystalline silicon terrestrial PV modules: design qualification and type approval as per the latest 2016 edition.
  (ii) IEC 61730 part 1 and 2: Module construction and PV module safety qualification as per the latest edition.
  (iii) IEC 62804: potential induced degradation.

    (iv) IEC 61701: marine environment: salt mist test.

    (v) IEC 62716: agricultural environment: ammonia resistance.

    (vi) IEC 61215-10.16: high snow and wind loads mechanical resistance.

    (vii) IEC 61215-10.12: cold environment: humidity freeze.

    (viii) IEC 60068-2-68: desertic conditions: dust and sand.

    (ix) Accelerated tests conducted on the modules and their details.

(c) Verification of all documents related to Factory certifications, specifications of all bill of materials (BoM), quality control plan, consistency and adequacy of the documentation related to quality, etc.

    (i) ISO 9001 certificate of the manufacturing unit.

    (ii) ISO 14001 certificate.

    (iii) OHSAS 18001 certificate.

    (iv) Product datasheet.

    (v) PAN files verification.

    (vi) Quality control plan.

    (vii) Production flow chart.

(d) Supervision of the production process

    The principle of production supervision is to ensure that the vendor uses the correct components and that the manufacturing processes implemented can guarantee the conditions to obtain the required level of quality for the products.

    (i) Material control auditing.

    (ii) Production processes auditing and production monitoring at the vendor's module production facility.

The supervision starts with the reception of the components intended to be used for the production lot until completion of the packing and shipment of the products from the factory.

    The material control shall cover:

- Actual use of agreed BoM material (solar cells, EVA, back sheet, solar ribbon, frames, junction box, glass, etc.).
- Control of the correct Incoming inspection of the materials, including supplier reports.

The process control should cover:

- Incoming inspection of materials.
- Proper storing of the materials in a warehouse.
- Module manufacturing (tabbing and stringing/layup/lamination/framing/junction box assembly/curing/cleaning).
- Performance tests, especially flash test (calibration, reference module).
- Reliability tests, especially electroluminescence test.
- Safety tests.
- Packing.

Online and offline process checks during module manufacturing, which include:

- Solder joint strength.
- EVA gel content test.
- EVA/glass and EVA/back sheet peel strengths.

- Frame adhesion check.
- Diode polarity check and Hi-pot test.

(e) Skills and competencies of operating personnel.

(f) Packaging inspection and handling of solar modules.

    Checking the documentation of qualified tests on packaging as per IEC.

(g) Preshipment/container loading inspection.

(h) Third-party testing of solar modules for flash test and electroluminescence test.

(i) The following are the tests to be conducted on the modules:

    (i) *I-V* curve testing; **(ii)** visual inspection; **(iii)** electroluminescence test; **(iv)** Hi-pot test; **(v)** dielectric withstanding test and wet leakage; **(vi)** dimensional check of the module; **(vii)** circuit dimensional check as per drawing of module; and **(viii)** anodization check of the frame. LeTID test results, low irradiance performance test results, potential-induced degradation, and dynamic mechanical load test results have to be checked.

    As part of accelerated tests, PVPL & DNV GL conducts tests on solar modules two to three times of the specification that is specified in IEC standards and provides solar reliability score card [3]. Thermal cycling test from –40°C to 8°C will be conducted for 800 cycles at interval of 200 thermal cycles, whereas as per IEC 61215 standard, it is 200 cycles only. The Damp Heat test will be conducted for 2000 hours instead of 1000 hours. UV exposure test will be for 90 kWh, which is double the UV energy specified in IEC standard. Dynamic mechanical load test will be conducted for 1000 cycles from –1000 to +1000 kPa. After that, 50 thermal cycles (TC50) will be conducted on the same module. After TC 50, the same module undergoes 30 cycles of humidity freeze test. The PID test will be conducted for a duration of 96 × 2 hours. Pre and post of every test under each stage of interval, visual inspection, *I-V* testing, EL testing, and wet leakage tests are conducted on the modules. These type of tests provide more confidence to the customer and the reliability of modules will be very high.

9.2.2 Inverters or power conditioning units

There are three types of inverters/PCUs that can be considered for grid-connected solar PV power plants.

1. Central inverter or power conditioning unit.

2. String inverter.

3. Bidirectional inverters with energy storage application.

9.2.2.1 Central inverters

Central inverters are single units or multiple units key to a grid-connected solar PV power plant. They can be at a single location or multiple locations in a grid-connected solar PV power plant. The central inverters are available from 500 W to 4.5 MW capacity with meeting the maximum DC voltage of 1000/1500 V.

    Some inverters are designed with air cooling with exhaust fans and some inverter components are cooled using a circulating coolant.

Some inverters will have the ability to monitor all the parameters in the display while some inverters will not have monitor display facility and the parameters can be monitored in the SCADA system.

Inverters are called power conditioning units. A PCU consists of an electronic inverter along with associated control, protection, and datalogging devices as well as remote monitoring hardware that is compatible with software used for string level monitoring.

The inverters will have a European efficiency >98% as per the IEC 61683 standard. The PCU is designed for indoor or outdoor installation, taking care of the different environmental conditions of the plant locations. The inverters shall have a minimum protection of IP 21 and Protection Class II. The voltage and frequency synchronization with the grid is achieved by sensing the grid voltage and phase and feeding this information to the feedback loop of the inverter. The inverters have self-commutated pulse width modulation (PWM) technology and control the output voltage and frequency. Figs. 9.2 and 9.3 show the outside and inside views of a 1 MW inverter manufactured by Fimer Group under ABB trademark licence [4].

The PCUs are capable of controlling the power factor dynamically by supporting the reactive power requirement. Central inverters are integrated with a single maximum power point tracker (MPPT). So, for example, for a 1000 kW inverter, more than 100 strings will be connected to the input of the inverter. The current mismatch with a large number of strings will be more compared to string inverters.

The system will be in automatic sleep mode at night. It automatically wakes up in the morning and begin to export power, provided there is sufficient solar energy and the grid voltage and frequency are in the required range. A self-diagnostic system check occurs upon start-up of the inverter. It checks the

FIG. 9.2

ABB central inverter of capacity 1 MW.

Courtesy: Fimer SPA.

FIG. 9.3

1 MW Fimer (ABB) central inverter—inside view.

Courtesy: Fimer SPA.

insulation resistance of the solar array and if the value is lower than the specified value, the inverter will not start its operation. During operation, if the insulation resistance comes down due to earth fault of the string, the inverter automatically gets tripped. The inverter will have a sinusoidal current modulation with excellent dynamic response. Comprehensive network management functions such as LVRT and the capability to inject reactive power to the grid are included in the inverter. Total harmonic distortion (THD) in current is <3% and it shall have minimum load loss in sleep mode. Power factor control range: 0.8–0.9 (lead-lag).

The central inverters are based on IGBT technology and have the following features:

- Earth fault monitoring.
- Grid monitoring.
- DC input monitoring through hall sensor or current transformer.
- Anti islanding.
- DC reverse polarity.
- Overcurrent protection (AC and DC side).
- Short circuit protection (AC and DC side).
- Undervoltage/overvoltage protection.
- Grid synchronization loss protection.
- Over temperature protection.
- kW, kVA, kVARH, kWH, PF, Hz, V, A, and kVARH monitoring and recording.
- Emergency stop switch.
- Microprocessor-based circuit breakers with plug settings in both DC and AC sections for short circuit, overload, and earth fault protection.

- Suitable contactor for grid connection with low voltage ride through (LVRT)/high voltage ride through (HVRT), arc fault features.
- Surge protection device.

Standards followed for inverters
  Efficiency measurement: IEC 61683
  Environmental testing: IEC 60068-2 or IEC 62093
  EMC, harmonics, etc.: IEC 61000 series, 6-2, 6-4
  Electrical safety: IEC 62109 (1 and 2), EN 50178
  PV—Utility interconnections: IEEE standard 929-000
  Protection against islanding of grid: IEEE1547/UL1741/IEC 62116
  Grid reliability test standard: IEC 62093

The inverter manufacturers provide protection for the basic parameters that include: DC overvoltage, AC overvoltage, AC short circuit, DC short circuit, frequency out of range, voltage out of range, DC inverse polarity, ground fault, overcurrent, grid synchronization loss, over temperature, DC bus overvoltage, cooling fan failure in short circuit, surges due to lightning, negative grounding (GFDI), insulation monitoring, HVRT, LVRT, antiislanding, and AC output breaker. The inverters support the reactive power whenever required.

In addition to those protections, the following technical aspects are very important to understand solar inverters for solar projects.

Maximum power point tracking voltage range

A good solar power inverter has a large MPPT operating voltage range to maximize the power extraction from the solar array. The maximum power point of an array should always be within the inverter's MPPT window through a wide variety of operating conditions. The solar array output voltage depends on the module temperature and the solar irradiance falling on the array. The PV output voltage is maximum during early morning in the winter season and lowest at noon on hot summer days. Very cold winter days may drive the system voltage above the inverter's maximum voltage rating, or hot summer days may leave the inverter unable to track down to the lowest peak power voltage. The expected solar PV array maximum power point voltage needs to be within the MPPT range of the inverter. In the event that the solar array voltage is outside the range of maximum power point voltage, the inverter is not capable of generating the maximum energy. This voltage window of the MPPT decides the minimum number of solar PV modules required in a string to be connected to the inverter.

Total harmonic distortion (THD): The total harmonic distortion (THD) is defined as the ratio of the sum of the power of all the signal harmonics to that of the power of the fundamental frequency. THD, always expressed as a percentage, describes the power quality entering the utility grid from an interconnected inverter.

There are three types of efficiency rankings used for inverters.

Peak efficiency is the maximum percentage of DC input power inverted to AC output power. It indicates the performance of the inverter at the optimal power output. Every inverter is tested at a range of input voltages and power levels. The results of these tests are often summarized as a single efficiency curve that is published on an inverter data sheet. Fig. 9.4 shows the efficiency versus percentage of loading as a function of operating DC input voltage for a typical inverter.

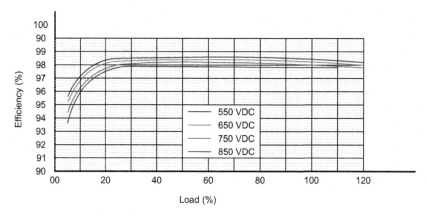

FIG. 9.4

Efficiency versus percentage of loading curve of an inverter as a function of voltage.

For two-stage inverters, the efficiency increases as the DC input voltage increases, whereas for single-stage inverters, the efficiency decreases with the increase of DC input voltage.

European efficiency is defined as an averaged operating efficiency over annual power distribution corresponding to central Europe climatic conditions. The Joint Research Center (JRC/Ispra) has developed this definition based on the climate of its test center at Ispra, Italy. This is used by all inverter suppliers. It is more useful than highest peak efficiency, as it demonstrates performance at different loading levels of the inverter during a typical day.

The empirical formula for European efficiency is

European efficiency $= 0.03 \times$ Eff at $5\% + 0.06 \times$ Eff at $10\% + 0.13 \times$ Eff at $20\% + 0.1 \times$ Eff at 30% $+ 0.48 \times$ Eff at $50\% + 0.2 \times$ Eff at 100%.

The above formula can be explained as follows. For example, $0.48 \times$ Eff at 50% is explained as the efficiency of the inverter at 50% of the nominal power of the inverter and the inverter operation time in this condition is 48% of the total inverter operation. This implies that the European efficiency considers the areas with medium solar irradiation levels. Considering different levels of solar irradiance and their occurrence as well as input power operating at different time periods, the weighted average efficiency shall be determined.

California Energy Commission (CEC) efficiency is also a weighed efficiency. This is very similar to the European efficiency. However, it has different assumptions on weighing factors. The CEC proposal includes considering climates of higher solar irradiance regions like the US southwest regions. This is now specified for various inverters installed in the United States.

California Energy Commission (CEC) efficiency is defined by the following empirical equations:

$$\textbf{CEC efficiency} = 0.04 \times \text{Eff at } 10\% + 0.05 \times \text{Eff at } 20\% + 0.12 \times \text{Eff at } 30\% + 0.21 \times \text{Eff at } 50\%$$
$$+ 0.53 \times \text{Eff at } 75\%. + 0.05 \times \text{Eff at } 100\%.$$

CEC efficiency considers a 53% operating time of the inverter with 75% of nominal power.

Each inverter is subjected to tests at six input power levels between 10% and 100% of the inverter rated power and at three DC input voltages such as low, nominal, and high DC voltage. Averaging of the test results takes into account the amount of time that an inverter is operational in each loading range of operation.

It is the effective efficiency of the inverter across its operating range of loads. In EU efficiency, the highest weightage is given for half load (that is, 48% weightage for 50% load of its rated capacity). Then, 20% weightage for full load efficiency, 10% weightage for one-third of the rated load efficiency, 13% for one-fifth load, 6% for one-10th load, and 3% for one-20th load. From a customer point of view, the higher the number in EU efficiency, the higher the efficiency at half load, but a relatively lower EU efficiency inverter might offer higher efficiency at higher loads.

Auxiliary consumption

For the control and operation of the inverter, power is required. This power is called auxiliary consumption, which is divided into three categories. The power consumption takes place to run the exhaust fans to cool the inverter or to drive the pump of the coolant, to power the control circuits for operation and standby, or night time on conditions. Lower auxiliary consumption will result in higher energy.

Temperature derating: The thermal behavior of inverter has to be checked by studying power output in kVA versus ambient temperature graph of the inverter. The typical temperature versus output kVA of the inverter graph is shown in Fig. 9.5. This will help to understand when the output of inverter will start derating and at what temperature.

Many central inverters control power fold-back by internally monitoring the temperatures of certain critical reliability components; others use ambient temperature and instantaneous power-level measurements to predict component temperatures. The inverter control system regulates these component temperatures, to ensure that overheating does not compromise the lifetime of the critical reliability components such as IGBTs or DC link capacitors. Power fold-back is a condition of state in which the inverter lowers its output power in response to abnormal conditions such as a higher

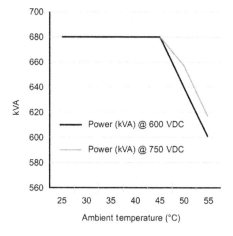

FIG. 9.5

Thermal behavior of a central inverter.

temperature/input power or other conditions. Typically, inverters do this power fold-back by shifting the solar array operating voltage from its maximum power point.

9.2.2.2 String inverters

In a solar PV power plant, the solar PV modules are connected in series and parallel to achieve the required kWp or MWp solar PV array capacity. Solar PV modules are installed in rows for ground-based systems. For example, if we have 60 solar PV modules, 10 are connected in series to form a string to get the desired input DC voltage. The number of parallel strings, in this case six, can be connected in multiple strings to a single string inverter. Multiple strings add up to a capacity or more than the string inverter DC rating. The string inverter converts the input DC power to AC power and is fed to the LT bus or stepped up using a transformer to transmit power to the nearest grid station.

First, string inverters were in the single-phase version around 2007/2008. These single-phase inverters were initially connected in a three-phase configuration. Subsequently, three-phase string inverters were designed and have become the first choice for large-scale MWp-level solar PV power plants without shading issues. Also, string inverters provide redundancy as their rating is around 150–200 kVA while central inverters come in multiples of 500 kVA or larger MVA capacity.

In the event of shading on one of the solar PV modules in a string, the reduced power generation is limited to the shaded string only while power generation remains intact for other strings that are not shaded. The technology of string inverters has become rugged, which helps in generating 2%–3% more energy while providing redundancy.

The use of an array junction box or a string combiner box can be avoided with a string inverter while the ACDB is a must to combine the AC output of all string inverters before connecting to the grid.

As regards comparable energy generation, string inverters generate more than 2%–3% energy as regards a central inverter while the cost of a string inverter is higher compared to a central inverter. Three-phase string inverters are available for 1000 and 1500 V system voltages. These will have more MPPTs so that the array circuits in different orientations and different string lengths can be combined with the modules used for the string inverter. With the use of string inverters, the current mismatch will be reduced and there will be an increase in the energy yield compared to a central inverter. The string combiner boxes are not required for string inverters, whereas AC combiner boxes to combine the string inverters will be required. As the inverter output voltage is higher, relatively higher size AC cables are required from inverter to AC combiner boxes and from AC combiner boxes to the LT panel, where the power is evacuated. The plant with string inverters will be costlier compared to the plant with central inverters. The BOS cost will increase with the string inverters.

With the usage of string inverters civil work is reduced compared to central inverters, as there is no requirement of construction of inverter room. The failure and maintenance loss is much higher for centralized inverters than string inverters. Even if a string inverter fails, you don't lose much generation. Repair of string inverters is also easier than central inverters. To manage the temperature of internal components in summer, the inverters come with a self-cooling mechanism, equipped with IP67 fans to avoid overheating. The lifespan of inverters covers around 25 years. Inverters will have integrated cloud-based communication platform system. The monitoring device collects all the data of individual inverters and sends it to a server and one can monitor more inverters through RS485 connection. The performance of the whole plant as well as that of individual inverters from the office or any remote location around the world can be checked. Some of features of the latest inverters are one-click smart *I-V* curve diagnosis across entire plant, grid side failure function recording, and online inverter support service. With string monitoring and *I-V* curve scanning, it is easier to recognize a fault in a string or a low performance string. It also enables total harmonic distortion voltage (THDv), ISO and grid line impedance analysis, and the wave record of the inverter. The DC isolation and AC line impedance show the system condition on the DC/AC side to let the O&M personnel know when, where, and why there may be a fault. Different aspects of string inverters are discussed in Chapter 8.

For string inverters, there is no provision for negative grounding. So, the solar array of the plant with a string inverter will be with floating ground. As the system voltage is increasing from 1000 to 1500 V, there is a greater probability of the PID effect on the solar modules. If there is a problem with a string inverter such as rodents eating cables or other issues, the inverter cannot be repaired on the plant premises and it has to be replaced. If problems arise with string inverters during their lifetime of operation, there is a possibility of replacement of all the string inverters within a period of 10 years. The cost of maintenance with string inverters is more compared to central inverters.

To meet grid code requirements, the inverter will has the provision of LVRT/HVRT and the support of reactive power. This aspect is easily met by the central inverters, whereas the old version string inverters may not be able to meet the LVRT/HVRT and reactive power support requirements. Nowadays the string inverters are having grid supporting functions such as active and reactive power management, low/high voltage ride through and Q at night function. Q at night function provides PV power plant operators with key advantages. For starters, it meets the reactive power needs of their PV power plant and supplies power to any installed local electrical appliances, eliminating the need to purchase reactive power or the expenses for a compensation plant.

As there is no galvanic isolation in the string inverters, the transient currents flowing from the transformers to the inverter cause the failure of the bypass diodes of nearby solar modules.

It is better to study the financial aspects, including the maintenance and energy generation aspects, and decide on the use of string inverters for large megawatt-level solar PV plants. But project developers and EPC players are using string inverters for ground-mounted solar plants of size 100 MW also.

9.2.3 DC cables and connectors

Every solar module comes with a single core 4 mm^2 copper cable of solar grade, meeting all the requirements laid down by the IEC standards. The cables also come with MC4 connectors. The solar modules are mounted on the module mounting structure and connected in series using the cables and connectors assembled with the modules. A number of modules are connected in series to form a string. Single strings or double strings connected via Y-connectors are combined in a string combiner box. The minimum guarantee expected for all the solar photovoltaic cables will be 25 years. So, the cables have to meet stringent requirements. Cables and associated connectors, which are used for the interconnection of solar arrays and all other BOS components, must be of solar grade and must withstand the harshest environmental conditions such as high temperature, snow, UV radiation, rain, humidity, dust, salt spray, invasion by fungi, etc., and must comply with the latest IEC standards.

Construction
- Conductors are electrolytic-grade high-conductivity annealed tinned copper.
- Conductors are multistranded, smooth, uniform in quality, and free from scale and other defects.
- Cables are essentially resistant to UV, ozone, and high-temperatures.

- The insulation, the inner sheath, and the outer sheath are of high-grade cross-linked compound (XLPE).

Cables are a flame-retardant, low-smoke (FRLS) type designed to withstand all mechanical, electrical, and thermal stresses.

For single-core armored cables, the armoring is of aluminum wires. For multicore armored cables, the armoring is of galvanized steel. The minimum area of coverage of armoring generally is 90%.

Typical technical specifications of DC cables for 1000/1500 V nominal system voltage:

Electrical characteristics
- Rated DC voltage: 1.5 kV.
- Maximum permitted DC voltage: 1.8 kV (conductor/conductor, nonearthed system, circuit not under load).
- Maximum permitted AC voltage: 0.7/1.2 kV.
- Working voltage: DC 1000 V/1500 V.
- Insulation resistance: 1000 mΩ/km.
- Spark test: 6000 VAC (8400 VDC).
- Voltage withstand: 6500 V as per EN 50395 for 5 min.
- Ampacity: according to requirements for cables for PV systems.

Thermal characteristics
- Ambient temperature: −40°C to 90°C.
- Maximum temperature at conductor: 120°C (20,000 h).
- Short circuit temperature: 200°C (at conductor max. 5 s).
- Thermal endurance test: according to EN 60216-2 (temperature index 120°C).
- High temperature pressure: test according to EN 60811-3-1.
- Damp-heat test: according to EN 60068-2-78.
- 1000 h at 90°C with 85% humidity.

Chemical characteristics
- Mineral oil resistance: according to EN 60811-2-1.
- Ozone resistance: according to EN 50396 part 8.1.3 method B.
- Weathering-UV resistance: according to HD 605/A1 or DIN 53367.
- Ammonia resistance: 30 days in saturated ammonia atmosphere (internal testing).
- Very good resistance to oils and chemicals.
- High wear and robust, abrasion-resistant.
- Acid and alkaline resistance: according to EN 60811-2-1.

Fire performance
- Flame-retardant according to IEC 60332-1-2.
- Low smoke emission <20% as per ASTM D-2843.
- Halogen-free according to EN 50267-2-1/-2, IEC 60754-2.
- Acid gas emission not more than 0.5% as per IEC 60754-1 pH minimum 4.3 as per IEC 60754-2 conductivity maximum 10 as per IEC 60754-2.
- Toxicity according to EN 50305, ITC-index <3.

The cable size has to be properly selected considering the current to be carried, the length of the cable, and other environmental factors such as temperature, cable bundling factor, etc., for the derating of the cable. The power loss due to voltage drop should not exceed 2%. For a better energy yield, a higher size of cable will help with reduced cable losses.

Connectors: The straight connectors suitable to mate with MC4 or equivalent connectors are used in solar PV plants to connect source circuits to string combiner boxes. To avoid DC arc faults in the solar plant, reliable connectors MC4 or equivalent are to be evaluated. The solar module connectors should be water tight sealing conforming to IP 68 and are supplied as male (plug) and female (socket) types. These are suitable for crimping $1C \times 4$ mm^2 and $1C \times 6$ mm^2 XLPE insulated copper cables. They are available for system voltages of 1000 and 1500 V. The temperature variations by environmental factors cause expansion and contraction of contact interfaces of connectors. The relative movement of the contact interfaces develops stress. If the temperature rise is more for any connector, with a period of time the connector dimension may change due to expansion and contraction of connectors and the connectors might get breakdown fast. Corrosion can happen through the ingress of contamination at the contact interface. Gradual age-related changes also change the stiffness and morphology of the connector material [5]. For electrical compatibility, the contact resistance of the connectors has to be measured as per IEC 60512 and temperature rise has to be measured as per EN 50521:2008. In addition, the mechanical compatibility has to be confirmed by insertion-retention force test. The water proof test, insulation resistance test, and wet leakage current tests are to be conducted as per EN 50529. The contact resistance should be <0.5 mΩ. The insertion force should be <50 N and the retention (withdrawal) force should be >50 N.

To reduce the quantity of source circuit cables, Y-connectors will be used. Two single solar array strings are connected parallelly to form two parallel configurations, these Y-connectors are useful. The Y-connectors on positive terminal will have a fuse and meet all the requirements that are mentioned for MC4 connectors.

9.2.4 Array junction box/DC string combiner box

All the strings are to be connected to the inverter. The strings are grouped and combined in a string combiner box (SCB). The SCB is also called the array junction and a string monitoring facility can be incorporated. These can be configured for many inputs and with a single output.

The SCB is a box with an enclosure made to be dustproof, verminproof, and waterproof. It is made of thermoplastic/metallic materials in compliance with IEC 62208. The junction box enclosure is sunlight/UV resistive as well as fire-retardant, suitable for outdoor installation. It will have an IP 65/IP 66 degree of protection. The junction boxes will have suitable cable entry points fitted with cable glands of appropriate sizes for both incoming and outgoing cables. The individual string terminal enters the junction box through the incoming gland and connects to the respective polarity terminals. There are two copper busbars of appropriate size, one for positive and the other one for negative terminals. The terminals are connected to the respective positive and negative copper busbars. Thus, the positive and negative terminals of the strings are separately combined in the busbars. The independent output cables of the positive and negative terminals are taken out from the outgoing glands. There is a DC disconnector switch of the appropriate current-carrying capacity integrated in the SCB.

There will be a surge protection device inside the SCB to suppress the surge currents arising due to lightning or any other transients. The ratings are 1 kV, 25 kA, and protection level type 2 for a 1000 V system voltage. Type I + II surge protection device is preferred.

The array junction box will have suitablly rated DC isolating miniature circuit breaker as the DC disconnector switch at the output side of the junction box. The fuse is provided at both the positive and negative terminals depending on the requirement. For standard p-type wafer-based Si solar PV modules, the fuse is provided at the positive terminal for negatively grounded solar array circuits. The fuse protects the modules from the reverse current overload.

Suitable spaces for workability and natural cooling are required in the combiner box. Otherwise, the heat will build up and burning takes place. The terminal screws should be properly tightened. Loose contacts cause arcing, which causes burning of the junction box.

To reduce the quantity of DC cable, Y-connectors are employed to parallel the two strings. Then, the input DC cable carries double the amount of the string current. The string fuses and the DC disconnector switch have to be designed as per the required currents. It is preferable to have a fuse in one of the Y-connectors.

The junction boxes will have a continuous gasket made of a polyurethane seamless gasket to provide water tightness and prevent the ingress of dust. Fuses are provided on the positive side of the string in the array junction box, if the array negative terminal is required to be grounded. The array junction box should undergo qualification test as per IEC 61439-2, fire resistance/flammability test as per UL 94V, mechanical impact resistance as per IEC 62262, and enclosure protection as per IEC 60529.

A SCADA communication device such as RS 485 will be integrated for string monitoring facility, if required. There will be a DC-DC power converter to provide the power to string monitoring unit by converting from the string directly. Hall sensor or shunt is used to sense the string current. Fig. 9.6 shows the image of a string combiner box. It does not have a string level monitoring facility.

FIG. 9.6

String combiner box—inside view with arrangement of input and output of strings.

9.2.5 **Module mounting structure and types**

The means on which the solar modules are anchored are referred to as the racking system or module mounting structure (MMS). It should carry the load of the modules, withstand all the wind loads, and work for 25 years. The base columns that are penetrated into the ground support the MMS. To mount the modules, a large area is required, which is created using purlins and rafters.

Based on the soil load-bearing capacity, the base columns are fixed to the ground using ramming or piling. For pilings, reinforced cement concrete (RCC) or plain cement concrete (PCC) is preferred.

If the base columns or vertical leg assemblies of the array structures are made up of mild steel (MS) hot dip galvanized (HDP) material, it will be corrosive-resistant and provide the required strength to meet all the mechanical loads. Presently, the rafters and purlins are used with pregalvanized mild steel or high strength galvaluminum or HDG MS to reduce the weight of the MMS.

Adequate galvanization thickness is required for the materials to take care of long-term corrosion issues.

- The module mounting structural material used has be resistant to corrosion while it should have electrolytical compatibility with materials such as the solar PV module frame, its fasteners, and the nut and bolts.
- Nut and bolts and the supporting structures, including module clamping means, should be corrosion resistant. SS-304/316 material is suitable for this.
- The support structure and foundation are designed to withstand the required wind speed of the location.
- The lower end of the solar module is usually maintained at a height greater than 0.5m above the ground.
- The design of MMS should meet the local codes of the country.
- The structures are designed to allow easy replacement of any module. The requirements of the MMS for a ground-mounted system and the design methodology are presented in Chapter 7, Section 7.1.2(v).

Fixed tilt

Solar modules are to be anchored to a structure exposing the solar cell sides toward the sun in the outside. The means to mount the solar modules is called a module mounting structure. To get the proper voltage within the MPPT window in the inverter, solar modules have to be connected in series to form a string. Based on the number of modules required for a string, the size of the mounting structure will be decided. For a 1000 V system with 72 cell modules, considering a normal inverter, the string length comes to 20. The MMS, which is synonymously called a module table, will have modules of 20 or multiples of 20. For a 20 module table, the modules can be arranged in a 2×10 matrix in portrait configuration or a 2×20 or 2×40 matrix type. For these arrangements a single pole type structure as shown in Fig. 9.7 can be used. This is the fixed tilt angle type. The tilt angle will be decided based on the latitude of the site and the energy yield analysis.

Fig. 9.7 shows the structure with fixed tilt angle configuration. It has a single vertical pole in situ fixed with the concrete pile made to the ground.

FIG. 9.7

Module mounting structure with fixed tilt configuration with a single pole.

FIG. 9.8

Module mounting structure with fixed tilt configuration with double pole.

If two modules of 72 cells are in portrait configuration, the vertical distance will be 4 m. If the vertical distance exceeds 4 m, a two-pole structure is required to support the modules and withstand the wind loads. A two-pole structure MMS is shown in Fig. 9.8.

The ballast type of MMS is also used for ground-mounted systems.

For the fixed tilted angle type, the angle will be decided based on the latitude of the location. The modules are to face south if the site location is in the northern hemisphere or toward the north if the site location is in the southern hemisphere.

For a fixed tilt angle, there is not much maintenance on the structure. During winter, the modules harvest more radiation compared to the summer. Depending on the geographical condition, the enhancement of incident irradiance on the plane of the array will vary.

Seasonal tilt

To get better radiation, the modules are tilted to different angles in different months. To accommodate this feature, the MMS should have a provision to tilt the modules in two or three positions according to the requirements of the season. Latitude +10 or 15 degrees can be used for winter months from October to February in the northern hemisphere. Latitude −10 or 15 degrees can be used for summer months from April to August in the northern hemisphere. For March and September, the latitude angle can be considered as the tilt angle. This type of MMS is called a seasonal tilt structure. As most of the plant employs manual labor during the tilting of the angle, there is a possibility for the generation of microcracks in the solar modules due to mechanical stresses and the components may undergo wear and tear. But the seasonal tilt configuration may provide an energy yield of 3%–6% depending on the geographical location of the site.

In the seasonal tilt configuration of MMS, in winter conditions, the module tilt angle will be higher. Higher tilt angle MMS has to support higher uplift wind pressure. So, the MMS has to be designed for a higher tilt angle to meet higher wind loads. So, the cost of MMS will be higher compared to an MMS of fixed tilt.

As seasonal tilted MMS is with a higher angle, more area has to be left between the adjacent rows of MMS to avoid casting shadows on the modules. For seasonal tilt, the MMS should have a minimum number of solar modules for easy tilting of the MMS. So, the area required to accommodate 1 MW with seasonal tilt MMS will increase compared to a fixed tilt-based structure. Fig. 9.9 shows the MMS with a seasonal tilt arrangement. The circular arrangement shown with a vertical pole has got holes to change the tilt angle whenever required.

A seasonal tilt MMS provides flexibility in changing the tilt angle of the MMS to harness the maximum energy yield corresponding to changes in the sun's path. The overall PV output can be increased by almost 3%–6% depending on the latitude of the site by adjusting the inclination angle (tilt) of the solar panel twice/thrice a year. With a marginal increase in capex, the seasonal tilt MMS promises a

FIG. 9.9

Module mounting structure with a seasonal tilt configuration.

higher yield and incremental benefits for the long term. Seasonal tilt MMS is economical, simpler, durable, and has lower maintenance requirements than a tracking system. The frequent change of tilt angle of the modules manually causes mechanical stress in the solar modules and becomes a cause for generation of microcracks in the solar cells. If a screw jack is used to change the tilt angle of MMS, microcrack generation in modules can be minimized.

Single-axis trackers

A single-axis solar tracker is self-powered, and hence doesn't require any additional cabling at the site. The entire system comes as a plug and play philosophy in mind, which allows easy installation without any welding requirements. There are two types of horizontal single axis solar trackers. One is independent row tracker and the other is block tracker. Horizontal single axis tracker (HSAT) will have minimum tracking range of ± 50 degrees in east-west direction with a tracking accuracy is ± 2 degrees. Tracker provides flexibility to utilize plot shape in most efficient manner. The structure has to be designed to withstand wind loads as per ASCE 7-10 or with local country standards. The design should take care of withstandability of seismic loads per ASCE 7-10, yet keep the structure weight to a minimum. Tracker is designed based on wind tunnel force coefficients specific to the static and dynamic conditions experienced during operation. The HSAT will have for stability with minimum deflection and sagging. Maximum permissible limit for deflection shall be in compliance to local Building Code. Expected deflection shall not exceed $L/240$ for beams and $L/150$ for columns, where L is length/span of the tracker structural member. This shall also be in compliance to the module manufacturer's installation guidelines. The tracker will have autostowing features and its mechanical components have to withstand environmental conditions for 25 years to ensure reliability. The wireless integration helps control and monitor the tracker's critical parameters with minimal cabling. The smart controller is compatible to site SCADA and provides real-time monitoring of the tracker system.

A single-axis solar tracker provides the lowest levelized cost of electricity (LCOE). These trackers help to achieve a 15%–20% energy generation increase, depending upon the site location, as compared to a fixed module mounting structure. Also, there is not as significant an increase in the initial cost, which is around 10% of the total cost of a solar PV power project. With continuous lowering of the feed-in tariff, a single-axis solar tracker provides a competitive edge over the fixed module mounting structure. Figs. 9.10 and 9.11 show the module mounting structure with a single-axis tracker. The modules are mounted facing east and tracking by east to west. The tracker uses servocontrolled-based motors to push and pull the structural yoke connected to the arrangement of the mounting systems. With this single-axis tracker, the energy yield may increase to 15%–20% depending on the location where it is installed.

Dual-axis tracker

Tracking both the sun's east-to-west azimuth and the elevation off the horizon maintains a more constant and accurate angle of incidence between the collector aperture and the sun. This results in a higher capacity factor and specific yield compared to fixed-tilt or single-axis tracker mounting, which may be desirable in certain flat-plate PV applications. In solar thermal power and concentrated photovoltaic applications, dual-axis tracking is a design requirement. Fig. 9.12 shows the modules mounted on a dual-axis tracker.

FIG. 9.10

Module mounting structure with a single-axis tracker.

FIG. 9.11

A single-axis tracker with a push-pull servomotor-based system.

The power curve for any PV array mounted on a tracker is broader than that for a fixed array, and thus is deemed to add better shoulders to the curve. Fig. 9.13 shows the relative power curves for flat-plate PV mounted at a fixed tilt, flat-plate PV mounted on single-axis trackers, and concentrated PV mounted on dual-axis trackers. A dual-axis tracker with concentrated PV modules gives maximum advantage in energy generation. But the dual-axis tracker is very complex for CPV modules and less complex for normal standard modules. The energy yield with a dual-axis tracker will be 5% more compared to a single-axis tracker.

FIG. 9.12

Solar PV modules arranged in a dual-axis tracker.

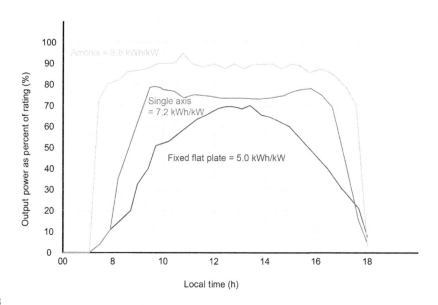

FIG. 9.13

Energy yield in a day for a fixed-tilt, single-axis tracker and a dual-axis tracker with concentrated PV [6].

Courtesy: Stephen, Solvida Energy Group Inc.

9.2.6 Civil foundations

To ascertain the soil parameters of the proposed site for construction of the control room, pilings for MMS, HT lines, array yards, and subsoil investigation will be carried out through a certified soil consultant.

The scope of the subsoil investigation covers:

- Execution of complete soil exploration including boring, drilling, and the collection of undisturbed soil sample where possible, otherwise disturbed soil samples.
- Conducting laboratory tests of samples to find out the various parameters mainly related to load-bearing capacity, ground water level, settlement, and subsoil condition.
- Submission of detailed reports along with recommendations regarding the suitable type of foundation for each bore hole along with recommendations for soil improvement where necessary.

The soil test also includes an analysis of a water sample.

Foundation: The part of the structure that lies below the ground level is referred to as the substructure or foundation. The purpose of the foundation is to effectively support the superstructure by transmitting the applied load effects (reactions in the form of vertical and horizontal forces and moments) to the soil below. Cement concrete and reinforced cement concrete are used for the foundation. Plain cement concrete (PCC) is defined as any solid mass made by the use of a cementing medium. PCC has cement, sand, gravel, and water as ingredients. Concrete is remarkably strong in compression, but it is equally weak in tension. Its tensile strength is approximately 1/10th of its compressive strength. Reinforced cement concrete is concrete with a steel bar embedded in it.

Any concrete foundation consists of PCC and RCC.

Cement: On the market, we have ordinary Portland cement (OPC) and the blended cement either as Portland Pozzalana cement (PPC) or slag cement. In addition, there are various other types of cements, such as:

- Rapid hardening Portland cement (RHPC).
- Hydrophobic Portland cement (HPC).
- Low heat Portland cement (LHPC).
- Sulfate resisting Portland cement (SRPC).
- Portland white cement (PWC).

Ordinary Portland cement (OPC)
- Most commonly used in general concrete construction.
- It has lime, silica, alumina, and iron oxide as ingredients.
- The cement is available in two different grades: 43 and 53. The number denotes the average compressive strength of at least three mortar cubes at 28 days.
- 43 Grade cement is used for precast concrete production, besides sleeper manufacturers and other building components.
- 53 Grade cement is used for high-strength concrete such as bridges, flyovers, large-span structures, and high-rise structures.

Portland pozzalana cement (PPC)/slag cement
- Produced by intergrinding OPC with pozzolana (fly ash) obtained as a byproduct from coal thermal power plants. It offers great resistance to aggressive water attack. It is useful in a marine environment.

Sand: According to standards, sands are those, most of which pass through 4.75 mm sieve. Sand is generally considered to have a lower limit of 75 µm.

Gravel: According to standards, gravels are those, most of which retained on 4.75 mm sieve. Gravel is generally considered to have a higher limit of 40 mm.

Water: Potable water is generally considered satisfactory for mixing concrete. Water should be free from oils, acids, alkalis, salts, sugars, and organic materials that may deleterious to concrete or steel. The pH valve of water shall not be less than 6.

For reinforced concrete, reinforcing bars of plain mild steel or high-strength deformed steel of nominal diameters ranging from 6 to 50 mm are used. Plain mild steel bars are less commonly used in reinforced concrete because they possess less strength (250 MPa yield strength). However, they are used in practice where nominal reinforcement is required. High-strength deformed steel bars with a yield strength of about 415 MPa are commonly used in RCC.

Concrete grades are termed M10, M15, M20, M25, M30, etc., where M refers to the mix and the number refers to the specified compressive strength of a 150 mm cube cured after 28 days, expressed in N/mm^2.

The cement:sand:gravel ratios of the concrete grades M10, M15, M20, and M25 are (1:3:6), (1:2:4), (1:1.5:3), and (1:1:2), respectively.

Based on the soil test report and the soil-bearing capacity, the foundation for the structure will be designed by taking all the loads into consideration.

The bore is drilled in the ground and a piling is built up with a PCC. This is called pile casting. The vertical leg of the structure is connected to the concrete piling using interfacing structural elements. For most MMS foundations, the vertical base frame of MMS is in-suite fixed while making the piling. The foundation for the MMS structure with in-suite piling is shown in Fig. 9.14.

Ramming type

Without any foundation, the vertical base frame of MMS is directly made to penetrate into the ground using special equipment, as shown in Fig. 9.15. This is suitable for silty sand soil. The machine applies force on the vertical leg and makes it penetrate into the ground. The ramming type of structure firmly fixed to the ground without any piling is shown in Fig. 9.16.

9.2.7 AC cables

Inverter to transformer: Requirements and design

Cables are to be compliant to most IEC standards such as IEC 60502. IEC 60502-2:2014 specifies the construction, dimensions, and test requirements of power cables with extruded solid insulation from 6 kV up to 30 kV for fixed installations such as distribution networks or industrial installations. XLPE insulated PVC sheathed Al or copper conductor cables are used for solar PV systems. For a 1500 V system voltage, the appropriate design and testing have to be followed. Copper and aluminum are

FIG. 9.14

Foundation with in-suite piling of the vertical leg of MMS.

FIG. 9.15

Equipment used for ramming the vertical frames of MMS.

FIG. 9.16

Vertical frame of MMS fixed to ground by ramming.

widely used as conductor materials in the cable industry. Conductors are classified in several ways as solid, stranded, flexible, and extra flexible. Insulation provided on the conductor are of different materials such as PVC, XLPE, and various types of rubber. Inner sheathing (beading) works as a binder for insulated conductors in multicore cables and it is mainly made up of PVC and rubber material. Armoring is a process to provide extra earthing shielding to current carrying conductors. It also provides extra protection and mechanical strength to cables. There are mainly G.I. wire armoring and G.I. steel strip armoring. The outermost cover of the cable is PVC outersheath made up of PVC or rubber material. It is provided over the armor layer for overall mechanical, weather, chemical, and electrical protection of the cable. Some of the standards followed for components of cables are given below.

IEC 60287: Recommended current ratings for cables.
IEC 60228: Conductors for insulated cables.
IEC 60502: Mild steel wires, strips, and tapes for armoring of cables.

- Cables sizes are selected considering the power loss, current carrying capacity, voltage drop, maximum short circuit duty, and the period of short circuit to meet the anticipated currents.
- Cables are of 1.1 kV grade, single/multicore, extruded XLPE insulated with an extruded PVC inner sheath.
- The conductor made of electrolytic grade aluminum will be smooth, uniform in quality, and free from scale or any defects.
- The maximum conductor temperature will not exceed 90°C during continuous operation at the full rated current. The temperature after a short circuit for 1.0s will not exceed 250°C with an initial conductor temperature of 90°C.
- Cables are usually armored with mild steel wires or strips for underground and in trench installations.

Transformer to substation: Requirements and design

The power cable consists of conductors covered by insulation layer, inner sheath, armor, and PVC outer sheath.

Conductor: It is the only element in the cable for current carrying path and it is usually made up of copper or aluminum material. The conductors are classified as solid, stranded, flexible, and ultraflexible. *Insulation*: Insulation layer is provided over the conductors to isolate them from other conductors and to protect them. It is made up of PVC, XLPE, or rubber.

Inner sheath (beading): It is generally used for multicore cables. It acts as binder for installed conductors together in a multicore power cable. PVC and different varieties of rubber materials are used as inner sheath material.

Armoring: It is mainly a process for providing earth shielding to the current carrying conductors. It provides extra mechanical strength and protection to the cables. There are mainly GI armoring or GI steel strip armoring.

Outer sheath: It is provided over the armor for overall mechanical, weather, chemical, and electrical protection to the cable. The material is same as that of used for inner sheath. Usually, it is made up of PVC or XLPE. Cables shall be compliant to most recent standards.

IEC: 60502 Cross-linked polyethylene insulated, PVC sheathed cables with working voltages from 3.3 kV up to and including 33 kV.
IEC: 60502 Two mild steel wires, strips, and tapes for armoring of cables.

- Cable sizes are selected considering the power loss, current carrying capacity, voltage drop, maximum short circuit duty, and the period of the short circuit to meet the anticipated currents.
- Cables are of flame-retardant, low-smoke (FRLS) type.
- Aluminum conductor used in power cables shall have a tensile strength of more than 100N/mm^2. Conductors shall be multistranded.
  - A high-quality XLPE insulating compound of natural color is used for insulation. Insulation is made by the triple extrusion process and it is chemically cross-linked in a continuous vulcanization process.
  - Cables come with conductor shielding as well as insulation shielding and consist of an extruded semiconducting compound. It will have an additional insulation shield with semiconducting and metallic tape shield over the insulation shield.
  - An inner sheath is applied over the laid up cores by wrapping of plastic tapes and armoring may be of galvanized steel wires or galvanized steel strips.
  - The outer sheath is of a suitable grade PVC compound applied by the extrusion process.
  - The cable shall withstand all mechanical and thermal stresses under steady-state and transient operating conditions.
  - A $3\text{C} \times 185$ or 240mm^2 Al armored cable with XLPE insulated cables is used from the transformer to the 33 kV switchgear based on the fault current and the time considered for interrupting the fault.

The following type tests and routine tests are conducted on the cables: Conductor resistance, high-voltage test at room temperature, test on armor wire/strip, tensile strength, percentage of elongation, torsion/winding, weight of zinc coating, dimensions, uniformity of coating, thickness of insulation and sheath, physical test for insulation, hot set test on insulation, aging in air oven, shrinkage test, water absorption (gravimetric) test, physical test on sheath, loss of mass, hot deformation and shrinkage, thermal stability and heat shock and insulation resistance (volume resistivity).

The following acceptance tests will be conducted on the cables: conductor resistance, thickness of insulation and sheath—as per data sheet, insulation resistance (volume resistivity) at room temperature and at maximum working temperature (90°C), tensile strength and % elongation for insulation and outer sheath, H.V. test on full drum (injection of 3 kV for 5 min), hot set test for insulation (at 200°C), armor dimension test, and overall diameter check of the cable.

9.2.8 Grounding of solar PV power plants—Grid-connected

Solar modules have grounding holes on their frames that can be used to connect to the frames of the module mounting structure. The piercing type of nuts and bolts is used in connecting to the structural frame to have good electrical continuity. All module mounting structures are connected and a grid is formed and connected to the earth pits. Module-to-module connections are also done.

The basic design calculation for the solar array and inverter earthing system is given below. This is a sample calculation considering IS standards with GI pipe. For maintenance-free earth pits containing chemical filling, the calculations will be different.

The solar PV plant capacity is 1 MW.
Maximum system voltage: 1500 V DC.
AC power evacuation voltage: 11 kV.

The earthing system consists of a main earth grid (strip), an earth electrode, and an earthing conductor.

The earth grid and earth electrodes are buried in soil in the solar array field or embedded in concrete inside the buildings to which all the electrical equipment and metallic structures are connected to have earth continuity for safety reasons.

The earth grid is a GI strip that will be buried 100 mm in depth.

The earth electrode is a GI pipe that will be installed at a 3000 mm depth from the ground level.

The earth conductor is a GI or copper wire used to connect between metallic structures.

Ultimately, the total earth resistance at any point of the earthing system shall not be more than 1 Ω.

The earthing systems are of two types: body earthing and neutral earthing.

All the inverters are connected to the GI strip to form an earth grid and directly connected to an earth electrode with a GI clamp, nut, and bolt.

Similarly, the metallic structures used for mounting the solar modules are looped with GI wire at the legs and connected to the earth electrode; ultimately, a grid formation will be done.

An example of earthing calculation is given below

There are 24 source circuit strings combined in a combiner box. The I_{sc} of the solar module is 9.28 A and 4 string combiner boxes are connected to a 1 MW inverter. There are 96 strings and the total current in the input of the inverter considering 25% extra is 1113.6 A. The power distribution system nominal fault level of the DC system can be considered as 1113.6 A. The duration of the fault level for the purpose of conductor sizing is 3 s. The soil resistivity is 100 Ω m. The voltage level of the DC array yard is 1500 V.

Step 1: To determine the minimum size of the earth conductor

The earthing conductor size can be calculated by

Cross-sectional area of the conductor $S = \dfrac{I \times \sqrt{t}}{K} = \dfrac{1113.6 \times \sqrt{3}}{80} = 24.11 \text{ mm}^2$

where

S = cross-sectional area of the earth strip in mm^2
I = fault current in amps = 1113.6 A
t = operating time of the disconnecting device in sec = 3 s
K = material factor = 80 for steel (as per IS 3043, p. 31)

As per the CBIP manual on substation publication number 223, corrosion allowance is recommended as per the following rule:

In case of a conductor to be laid in soil with soil resistivity greater than 100 Ω m—no allowance

In case of a conductor to be laid in soil with soil resistivity from 25 to 100 Ω m—15% allowance

In case of a conductor to be laid in soil with soil resistivity lower than 25 Ω m—30% allowance

The cross-sectional area considering a 15% corrosion factor is 27.7 mm^2

The conductor chosen is 25×3 is 75 mm^2

Maximum permissible current density (I_p) (according to p. 27, IS:3043)

Maximum permissible current density $I_p = \dfrac{7.57 \times 1000}{\sqrt{\rho} \times \sqrt{t}}$ is 437.054 A/m^2

Considered earth strip length (l)—650 mm
Considered earth strip width (w)—25 mm
Considered earth strip height (b)—3 mm
Surface area of selected earth strip per 1 m—0.056 mm^2
Maximum current dissipated by earth strip per running meter—$437.054 \times 0.056 = 24.541$ A
Dissipated current through earth strip—$650 \times 24.541 = 15951.384$

So the fault current can easily dissipate in the earth strip. Thus, the selected size of the earthing conductor meets the requirements.

Step 2: Calculation of resistance of earth electrode for DC side

Resistivity of the soil (ρ)—100 Ω m
Length of the rod (L)—300 cm
Diameter of the rod (d)—1.72 cm
Resistance of the rod electrode $\text{Re} = \dfrac{100\rho}{2\pi L} \times \log e \dfrac{2L}{d}$ is 31.075 Ω

Step 3: Calculate net resistance of earth electrode

Number of electrical earth electrodes (earth pits) considered (N) is 12

The resistance of earth pit $R_{earth} = \dfrac{\text{Re}}{N}$ is 2.59 Ω

where R_{earth} is the net resistance of earth pits.

Step 4: Calculate resistance of earthing strip

Resistance of the earth strip $R_s = \dfrac{100\rho}{2\pi L} \times \log e \dfrac{4L}{t}$ is 0.283 Ω

Step 5: Calculate total grid resistance

Total grid resistance is $\dfrac{R_{earth} \times R_s}{R_{earth} + R_s}$ is 0.255 Ω

Main earth grid conductor size (DC side) is 25 × 3 mm GI strip

Number of earth pits considered (DC side) is 12

Earthing system resistance (DC Side) 0.255 < 1 Ω.

Hence, as per the design calculations, the effective resistance of the earthing system is 0.255 Ω and it is a safe resistance value.

The AC earthing can also be calculated in a similar way.

The fault current is 25 kA for 1 s

AC voltage level of HT side—11 kV

AC voltage level of LV side—800 V

Cross-sectional area of the conductor $S = \dfrac{I \times \sqrt{t}}{K} = \dfrac{25{,}000 \times \sqrt{1}}{80} = 312.5$ mm^2

Considering a corrosion factor of 15%, the cross-sectional area S is 359.375 mm^2

The conductor considered is 50 × 8 mm.

Maximum permissible current density $I_p = \dfrac{7.57 \times 1000}{\sqrt{\rho} \times \sqrt{t}}$ is 757 A/m^2

Considered earth strip length (l)—1000 mm

Considered earth strip width (w)—50 mm

Considered earth strip height (b)—8 mm

Surface area of selected earth strip per 1 m—0.117 m^2

Maximum current dissipated by earth strip per running meter—757 × 0.117 = 88.418 A

Dissipated current through 100 m earth strip—100 × 88.417 = 8841.7 A

Calculation of resistance of earth electrode for AC side

Resistivity of the soil (ρ)—100 Ω m

Length of the rod (L)—300 cm

Diameter of the rod (d)—0.866 cm

Surface area of selected earth rod—123.864 mm^2

Resistance of the rod electrode $\mathrm{Re} = \dfrac{100\rho}{2\pi L} \times \log e \, \dfrac{2L}{d}$ is 31.075 Ω

Number of electrical earth electrodes (earth pits) considered (N) is 10

The resistance of earth pit $R_{earth} = \dfrac{\mathrm{Re}}{N}$ is 3.108 Ω

Resistance of the rod electrode $R_s = \dfrac{100\rho}{2\pi L} \times \log e \, \dfrac{4L}{t}$ is 69.77 Ω

Total grid resistance is $\dfrac{R_{earth} \times R_s}{R_{earth} + R_s}$ is 2.975 Ω

Main earth grid conductor size (AC side) is 50 × 8 mm GI strip

Number of earth pits considered (AC side) is 10

Earthing system resistance (AC side) 2.975 < 5 Ω.

9.3 Design of grid-connected solar PV power systems

The following parameters are required to carry out the design of a solar PV system:

Site location: latitude and longitude of the place.

Total area available to set up the power plant. Capacity of the plant required. Generation requirements and PLF/CUF values as per power purchase agreement (PPA). The power evacuation voltage should be known. The distance of the transmission line from the plant premises to the point of power evacuation. Once these parameters are known, a site survey can be conducted and a feasibility report can be prepared for setting up the solar PV plant in the said location.

Once the site location is known, the solar irradiance assessment and wind load study can be carried out. A feasibility study has to be conducted.

Site survey

A number of factors related to economic feasibility, power evacuation, and environmental aspects are involved in site selection. The important factors that influence the project site selection are given below:

- Technoeconomic.
- Infrastructure logistics.
- Environmental.

Technoeconomic

The technoeconomic considerations are as detailed below:

- Availability of a large area of waste and unused land.
- Availability of solar insolation throughout the year.
- Availability of adequate quantity of water year-round within a reasonable distance from the site for cleaning modules.

Infrastructure logistics

- Availability of infrastructure facilities such as ports, railways, and road access to the site for ease of transportation of plant equipment, etc.
- Facility for interconnection with transmission and distribution system for evacuation of power.
- Minimum investment requirement for development of required infrastructure.
- Availability of facilities such as medical, education, market, and railway station within a reasonable distance.

    The environmental considerations critical to the selection of a site are listed below:

- Avoidance of use of forest land.
- Minimum use of agricultural land.
- Minimum requirement of felling of trees.
- Minimum displacement of people.

To capture all the above factors that are to be considered in site selection, a format is given in Table 9.1.

Substation details regard the capacity, grid outage aspects, and power evacuation arrangement, Transmission line distance and difficulties for right of way for transmission line pole erection can also be assessed. The soil type can be assessed to know regarding the suitability of piling or ramming for the MMS structure foundation. The boundaries of the area can be taken and array capacity that can be accommodated in that site can be assessed.

Table 9.1 Selection criteria for site evaluation.

Category	Criteria	Unit
Location	Site name	
	Project developer	
	Local contact (name, telephone)	
	Region/municipality	
	Address	
	Latitude	N degree
	Longitude	E degree
	Elevation	m
	Time zone	Hours $\pm$ GMT
Meteorology	Mean annual GHI	kWh/m^2/annum
	Mean annual DHI	kWh/m^2/annum
	Maximum wind speed	m/s
	Average wind speed	m/s
	Main wind direction	
	Seasonal winds are sandy/dusty	Yes/no
	Microclimatic impact on radiation	
	Extreme climate (heavy snow/temperature/fog)	Yes/no
	Annual sum of rainfall	mm
	Seasonal distribution of rainfall	
	Average relative humidity	%
Land characteristics	Size of land	m × m (orientation)
	N-S oriented rectangular shape size	m × m (N × E)
	Use of land	
	Land ownership	
	Land cover (type of vegetation)	
	Surface profile	
	Slope	

Table 9.1 Selection criteria for site evaluation—cont'd

Category	Criteria	Unit
	Type of near shading obstacles	Yes/no
	Distance to these obstacles	m
	Type of distant shading obstacles	
	Shading angle	
	Distance from the coast	km
	Distance from greater water bodies	km
	Surface soil composition (if available)	
	Ground composition (if available)	
	Earthquake frequency and strength)	
	Flooding risk	Yes/no
	Site protection	Yes/no
	Fire risk	Yes/no
	Armed conflict in region of site	
	Comments to site	
	Comments to site	
	Quality of surface water	
	Quantity limits	km
	Depth of groundwater	ft
Infrastructure	Road available to access the site	
	Road/railway to be constructed	km
	Distance from closest highway	km
	Distance from closest seaport	km
	Distance from closest city/town	km
	Distance from closest airport	km
	Distance to MV grid	Yes/no
	Voltage level	kV
	Distance to nearest MV substation	km
	Substation voltage level	kV
	Distance to nearest HV substation	km
	Substation voltage level	
	Telecom available	Yes/no

Solar resource data and irradiance assessment

Solar resource and irradiance assessment refers to the analysis of a prospective solar energy production site with the end goal being an accurate estimate of PV system's annual energy production. There are a variety of possible solar irradiation data sources that may be accessed. The datasets either make use of ground-based measurements at well controlled meteorological stations or use processed satellite imagery. Installation of ground-based instruments such as pyranometers or pyrheliometers at site and monitoring the data for longer period is used to obtain high-quality solar radiation and other meteorological data with a much lower uncertainty than satellite-derived data.

- NASA's Surface Meteorology and Solar Energy dataset holds satellite-derived monthly data for a grid of $1° \times 1°$ covering the globe for a 22-year period (1984–2005). The data are suitable for feasibility studies of solar energy projects.
- SWERA obtains primary inputs into its models from geostationary satellites. The satellites provide information on the reflection of the Earth-atmosphere system and the surface and atmosphere temperatures, which are useful in determining the cloud cover. SWERA also uses data such as elevation, ozone, water vapor, snow cover, etc., to attain final results. Model outputs are verified with ground-based data to ensure the quality of the measurements.
- The Meteonorm global climatological database and synthetic weather generator contains a database of ground station measurements of irradiation and temperature. When a site is more than 20 km from the nearest measurement station, it outputs climatologic averages estimated using interpolation algorithms. When no radiation measurement station within 300 km of the site, satellite information is used. If the site is between 50 and 300 km from a measurement station, a mixture of ground and satellite information is used. The accuracy of irradiation figures close to measurement stations is within a few percent. The interpolated global irradiation figures for India are given with an uncertainty of 5%–7.5% for yearly values. The uncertainty increases with the distance between the site and the measurement station, especially in hilly and mountainous terrain.
- Data from total meteorological year

Correlation to the resource data may be done with detailed designs with datasets derived from satellite imagery.

The uncertainty analysis for the radiation data has to be done. From this data, the monthwise GHI and ambient temperature values are used as inputs to the energy yield estimation software.

Selection of module and inverter

The crystalline and thin film-based PV technologies are used globally for utility-scale power generation. Thin-film modules give maximum specific production with a significantly high performance ratio in high temperature and diffused irradiation zones. Though thin film modules give maximum specific production at sites of higher irradiance due to their lower temperature coefficient, long-term performance records are not available. The plant with thin-film modules requires more area (5–6 acres/MW) and requires more structures and cabling. Crystalline modules suffer from low annual degradation compared to thin-film technologies and hence provide high energy yields in the long run. Crystalline Si has been the workhorse since the inception of the technology and has a proven record of performance for more than 25 years.

Electrical, mechanical, financial, environmental, and customer-related factors are the main criteria that are being often followed in evaluation of various investment projects for making a decision. The cost criterion can be subdivided into variable cost and total investment cost. The environmental criteria include required land area to install the panels and material manufacturing effect. The financial criteria for the solar modules are cost per watt, total cost of investment, and rate of return on investment. Customer service, spare parts availability, and the reliability of the company are the factors considered for customer satisfaction toward the solar panels.

A higher power rating does not mean that the panels are more effective at producing power. With the increased efficiency and power, the required area to accommodate the modules will reduce. A positive power tolerance with narrower range is preferable to a wider one, because it represents more certainty and reduces the mismatch. The modules with lower temperature coefficient of power and better low light performance are preferable. Higher warranty terms and lower power degradation are also the factors to be considered in selecting the modules.

Based on the cost considerations, reliability, and performance, a solar module can be identified.

Accordingly, the inverter can be finalized based on the financial analysis between the cost and the generation between the central and string inverters. Among the different brands, based on reliable performance and the service at previous plants, a vendor can be selected. The capacity of the inverter can be decided based on the system voltage, the requirement of system redundancy, and the effect on the cost of BOS elements and the financial analysis.

String length design

A sufficient number of modules are required to connect in series to get voltage that will be always within the MPPT window of the inverter. The number of modules in a string is called the string length. The parameters required for the estimation of string length are:

Module data such as V_{oc}, V_{mp}, I_{sc}, and I_{mp} and their temperature coefficients as well as NOCT/NMOT and yearly degradation characteristics.
The system voltage or the maximum DC voltage the inverter can withstand.
The MPPT voltage window.
The ambient temperature of the site during winter, the coldest day.
The ambient temperature of the site during summer, the hottest day.

For a lowest ambient temperature, the V_{oc} of the module will be estimated. Dividing the system voltage with the V_{oc} of the module at the lowest temperature gives the maximum number of modules required to be in a string. For the highest ambient temperature, the V_{mp} of the module is estimated.

The minimum MPPT voltage is divided with the V_{mp} at the highest temperature, giving a minimum number of modules for a string. To take the module degradation aspect and the life of the PV plant into consideration, a 15% correction factor can be applied to get the string length. The calculations are presented in Chapter 8.

For a system voltage of 1000 V for a normal central inverter, the string length is generally 20 with the consideration of a 72 cell module. For a 1500 V inverter, the string length becomes 30.

Module mounting structure design

Considering the wind speed, the wind load can be calculated. Considering the area of the module, the size of the mounting structure, and the inclined angle of the MMS, the upward and downward thrusts will be estimated. The structural elements which are vertical columns, Purlin, rafter are designed considering their mechanical properties to withstand the upward thrust. The drawing of the module structure is made and a STAAD file will be prepared, dividing the structure into different nodes. The

STAAD file will be fed to the STAAD Pro software and the acting load simulation is done for different loads and safety factor options. The deflection response of the structural elements will be obtained. By suitably iterating the MMS parameters using a design simulation program, the MMS is finalized.

A variable module mounting system with different tilt angles in different seasons harvests the maximum energy yield over a year. The mounting structure will be restricted if the seasonal tilt is considered. The mounting structures to be selected shall comply with the appropriate industrial standards and shall be capable of withstanding onsite loading and climatic/wind speed conditions.

The materials to be used for the mounting structures are hot-dipped galvanized mild steel, pregalvanized mild steel, galvaluminum sheets, or aluminum alloys.

During the detailed design stage, a geotechnical analysis of the soil has be done to determine the profile, size, and class of the grounding piles or foundations required. The analysis will also determine the installation method. The load-bearing capacity tests along with soil sampling and analysis will be done to ascertain the soil profile.

The design aspects of MMS are discussed in Chapter 8.

Tilt angle

Reducing the tilt angle can be beneficial in several ways: it allows better row spacing without raising the row-on-row shading. It also decreases the top module height and reduces the wind loading on the frame, which could lead to cost savings in the mounting system and support the posts and footings of the mounting system.

Fig. 9.17 shows a module mounting structure with a fixed tilt. It has two modules mounted in portrait configuration in vertical condition. The tilt angle is 10 degrees.

DC to AC ratio

Solar PV designers as well as solar power developers face challenges to remain in competition for developing large MWp-level solar PV power plants. With the reduction of prices of solar modules, solar system designers are finding the financial benefits by changing the solar PV array-to-inverter ratio greater than 1.

FIG. 9.17

Module mounting structure with a fixed tilt.

The solar PV *array-to-inverter power ratio* is defined as the ratio between the solar PV array capacity (DC MWp) and the inverter capacity (AC MW). The solar PV array capacity is a summation of all the solar PV modules' wattage (solar PV module rating under standard test conditions (STC), meaning at $1000\,W/m^2$, 25°C cell temperature, and air mass 1.5). The combined inverter rating is the maximum output power rating of all the inverters, which is called AC capacity of the solar plant. As an example, if a solar PV array has an aggregate capacity of 1250 kWDC to the combined inverter rating of 1000 kWAC, the solar PV array-to-inverter ratio is 1.25, or 125%. Other used definitions of the solar PV array-to-inverter ratio are also known as the DC load ratio, the DC-to-AC ratio, the over-sizing ratio, and the overloading ratio.

When solar PV module prices were higher, it was the main goal of the PV system designer to define the solar PV array-to-inverter ratio, which could ensure that the power generated by the solar PV array was not allowed to be wasted. PV system designers specifically wanted to avoid inverter overloading, which occurs whenever the solar PV array is capable of generating more power than the inverter can handle. A decision for the optimal solar PV array-to-inverter sizing ratio is done by the analysis of the annual energy generation per kilowatt of the solar PV array capacity at different ratios. This kWh/kWp ratio is known as the specific yield and is a measure of solar PV array production efficiency.

Because solar PV module prices have decreased consistently over the years, the criteria for solar PV design has changed. In lieu of focusing on solar PV array production efficiency and maximizing the output of each solar PV module, solar PV designers have started designing for maximum financial efficiency at the system level.

The incremental cost to enhance the solar PV array capacity is insignificant compared to the energy production gains. This helps solar PV system designers capitalize on higher solar PV array-to-inverter ratios up to 1.5 and higher, despite the risk of solar PV power generation exceeding the inverter capacity rating during peak sunshine hours.

Higher DC load ratios allow designers to get more value from fixed development costs. They also allow designers to capitalize on high-value energy rates, or time-of-delivery or time-of-day (TOD) rate structures that incentivize summer production. Increasing the DC load ratio is a compelling design approach when there is a limit on the AC system size but no corresponding limit on the DC system size. It allows solar PV system designers to increase solar array production due to cloudy conditions or solar array degradation over a long period.

When system designers increase the DC-to-AC load ratio above 1.0, the total system cost increase is not directly proportional to the increase in solar PV array capacity, as other project fixed costs (legal fees, permitting costs, inverter and interconnection costs on the AC side, etc.) remain the same. Hence, the cost for this additional solar PV array capacity is limited to the array string combiner box, MMS, the DC cables and their interconnection, labor and material costs relating to the addition of solar PV modules, and associated DC components. By increasing the solar PV array-to-inverter ratio, solar PV system designers/developers are able to take advantage of the fixed development and structural costs.

Solar developers and PV system designers have more incentive to increase solar PV array-to-inverter ratios when the price for the generated energy is high, and increasing DC load ratios allows designers to deliver more high-value electricity.

In desert and clear sky areaa (mountaintops and high elevation areas such as Ladakh in india), solar PV modules receive high solar irradiance, at times greater than $1000\,W/m^2$. High solar irradiance conditions are not so common in Europe, some parts of North America, North and East India, etc. For example, in Northern and Northeastern India, the solar irradiance on a tilted plane of the solar array on a typical clear sunny day may be $800\,W/m^2$. In such locations, it makes economical sense to increase the solar PV array-to-inverter capacity ratio.

Solar PV system designers can also decide on a higher solar array-to-inverter ratio to compensate for the unavoidable solar PV array degradation over the lifetime. With the aging of solar PV arrays, solar PV module performance degrades between 0.5% and 0.7% average on an annual basis. In order to ensure that the solar PV array's power output loads 100% capacity of the inverter after 10 or 15 years, it is suggested to use higher solar PV array-to-inverter ratios than required to meet 100% loading of the inverter in the first year.

When the DC/AC ratio of a solar system is too high, the likelihood of the PV array producing more power than the inverter can handle is increases. In the event that the PV array outputs more energy than the inverter can handle, the inverter will reduce the voltage of the electricity and drop the power output. This loss in power is known as "clipping." For example, a DC/AC ratio of 1.5 will likely see clipping losses of 2%–5%. Not as major as other losses, but still a noticeable effect. The inverter operates at higher voltage and there is no heating effect on the inverter components and inverter active cooling process takes care of thermal issues.

When the solar PV array power exceeds the inverter maximum power rating, the control logic in the inverter responds by shifting the operating voltage of solar PV array toward V_{oc} side. By limiting power, it is ensured that excess power is not dissipated as heat in the inverter. Hence, the inverter components are not stressed by this excess power under normal operating conditions. Optimal cooling provisions can help in controlling this effect. Active cooling with coolants are used in the inverters.

The most critical factor is withstanding the short-circuit current by the internal components such as busbars and disconnect switches during a fault on the input DC section of the inverter. In the event that the inverter fails to limit the input current from the solar PV array, all inverter components must be able to withstand the full short-circuit current of the solar PV array for the duration of the fault, without compromising the safety of the inverter. Fig. 9.18 shows inverter clipping losses due to a high DC-to-AC ratio [7].

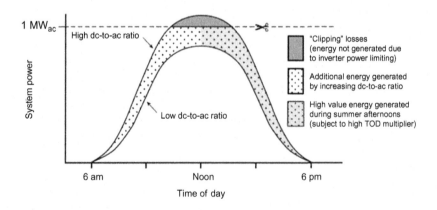

FIG. 9.18

Inverter clipping losses due to a high DC-to-AC ratio [7].

Interrow spacing

An optimal fixed-tilt system with a high DC-to-AC ratio for a particular project will yield better energy generation. The interrow spacing has to be maintained such that minimum horizon shading will occur. It also has to be sufficient for the movement of utility vehicles with trailers to drive between the rows to deliver modules, materials, and persons. For construction and O&M purposes, spacing should be required for the movement of vehicles carrying water for module cleaning. Although row-on-row shading may be a problem, this typically occurs in the early morning and late afternoon with the angle of the sun to the panels, and therefore their power output is still very small. However, more shading occurs in the winter; the effect of lost energy is negligible due to lower irradiance at that time.

Shadow analysis

Shadow calculations for the effect of interrow spacing and for the nearby control room buildings and trees are to be conducted. PVSYST or Google Sketch Up software can be used to estimate the effect of shadows. Iso shading curves for the particular location can be used to estimate the sun's altitude. The shadow has to be checked for winter solstice conditions.

Making of layout

As space is not a major constraint at the location chosen, the layout of the plant has to be chosen to maximize the annual energy yield. The distance between the rows of the mounting structures has to be chosen such that there is minimal interrow shading at the maximum sun angle on the winter solstice and adequate distance for maintenance purposes.

DC cable design

The size of the cable shall meet the following requirements:

The cable should withstand the short circuit time fault current in the circuit for the duration of the circuit interrupting device.

It should be able to carry a full load current after considering the appropriate derating factor for ambient temperature variation, method of laying, and bundling factor.

The voltage drop and power loss due to the DC cable should be 1.5% to get better energy generation. Current ratings and the derating factor can be taken for DC XLPE insulated, armored, copper/aluminum cable, from the cable manufacturer's catalogue.

The maximum fault current for the DC side can be considered as the short circuit current of the string or combined strings. To calculate the voltage drop in a conductor, the resistance of the conductor, the length of the circuit conductor, and the current flowing through the conductor are required. The voltage drop is calculated using the ohms law. For estimating the voltage drop due to the cable, the operating voltage of the source circuit is required. Usually, the V_{mp} of the module at STC is multiplied with the number of modules to get the operating voltage.

The voltage drop due to cable size affects energy generation. The cable resistance is directly proportional to its length. Depending on the distance from the string combiner box to the inverter, the size of the DC is cable is decided by maintaining the voltage drop <1.5%. Y-connectors are used to combine two strings as a solution for reducing the overall DC cable length. The average length of the string and the main DC cables used in a typical solar project in India has come down from 15 km/MW and 3–4 km/MW in 2011 to about 8 and 2 km/MW, respectively [8].

With large-scale projects shifting toward a 1500 V operating voltage for DC systems, the cable requirement per MW will be further lowered. An example of DC cable design and voltage calculation for 1 MW solar PV plant considering 405 W module is given below

Design inputs				Cable data				
Load description	String to SMB			Load description		String to SMB	SMB to inverter	Unit
Fed by	Series strings			Voltage grade	V_c	1.8	1.5	kV
Inputs	1			No. of cores		1	1	No
Module voltage	V_{mp}	40.5	V	Cross section area	A	4	300	mm²
Modules in series		30	Nos	Conductor material (copper/aluminium)		Copper	Aluminium	
System voltage	V_{mp}	1215	V	Insulation		XLPO	XLPE	
Current	I_{mp}	10	A	Current carrying capacity	Ic-Gd	44	390	A
No. of strings		1						
Full load current	I_L	10	A	Type of laying		In ground		
				Ground temperature		35	35	°C
Load description	SMB to inverter			Depth of laying		1	0.9	m
Fed by	SMB			Formation (horizontal / vertical)		Horizontal	Horizontal	
Inputs	4			Touching / spacing / trefoil spacing		Touching	Touching	
Module voltage	V_{mp}	40.5	V	No. of cables		12	4	Nos
Modules in series		30	Nos					
System voltage	V_{mp}	1215	V	**Thermal ampacity**				
Current	I_{mp}	10	A	Calculation of derating factor for laying in ground				
No. of strings		21		Derating factor for variation in ground temperature	G1	0.93	0.96	
				Derating factor for depth of laying	G2	0.98	0.98	
Full load current	I_L	210	A	Derating factor for grouping and touching (laid in ducts)	G3	0.48	0.64	
				Overall derating factor for ground	K-Gd	G1 x G2 x G3	G1 x G2 x G3	
						0.44	0.60	
				Derated current carrying Capacity of cable in ground		Ic-DRG K-Gd x Ic-Gd	K-Gd x Ic-Gd	
						19.25	234.82368	A
				Full load current	I_L	10	210	A

The current carrying capacity of the selected cable (4 mm²) after deration is 19.25 A, which is greater than required load current (10 A). Hence the cable is safe. The current carrying capacity of the selected 300 mm² cable after deration is 234.8 A, which is greater than required load current of 210 A. Hence the cable is safe.

Current in amps (I): 210 A; V_{mp} of the module at 25°C: 40.8 V

The output voltage of SMB at 25°C: 1215 V; cable considered: 1C × 300 mm², Al

Resistance of the cable at 20°C in Ω/km (R_0): 0.125 Ω

Resistance at temperature of 50°C (R_{Temp}): 0.451 Ω; $R_{Temp} = R_0 \times [1 + \alpha * (50 - 20)]$

Max. amps in ground/air/duct: 390 A

Voltage drop = 2 L * I * R_{temp}/1000 = 5.51 Ω (the cable length L is 93 m)

Percentage of voltage drop = (V_{drop}/string V_{mp}) * 100 = 0.45%

Power loss = (I * string V_{mp}) * V_{drop}%)/100)/1000 is 1.16 kW

For AC cable voltage drop calculation, the following formula is used:

$$V3\varphi = \frac{\sqrt{3}\,I(R_c \cos\varphi + X_c \sin\varphi)L}{1000}$$

$V3\varphi$ is the three-phase voltage, I is the nominal full load or starting current in A, R_c is the AC resistance of the cable (Ω/km), X_c is the AC reactance of the cable (Ω/km), $\cos\varphi$ is the load power factor, and L is the length of the cable in meters.

AC distribution board or LT panel

The AC low voltage switchboard or AC combiner box is suitable for operation with 415 V or inverter output voltage, 50/60 Hz, and a four-wire system. This is used to combine the string inverters in a power plant. The cable entry can be from the bottom or top side as required. It will have built-in overload and short circuit protection with a three-phase voltage meter of Class 0.5 accuracy. It serves as the connecting and disconnecting device for the three-phase AC output rated for suitable breaking capacity. It monitors functions of three-phase voltage meter. The system operates by a manual on/off switch to connect and disconnect the supply. A three-phase voltage meter can be used to see whether the output from the inverter and the MCCB is coming. To see the various voltages, the meter will have a seven-segment display with feather touch up/down keys. It will have overload protection and short circuit protection features.

The AC switchboard can be the indoor type and dust- and verminproof maintaining the IP 42 degree of protection.

The switchboard will be free-standing type and floor mountable. Doors are usually pad locking type, and all hinged doors are earthed through a 2.5 mm^2 tinned copper wire. It is usually made such that the isolator compartments should be openable only in the OFF mode. All the busbars and links shall be made of copper with color-coded heat-shrinkable insulating sleeves. Separate earth busbars are provided along the length of the switchboards. It will have meters for measuring different parameters; these will be digital electronic with LED displays. It will have LT switchgear having suitably rated circuit breakers, isolators, and multifunction meters.

Energy yield estimation

PVSYST and PVSOL are used in the solar industry for energy yield generation for the solar PV plant. The following are the input parameters required for energy yield estimation:

Module electrical data—PAN file.
Inverter data—OND file.
Latitude and longitude of site.
Elevation of site.
Time with respect to GMT.
Ambient temperature of the site in different seasons.
Weather data—GHI and ambient temperature of site—Meteonorm or any other software compatible database.
Row to row distance.
Tilt angle—fixed tilt, variable tilt, or tracker.

Table 9.2 Loss factors usually considered in energy yield estimation.

Shadow loss	1.20
IAM factor on global	2.60
Soiling loss factor	2.00
Loss due to low irradiation level	0.10
Loss due to temperature (it depends on the ambient temperature of site)	8.10
Module quality	0.00
Module array mismatch	1.00
Average DC ohmic losses	1.20
Inverter losses	1.60
Transformer losses	1.00
AC cable loss	0.30
Auxiliary power loss	0.70
Transmission line losses	0.00
LID of module for first year	2–2.5
Degradation of module in first year except LID	1.00
Yearly degradation of module from second year	0.70
Plant downtime	1.00

The energy loss factors. Typical values that can be considered are given in Table 9.2.

The software imports the meteorological data from a database such as Meteonorm and creates a met file.

The solar module and inverter parameters are considered based on the module PAN file and the inverter OND file, respectively. Considering the temperatures and module and inverter parameters, system sizing is done. PVSYST uses one diode-based model for modeling the solar module. Using the Perez or the Hay and Davis model, the irradiance values for the plane of the array will be estimated. By applying all the loss factors, the PVSYST estimates the PV plant's energy yield. The loss diagram looks as shown in Fig. 9.19 for a particular project in India.

Uncertainties in solar PV energy yield estimation

The uncertainty in energy prediction is challenging to quantify, as it is a function of many independent factors. Solar radiation is one of the core parameters that creates maximum uncertainty in the measurement. Meteonorm data has been used for the energy yield calculation, which has been generated through a synthetic weather generator.

For example, the following uncertainty factors are considered in energy yield estimation:

Uncertainty in PVsyst: 3%.
Uncertainty in radiation model: 4%.

The standard deviation of uncertainty σ is the square root of the sum of the squares of uncertainties. So,

$$\text{Total uncertainty} = \sqrt{(4)^2 + (3)^2} = 5\%$$

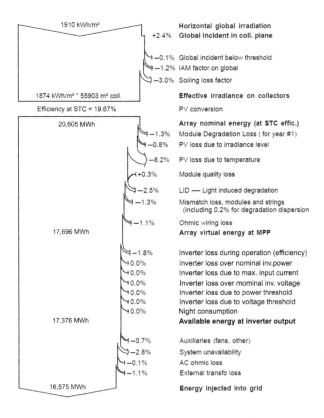

FIG. 9.19

Loss diagram from PVSYST for a project in India.

The combined standard deviation is established with the rule of squares. The factors for P75, P50, and P90 can be taken from the cumulative distribution function curve, as shown in Fig. 9.20. The zero shows P50, which is the energy yield value obtained from the PVSYST software.

P75 is 0.66σ times P50 and P90 is 1.28σ times P50.

The P75 level (the energy production exceeded 75% of the time) can be estimated as 0.67 times the standard deviation of uncertainty σ (σ is 5% in the example), that is, 3.3%; the P90 value becomes 1.28 times the standard deviation, that is, 6.4%, and the P99 value becomes 2.33 times the standard deviation, that is, 11.65%.

For a 5 MW solar PV system, the annual energy generation obtained is 7,964,852. This figure is considered as P50 and the estimated P75, P90, and P99 figures are tabulated in Table 9.3.

Earthing design

A key aspect of the PV plant electrical circuit design concerns the grounding of the PV arrays, electrical equipment, and steel structures. The PV arrays at most of the pilot plants have used floating DC ground, that is, a PV power circuit, and the DC bus are not earthed during operation. A safety earthing system

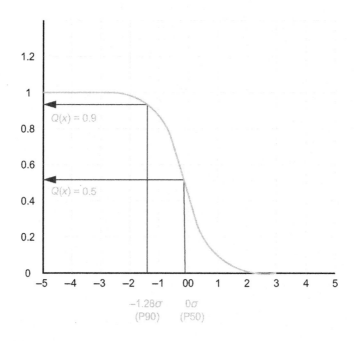

FIG. 9.20

Cumulative distribution function for uncertainty analysis.

Table 9.3 Energy yield and P50, P75, P90, and P99 values for a 5 MW project.				
Year	P50	P75	P90	P99
		3.3%	6.4%	11.65%
1	7,964,852	7,702,012	7,455,102	7,036,947

consisting of a buried GI flat conductor earthing grid will be provided for the switch gear/switchyard elements. The earthing system will be formed to limit the grid resistance to below 1 Ω. In the switchyard area, the touch potential and step potential will be limited to the safe values. The buried earthing grid will be connected to earthing electrodes buried underground. The neutral point of the transformer, the noncurrent carrying parts of the equipment, the lightning arrestors, the fence, etc., will be earthed rigidly. The following factors shall be considered for earthing system design:

- Magnitude of fault current.
- Duration of fault.
- Soil resistivity.
- Resistivity of surface material.
- Material of earth conductor.
- Earth mat grid geometry.

Lightning arrester design and connection

Lightning arrestors of adequate capacity shall be provided for an SPV plant.

The protection system is based on early streamer emission lightning conductor with air terminals being used for ground mounted solar PV plants. The air terminals shall provide umbrella protection against direct lightning strikes covering a radial distance of 100 m. The air terminal will be capable of handling multiple lightning strikes and should be maintenance-free after installation. Arrestors will be complete with an insulating base, self-contained discharge counters, and suitable milli-ammeters.

The lightning arrestors are to be installed in the solar field area.

These air terminals will be connected to respective earthing stations.

The earthing stations for the lighting discharges will be provided with test links of phosphorus bronze and located at 150 mm above ground level in an easily accessible position for testing.

Auxiliary power supply—Transformers

There is requirement of power at 415 V and three-phase for inverter operation; it requires power in standby mode at night. The inverters will be consuming power depending on their size. To meet all the loads, including SCADA, control room, office and administration, and operations and maintenance of the required plant, power either from the grid or tapping from the inverter output will be utilized using a suitable rated auxiliary transformer. The auxiliary transformers and accessories will have to conform to applicable standards. In the inverter room, an auxiliary transformer is installed. It is connected to the output of the inverter. The inverter output voltage will be stepped down to 415 V.

Grid connection

It is important that the SPV power plant is designed to operate satisfactorily in parallel with the grid under extremely high voltage and frequency fluctuation conditions, so as to export the maximum possible units to the grid. It is also extremely important to safeguard the system during major disturbances such as tripping/pulling-out of big generating stations and sudden overloading during falling of a portion of the grid loads on the power plant unit in island mode under fault/feeder tripping conditions.

The indicative selected PCU has the capability to monitor the grid parameters (voltage, frequency, and harmonics) and get the inverter output synchronized with the grid on a real-time basis. However, if the grid conditions vary beyond the preset window, this will ensure that the inverter is disconnected from the gird and is operated in sleep mode.

- The solar PV-based power plant envisages a power export during normal operating conditions. The grid connections will be at 11, 22, 33, 66, or 132 kV.

For 11, 22, and 33 kV levels of power evacuation, the inverter output can be directly stepped up to the required voltage level using an inverter duty transformer. The outputs of different transformers are combined in a switch gear. From the switch gear, the power can be evacuated to the grid at the nearby substation. The power is carried from the solar plant premises to the substation through a transmission line. At the solar plant end from the switchgear to the power connection point, a switchyard is required.

For a 66 kV voltage level, a two-step process is used. First, the inverter output is stepped up to 11 kV using a transformer. The output of all the 11 kV transformers is combined in a 11 kV switchgear. The output from the 11 kV switchgear is stepped up to 66 kV using a 66/11 kV transformer. A 66 kV switchyard is required.

A two-step process is followed for the evacuation of power at the 132 kV voltage level. First, the inverter output is stepped up to either 11 or 33 kV using the transformer. The output of all the 11 or 33 kV transformers is combined in an 11 or 33 kV switchgear. The output from the 11 or 33 kV switchgear is stepped up to 132 kV using a 132/11 kV or 132/33 kV transformer. A 132 kV switchyard is required.

A single- or double-bus arrangement will be made in the switchyard and substation based on the capacity of the power plant and the redundancy requirements.

Transformer

The transformer conforming to applicable standards such as IEC 60076 will be complete with the fittings and accessories such as the conservator, the breather, the Buchholz relay with contacts for alarm and trip, pressure relief devices, temperature sensor pockets, OTI and WTI, valves, earthing terminals, cooling accessories, bidirectional flanged rollers with locking and bolting devices for mounting on rails, air release devices, the inspection cover, the off-load tap changer (OLTC), the marshalling box, etc.

A typical photograph of a 5MVA 33/0.6-0.6-0.6-0.6 V inverter duty transformer is shown in Fig. 9.21. This is a five winding transformer, where four inverters are connected to its four primary windings.

FIG. 9.21

Typical photograph of an inverter duty transformer.

HT panel

The high-voltage switchgears such as 11, 22, 33, and 66 kV are called medium voltage switchgears. A 110 or 132 kV switchgear is called a high-voltage switchgear. The switchgears are used to combine the output from the transformers and provide protection to the system and supply the power to the switchyard for evacuation to the respective voltage level transmission line. The switchgears are equipped with protections for transformers and other equipment using circuit breakers, isolators, instrument transformers, and relays.

Circuit breakers

Circuit breakers of a suitable type are provided in the switchgear and in solar PV plant switchyard as well as in the substation of each transformer feeder. The circuit breaker and accessories are to conform to IEC standards.

The circuit breaker are totally restrike-free under all duty conditions and are capable of breaking the magnetizing current of the transformer and the capacitive current of the unloaded overhead lines without causing overvoltages of abnormal magnitudes.

The circuit breakers will be suitable for use in the switchgear under the operating conditions.

The closing coil will be operated suitably between 85% and 110% of the rated voltage.

The shunt trip will operate properly in all operating conditions of the circuit breaker and between 70% and 110% of the rated voltage.

Protection, metering, and control cubicles

The transformer will have the following minimum protections such as overcurrent and earth fault relays on the HV and LV sides and protection from the grid side, in addition to the built-in protections (Buchholz relay, oil and winding temperature relays, magnetic oil level gauge) to isolate the equipment during fault conditions.

The feeders linking the plant substation and the EB substation will be protected with directional as well as nondirectional overcurrent and earth fault relays. A rate of change of frequency (df/dt) relay with underfrequency protection and a vector surge protective relay will also be provided to isolate the generating system during grid disturbances/overloading conditions.

A MIMIC for a complete power evacuation system will be made in SCADA. To ensure safe operation of the system, electrical interlocking between breakers/isolators/earth switches is done.

All the protection, control cubicles, and panels shall be housed in the SPV plant's control room.

In addition to the metering and monitoring arrangement in the junction box and inverter, monitoring of the voltage, current, and energy will be provided at the medium voltage switchboards. These meters will be digital with an RS 485 port for remote monitoring. They will have accuracy class of 0.5 or 1.0.

Similarly, the HV side is also equipped with voltage, current, power, and energy meters in order to correlate the energy generation and losses. Further metering for the utility shall be main and check meters of 0.2S accuracy class or as required and specified by the utility authorities.

Lightning arrestors

Lightning arrestors are required to protect the switchyard/substation from lightning. The lightning arrestor will be heavy duty station class type, discharge class III, conforming to IEC specifications.

Isolators and insulators

Isolators will be complete with an earth switch (wherever necessary), a galvanized steel base provided with holes, solid core post insulators with adequate creepage distance, blades made of nonrusting material, and an operating mechanism (gang operated, manual charging mechanism). They will be the center post rotating horizontal double break type and consist of three poles. Solid core type post insulators of adequate creepage distances (suitable for a very high pollution category) will be provided for insulation and support in the switchyard at the plant/substation side.

Weather monitoring station

The weather monitoring station includes pyranometers to measure the global horizontal irradiation and irradiation at the inclined plane. It will have wind speed and direction sensors, a temperature sensor, a barometer, a rain gauge, an anemometer tripod, etc., to measure weather data and a datalogger to record all the data. This data will help in correlating the plant performance with the actual irradiation and weather conditions.

9.4 Simulation tools for design of grid-connected solar PV power system

Google Sketch Up
PVSYST
PV Design Pro
RetScreen
Google Sketch
PVwise
Scalian
PV Sol
PV Watts
Etap
STAAD Pro
Meteonorm
Solar Advisor Model

Google Sketch Up:

This software is used for shadow analysis of the ground-mounted as well as rooftop solar PV power plants, taking coordinates of the site location.

PVSYST:

This is the solar simulation software predominantly used by all solar PV power plant developers, EPC companies, designers, and researchers. This was developed by the University of Geneva. This uses met data from TMY, Meteonorm, NASA, Solar GIS, and other sources. This uses the Hay and Davis model as well as the Perez model with options. This can be used also for single- and dual-axis trackers. This also helps to use data of various technology-based solar PV modules such as HIT, etc. It also performs the shadow calculation for the solar arrays as well as the surrounding area. This software can be used for both crystalline (for bifacial modules also) and thin-film solar PV modules.

PV Design Pro:

This software was developed by Maui Solar along with Sandia National Laboratories. It uses a total meteorological year (TMY data base, GIS, and Meteonorm) for solar energy yield calculations of grid-connected solar PV power plants. It incorporates the Perez model for modeling solar irradiance.

RetScreen:

This solar simulation software is used for energy prediction, and was developed by Natural resources, Canada. It uses NASA meteorological data and other met data. It uses the isotropic sky model for solar irradiance calculations in the inclined plane.

PVSOL:

This is solar PV simulation software for the energy prediction of solar power plants. It was developed by Valentine Software. It uses weather data from MeteoSYN, Meteonorm, SWERA, PV GIS, and NASA Platform. It incorporates the Hay and Davis model as the solar irradiance model. This software can be used for both crystalline as well as thin-film solar PV modules.

PV Watts:

This is solar PV power plant simulation software developed by NREL. It uses the Perez model for solar irradiance simulation. It does so only for crystalline silicon-based solar PV modules and can be used to simulate plant performance for single- and dual-axis trackers. It uses met data from different American and Canadian agencies.

Solar Advisor Model (SAM):

This model was developed by NREL and Sandia National Laboratory, and helps to estimate the solar energy yield. It also calculates the financial parameters such as LCOE for specific site locations. It uses the Sandia PV array performance model and the Sandia inverter performance model. It also helps in doing detailed financial analysis, optimization, sensitivity, and statistical analysis. This software can be used for both crystalline silicon, thin-film solar PV, and concentrator PV.

Power Analytics:

This software is used for the power system analysis of centralized solar PV power plants. It mainly determines the financial calculation involving energy audits, project design support, and real-time plant operation analysis. It does feasibility studies for distributed renewable energy generation plant studies and microgrid islanding capability.

Easypower:

This software was developed by Easy Power. It does arc flash and harmonic analysis as well as the sizing of overcurrent protection devices.

ETAP:

This software is used for power system analysis of centralized solar PV power plants such as load flow and fault current study.

STAAD pro:

This software is related to the design of solar PV module structures, civil foundations, and inverter/control rooms. This does a detailed structural analysis for the foundations meant for module mounting structures, etc.

Meteonorm:

This software provides weather data as input for PVSYST simulation.

9.5 Typical schematic diagram and field application examples of grid-connected solar PV power systems

SLD of 5 MW system is shown in Fig. 9.22. String inverters are used as PCUs. The solar array strings (source circuits) are directly connected to string inverters. String inverters are combined in ACDB with maintaining proper protection. ACDBs are combined in an LT switchgear and stepped up to 11 kV using a transformer.

SLD of 50 MW system is shown in Figs. 9.23 and 9.24. A 50 MW solar plant can be divided into two nos of 25 MW blocks. Each 25 MW unit is divided into five nos of 5 MW blocks. At present, as the inverters of sizes 2 and 2.5 MW are available, one block can be 12.5 MW capacity or it can have different configurations with the usage of five winding transformers. The inverter duty transformers step up the inverter voltage to 33 kV. The transformers output will be brought to 33 kV switchgear using ring main units or vacuum circuit breakers. If the design of switchyard is with double bay, the plant will have better reliability.

1 MW solar PV power plant— 11 kV evacuation

Capacity: 1 MWp

Total area: Five acres ground mounted type

Total number of solar PV modules: 3229 (340 Wp)

Technology: Polycrystalline Si solar modules with fixed tilt

Number of solar PV modules in series per string: 32

Mounting arrangement: Fixed tilt of 9 degrees

Tilt angle: 9 degrees from horizontal

Inverter/PCU: three-phase string inverters 1000 kW, 660 V, three-phase, 50 Hz

Grid connection: 11 kV, 2 pole structure

Annual energy generation: 1.5 million units per annum

5 MW solar PV power plant—33 kV in Rajasthan, India

Capacity: 5 MWp

Fig. 9.25 shows the 5 MW solar PV plant with a solar array, inverters, a transformer, switchgear, and a transmission line.

Location: Khimshar, Nagaur District, Rajasthan, India (23.9°N, 73.4°E)

Total area: 35 acres

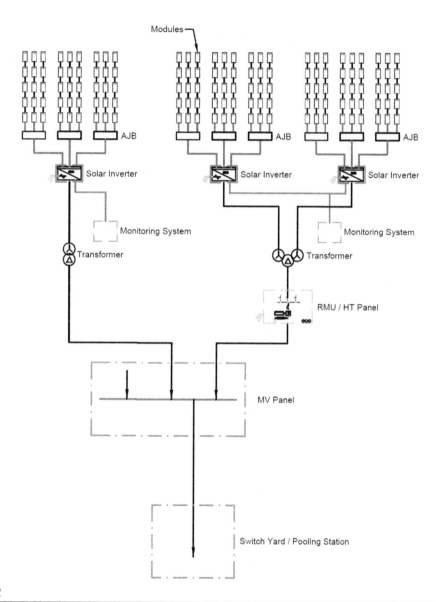

FIG. 9.22

Single line diagram of a 5 MW solar PV system.

Total solar PV modules: 24,841 (Kyocera/Mitsubishi/Sharp Solar)
Technology: Fixed crystalline (95.3%), thin film (0.48%), single-axis seasonal tracking (2.58%), dual-axis tracking (0.63%), CPV (1.0%)
Solar PV modules in series per string: 18/20
Mounting arrangement: Fixed ground mount

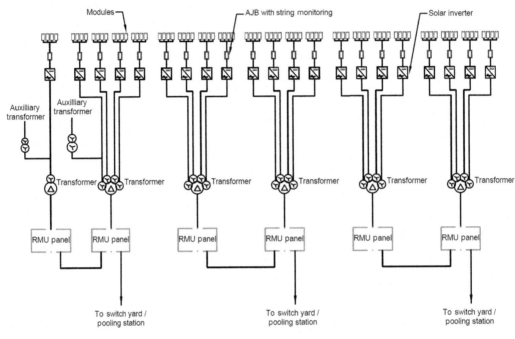

FIG. 9.23

Single line diagram of a 25 MW block of a 50 MW solar PV system.

Tilt angle: 26 degrees from horizontal
Inverter/PCU: 432 SMA string inverters, 11 kVA, 230 V, single-phase, 50 Hz
Grid connection: 33 kV, four pole substation
Annual energy generation: 7.5 million units per annum

10 MW solar PV power plant—66 kV—Charanka solar park, Gujarat

The power plant has an overall capacity of 10 MW connected to a 66 kV grid. Each string is designed to have 22 modules connected in series. There are 11 or 12 strings are connected one string combiner box. There are nine string combiner boxes, whose outputs are connected to one inverter. Two Inverters are connected to one 1.25 MVA, 300 V/11 kV Transformer. There are ten 1.25 MVA ONAN transformers in the compact substation and each is connected to an 11 kV HT panel via 11 kV ring main unit panel. The outputs are combined in a 11 kV switchgear and the 11 kV output is stepped up to 66 kV using a 12.5 MVA 11/66 kV transformer. The power is evacuated to 66 kV level. The configuration and connection schematic of 10 MW solar PV system is shown in Figs. 9.26–9.28.

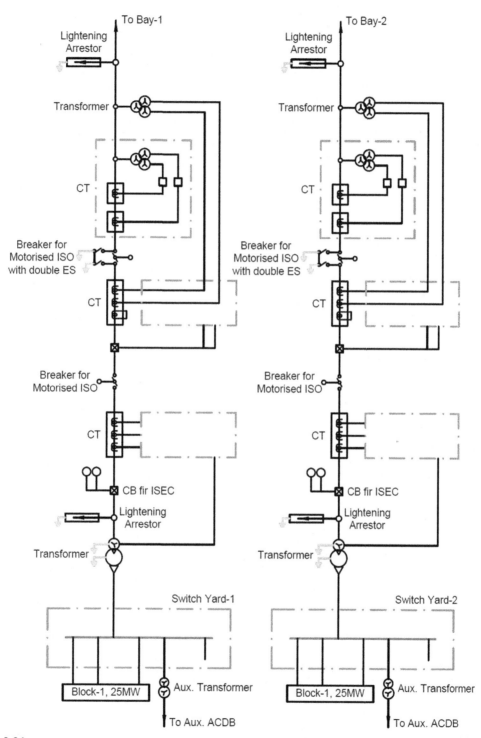

FIG. 9.24

AC single line diagram of a 50 MW solar PV system.

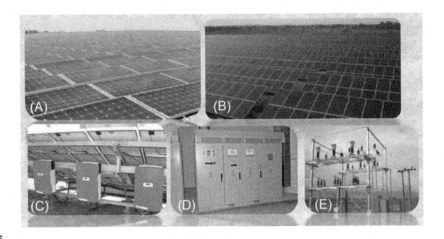

FIG. 9.25

A 5 MW solar PV system: (A) solar array, (B) string inverters mounted on the rear side of the MMS, (C) transformer, (D) LT panel combining all the string inverters, and (E) mini-switchyard with transmission line.

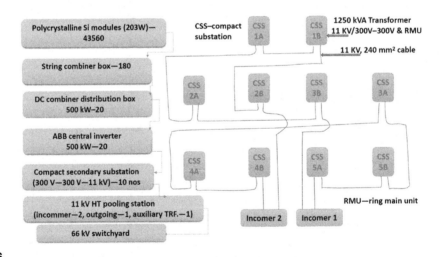

FIG. 9.26

Configuration of a 10 MW solar PV system. Thirteen to fourteen strings are connected to one string combiner box. Nine combiner boxes are combined in a DC distribution box. The output of DC distribution box, i.e., with 500 KWp capacity is connected to one 500 KW inverter. Two inverters are connected to three winding 1.2 MVA 0.300–0.300/11 KV inverter duty transformer. The output of the transformer goes to 11 KV ring main unit. The transformer and ring main unit are assembled in a compact substation (CSS). There are 10 numbers of CSS and their outputs are taken through 1 C × 240 mm2 11 KV cable as incomer 1 and incomer 2–11 KV switchgear. The *right side* of figure shows the combination of CSSs for incomer 1 and incomer 2 inputs.

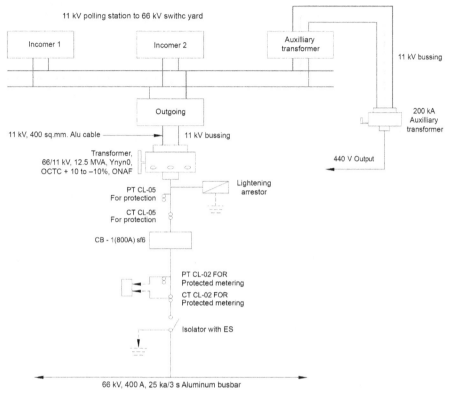

FIG. 9.27

AC schematic of a 10 MW solar PV system with 66 KV switchyard. Incomer 1 and Incomer 2 and auxiliary transformer feeders are inputs of 11 KV switchgear. A 11 KV switchgear output is stepped up to 66 KV using 12.5 MVA 11/66 KV power transformer and the output of transformer is connected to 66 KV switchyard. A 66 KV switchyard consists of protection and metering CT & PT, SF6 circuit breaker, main & check ABT meter, isolator with earth switch, and 66 KV lightning arrester.

9.6 Bifacial solar PV systems

It is a major challenge for solar PV power plant developers to increase the energy yield to meet power tariff rates, which are coming down. Grid parity has been achieved in different countries. As scientists and engineers continue to improve their solar panel designs to increase efficiency, the yield of solar PV power plants can be further increased by using modules with bifacial solar cells.

Bifacial solar PV modules absorb and convert solar irradiance received on the front and back sides as compared to standard monofacial solar modules, which only convert on front side. The back side of the solar PV modules receive reflected or diffused light from the ground or surface on which the solar PV modules are mounted, contributing to increased power generation. The reflected or diffused solar irradiance coming from the ground surface is called the albedo. The albedo factor

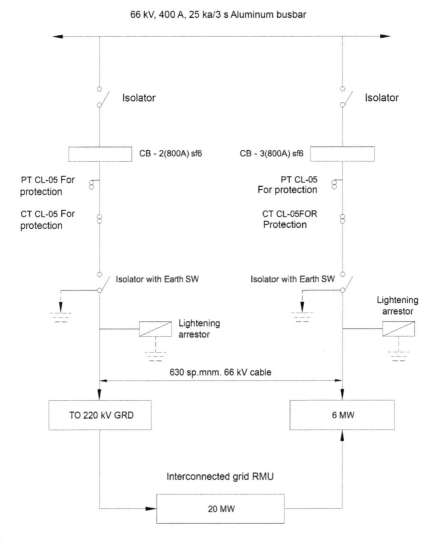

FIG. 9.28

A 66 kV pooling substation schematic for a 10 MW solar PV system. 10 MW is pooled with 20 MW and 6 MW systems and transmitted to 220 KV grid. Two bays are shown with SF6 circuit breakers and protection systems.

is defined as the ratio of the amount of light reflected back off the surface to the amount of the light incident on the surface.

The exact amount of albedo is solely dependent upon the surface where the solar PV modules are installed. The front side of the solar PV module absorbs most of the solar irradiance while the back side can absorb <5% or less compared to the front surface. The reflected diffused sunlight from the clouds, the diffused sunlight at rear side, the reflected direct sunlight from the ground and near by objects contribute to the albedo. The bifacial absorption is explained in Fig. 9.29.

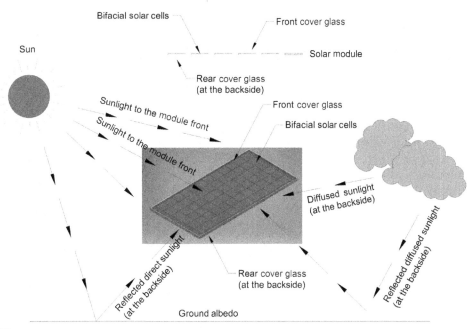

FIG. 9.29

Bifacial module light absorption. The reflected light from the ground, the reflected and diffused light from clouds, and reflected light from the neighboring objects reach the back side of the solar modules and get absorbed.

Bifacial solar PV modules have a few different designs. One of the bifacial solar PV module designs comes with a frame and an alternate design is frameless. A design uses a glass-glass bifacial solar PV module while another design incorporates clear transparent back sheet as a rear cover. The majority of the bifacial solar PV modules use p-type mono-PERC solar cells while some module manufacturers use n-type PERT and heterojunction solar cells. Most of the bifacial module designs are with half-cut cell based and with frame. All the bifacial solar PV module designs aim to produce power from both sides.

Back ground

The concept of a bifacial solar PV module has existed for many decades. As compared to any new technology, innovations at the right time can be converted into mass production at a competitive cost. Research on bifacial PV cells dates back to the dawn of the solar industry, according to Andrés Cuevas' oft-cited article, "The Early History of Bifacial Solar Cells." The first use of bifacial solar PV modules was in the Russian space program during the 1970s.

A significant development was initiated by Cuevas and his Spanish colleagues in 1980 by demonstrating and documenting a 50% output increase in their experiment in which the light colored surfaces directed reflected light (albedo) to the back of the bifacisal solar PV cell. As the costs were initially high in early 1980 to produce bifacial solar PV cells, initial bifacial solar PV modules for ground applications were delayed.

In 1997, the first field installation of a 10 kWp bifacial solar PV module was done in a noise barrier with a north-south orientation. Subsequently, after 10 years, the first commercialization of an HIT double bifacial solar PV module (UL listed) by Sanyo happened in North America.

The presence of metallic contacts on both sides of bifacial solar cells helps to produce additional power as compared to normal solar cells under the same environmental conditions. The additional gain in energy generation, which is dependent upon the solar cell design, solar module design, albedo factor, module mounting conditions, and other design factors such as module height, interrow distance, and usage of tracker can be between 3% and 30%.

Currently, there are four solar cell technologies available such as heterojunction technology (HJT), PERC, IBC, and PERT. Each of these cell architectures has varied levels of bifaciality. HJT and PERT have demonstrated 90% bifaciality while in the next range below at 80% is the IBC cell architecture followed by 70% bifaciality for PERC. PERC has the lowest bifaciality factor among all cell architectures.

Traditional solar PV modules use an opaque back sheet as a rear cover, whereas bifacial solar PV modules use either a clear back sheet or a glass on rear side. There is a reduced chance of potential-induced degradation (PID) effect for frame less bifacial solar modules, but there is a possibility of breakages during the stages of handling, transportation, installation, and working.

In order to covert sunlight into DC electricity on both sides of bifacial solar PV modules, the solar cells use selective area metallization so as to enable sunlight to pass between the metallized areas.

The following are the advantages of bifacial solar PV modules

- There is an improved energy generation due to bifacial energy gains and reliable performance over the life of the solar PV system due to improved durability. In view of the high conversion efficiency of bifacial solar PV modules, they help in lowering the BOS cost, which aids in mitigating the high initial cost of the total system.
- Unlike PV systems deployed with monofacial modules, bifacial PV systems can convert light that shines off the back of the solar PV module into electricity. This additional back-side production increases energy generation over the life of the system. Ongoing research and side-by-side testing suggest that a bifacial PV system could generate 5%–30% more energy than an equivalent monofacial system, depending on how and where the modules are installed. Moreover, the manufacturers' linear performance warranties for bifacial solar PV modules for 30 years are some of the best in the solar PV industry.
- In order to receive the reflected solar irradiance at the back-side of a bifacial solar cell, solar PV module manufacturers make use of either a UV-resistant transparent back sheet or another glass at the back. This will help to improve water ingress properties as well as durability.
- The bifacial module made from a double-glass solar PV module will give more rigidity and help to reduce mechanical stress in handling, transportation, and installation; it also does not allow more water ingress. In addition, there is a reduction in the power degradation rates year by year.
- Higher-efficiency modules not only reduce the area of the mounting system on a per kW basis, but also allow a developer to increase system capacity and energy harvest at a given site with fixed development costs. Due to this, there is a cost saving in land and the quantity of string combiner

boxes, DC cables, and module mounting structure will reduce causing the reduction in BoS cost of the system.
- The LCOE for a power generation asset is found by dividing the total lifecycle costs—both the upfront construction costs and the operational costs over time—by the total lifetime energy production. In the field, bifacial PV modules outperform their nominal power and efficiency ratings, which addresses the energy-generation side of the LCOE calculation. Bifacial solar PV modules will reduce the LCOE of the solar PV system.
- As the glass used in solar PV modules is getting cheaper due to high volume, double-glass solar PV modules are getting popular and have enabled the PV module manufacturer to extend module warranties up to 30 years.
- The back side generation of the bifacial solar PV module is dependent on the albedo at the site location. This calls for the appropriate site selection to achieve a high albedo. The albedo can be enhanced by adding material with good reflectance such as gravel, sand, white paint, etc. to the ground.

Basics of bifacial technology

Bifacial PV technology deals with making solar cells to generate power on both sides with solar irradiance. Recent advanced solar cell architectures are bifacial by nature. However, the optimization needed to make the solar cell rear side receptive to sunlight absorption is very minor. It deals with printing metallization pattern and antireflection coating that are similar to the front side. The bifacial concept requires some changes to the solar PV module design. An important attribute of the bifacial cell and module is described as bifaciality factor, which is the ratio of front side to rear side efficiency.

At the system level, there are several ways to improve the yield of a bifacial solar installation. As the output of the rear side of the modules relies to a large extent on the site albedo, developers need to select installation sites with a natural high albedo or increase the albedo artificially. East-west installation is another option.

The attribute of this technology is bifacial gain, which is calculated using the equation.

$$\text{Bifacial gain} = \frac{(Y_b - Y_m)}{Y_m}$$

where Y_b is the specific energy yield (kWh/kWp) of the bifacial PV and Y_m is the specific energy yield (kWh/kWp) of the standard monofacial PV.

The bifacial modules also offer the opportunity for vertical installations. Such installations can be considered for applications such as ground-mounted solar PV power plants (with space limitations) and sound barriers on highways and railways.

Issues to be sorted out

For the commercialization of bifacial technology, the issues of module bankability, a standardized method for power measurement, and development of a standard for testing need to be resolved.

Bifacial solar PV module technology has been facing difficulties while good progress has been observed for the various aspects. Bankability is improving with increased solar PV power system installations using bifacial PV modules. Many tier 1 solar PV module manufacturers have now entered this field. Meanwhile, the much-awaited IEC standard for measurement of the *I-V* characteristics of bifacial solar PV modules is in the final approval process. In addition, leading solar sun simulator manufacturers have started to supply sun simulators, which is now allowing manufacturers of bifacial solar cells and modules to use them in large solar PV module production facilities.

Fig. 9.30 shows the horizontal single axis tracker mounted with bifacial solar PV modules.

The gain in energy due to bifaciality is dependent upon the following factors

1. The type of solar cell used for the solar PV module and the bifaciality factor. Bifaciality factor can be defined as

$$\text{Bifaciality factor} = \frac{\text{Power of Rear side of module at STC}}{\text{Power of Front side of module at STC}}$$

The bifaciality factor is different for bifacial solar modules of different structures. PERC+ and IBC technology-based modules give a BF of >70%, whereas PERT and HIT-based modules give BF values of >90% and >95%, respectively. The half-cut solar cell technology-based bifacial modules generate more power due to reduction of series resistance of the cell and improved incidence of light reflected between the gaps of the solar cells.

2. **Bifaciality factor**: A higher bifaciality factor will increase the energy yield.
3. **Solar irradiance**: The exact site location and amount of diffused and direct radiation have a direct effect on the energy yield. The actual tilt angle of the solar PV module mounting system is determined by the latitude of the location. The more diffused radiation, the greater the energy

FIG. 9.30

Tracker-based bifacial solar PV system. Two modules in portrait configuration are arranged. Inter-row gap is maintained more and height of structure is more compared to fixed tilt system.

Table 9.4 Percentage of albedo for different surfaces.

Surface type	Albedo
Green field grass	23%
Concrete	16%
White painted concrete	60%–80%
White gravel	27%
White roofing metal	56%
Roofing membrane light gray	62%
Roofing membrane white	>80%

Data source: *https://solarkingmi.com/assets/How-to-Maximize-Energy-Yield-with-Bifacial-Solar-Technology-SW9001US.pdf.*

generation. Under same albedo conditions, if diffused radiance is higher, the bifacial gain will be higher. In the areas of high diffused radiation, higher bifacial gain can be achieved by increasing module height.

4. **Tilt angle**: An increase in tilt angle will result in increased reflected light, which will help in increased energy generation.
5. **Type of horizontal surface** from which the reflection is coming and its reflectivity (albedo factor): Table 9.4 gives the albedo factor for different materials covering the ground surface. Bifacial gain increases and almost linear with the increase of albedo factor. The albedo is an important aspect when selecting an installation site for bifacial PV systems. The reflected light from the ground varies according to the surface type. While water shows the lowest albedo of about 8%, fresh snow can reflect 90% of the sunlight.
6. **Row to row distance**: If the interrow space is increased, the reflected light will increase on the rear side and hence contribute to the energy yield. This can be described in terms of Ground Covering Ratio (GCR). GCR is defined as the ratio of module length in vertical direction to inter row spacing. If GCR increases, the bifacial gain will reduce due to lower inter row spacing. The suggested optimum GCR for energy gain is 40%–50%. If GCR reduces, the solar field area increases causing increased cost for the land.

7. **Elevation and height of structure:**

- Flush mounting of the solar PV module obstructs the reflected solar irradiance on the back side.
- Lower height of the solar PV module mounting structure shades the back side of the solar PV module.
- Increased structure height >1 m will assist in increasing the reflected solar irradiance and in turn yield higher energy. If the height of the module increases, there will be an improvement in the uniformity of back side irradiance. The optimum module height for improved bifacial gain is 1.2–1.3 m. Bifacial gain will improve with module height, but the wind loads also will increase causing increase in the cost of module mounting structure.

8. The use of a single-axis tracker enhances the reflected radiation on the rear side and boosts the energy up to 15% compared to monofacial modules mounted on a tracker.

Table 9.5 Gain in energy yield for different surfaces with and without a tracker.

Surface type	Gain in energy yield	
	Without tracker	With tracker
Grass	5.20%	10.57%
Sand	10.79%	24.42%
White-painted	21.90%	33.20%

To demonstrate the increased energy production from bifacial solar PV modules in actual field conditions, Trina Solar set up a test bed facility in Changzhou, China, to find out how much more energy bifacial modules generate compared to traditional solar PV modules in local environmental conditions. Field tests were conducted on bifacial solar PV module power along with a solar PV tracker on three different surfaces: grass, sand, and a white-painted surface.

Using monofacial solar PV modules as the base performance parameter, the various test results in energy gain are given in Table 9.5.

As the initial test results show, bifacial solar PV modules exceeded the performance of standard solar PV modules (monofacial) in various surface conditions. Also, the preliminary test data show that trackers add even greater energy gains to a bifacial solar PV module. In addition, when combined with a tracker and placed on a white-painted surface, bifacial modules outperform the standard monofacial solar PV modules by more than 30%.

9. Vertical mounting in the east-west direction generates higher energy yield for higher latitudes (NREL study).

Sandia Laboratories set up a test bed at the New Mexico Regional Test Center and measured the performance of bifacial systems over a period of 6 months. The study concluded that the bifacial gain varies throughout the day and the bifacial contribution is more in the morning, evening, and during cloudy conditions. Bifacial modules outperformed monofacial modules by 18%–136% [9]

As per a study by Solar World, the bifacial gain in energy (BGE) increases with the tilt angle of the modules (θ), the height above the ground (h), and the albedo factor (α) of the ground, as per the equation below:

$$\text{BGE}\,(\%) = A^*(\theta) + B^*(h) + C^*(\alpha),$$

where A, B, and C are numerical coefficients. The study confirms the following:

- High albedo gives higher energy yield.
- The higher the installation height of the module, the higher the energy yield.
- The bifacial boost is greater on days with lower insolation.
- The bifacial modules provide a considerable performance boost even in less than ideal circumstances.

Libal of Konstanz studied the LCOE with bifacial PV systems. The LOCE is reduced by more than 50% from a region with low irradiance in the north of Germany to a region with the highest irradiance in the Atacama desert in Chile. The bifacial PV system has strong potential to reduce the LCOE of solar PV generated electricity. In order to reduce the LCOE of the bifacial systems below the monofacial systems, a 10% premium on the system capex is required to exceed 10% in the bifacial energy yield gain [10]. Longi Solar also predicts that the LCOE is likely to be 5 US cents/kWh or less in all areas [11].

The bifacial solar PV module gain in energy depends on the module bifaciality factor, the albedo factor and the height of the module. The bifacial performance is quite sensitive to the enhanced albedo of the ground surface. E-W bifacial vertical modules can outperform optimally oriented monofacial modules, especially with an enhanced albedo. Vertical E-W bifacial solar modules generate higher energy in the morning and evening of the day than south-facing solar PV arrays. Bifacial solar PV modules significantly outperform monofacial solar PV modules in standard solar PV power system design. Better performance from bifacial solar modules is possible with optimized system designs that enhance the albedo, avoid backside obstructions, and minimize ground shading under the solar PV modules or array. A suitable design and choice of both solar PV modules and ground installation are likely to ensure a bifaciality gain of 10%–30%.

9.7 **Floating solar PV systems**

The solar PV panels designed and installed to float on water bodies and generates power are called floating solar PV (FSPV) systems. The water bodies such as reservoirs, hydroelectric dams, industrial ponds, water treatment ponds, mining ponds, lakes, and lagoons can be used for setting up the FSPV systems. In FSPV systems, solar panels, inverters, cables, and lightning arresters are mounted upon a pontoon-based floating platform, and the floating structure is anchored and moored.

Solar PV power plants require large quantities of land area compared to other power generating modes. Availability of land is a big problem. This is one of the primary reasons why different countries are interested in the floating PV technology. There are innumerable water bodies, large and small, available throughout the world suitable for setting up of solar PV plants. The following are the benefits of floating solar PV systems.

Improved energy yield performance: The evaporative cooling effect of water results in lower operating temperatures of the PV modules. The wind blowing on the water surface reduces the soiling on the modules. Combined with advantages in terms of reduced shading and soiling, the lower operating temperatures of a floating system increase its energy generation capacity compared to a land-based installation [12].

No loss of land space: Since FSPV plants are installed on water surfaces, the land requirement is greatly reduced and valuable land can be saved. Right of way issues for transmission line will also be reduced.

Water evaporation control: The FSPV plants deployed on the water bodies, causes shading of water surfaces, reducing the amount of light and wind in contact with the water surface causing the amount of water lost through evaporation is reduced. The rate of evaporation is directly linked to the size of area covered by the floating platform. With floating solar, around 70% of the evaporation could be prevented which would in turn help in the retaining sufficient amount of waters in the canals and small river bodies [13]. The ecology of the water body is not likely to be affected much and it will also reduce evaporation, thus helping preserve water levels during extreme summer.

Restriction of algae growth: Since FSPV plants provide shade to the water surface, they reduce the amount of sunlight reaching the water surface, which may cause a reduction in algae growth. This makes water less contaminated and helps in production of oxygen necessary for the aquatic life in sustaining and minimizing the associated water treatment and labor costs.

Ease of cleaning: In case of FSPV plants deployed on inland water bodies, water is readily available for cleaning purposes. Typically, areas with high solar energy potential tend to be dusty and arid, so in comparison to their ground mounted counterparts, floating PV system not only have to perform in a low dust environment, they can always use a sprinkler to bathe themselves clean. However, the quality of the water needs to be checked and should be used as per the guidelines of PV module manufacturers. In dam reservoir-based hydro power plants, solar power can substitute hydro-based generation during day time when sun is available. In such a case, the stored water in the reservoir will serve as an effective energy storage system. By deploying FSPV plants on reservoirs of hydroelectric plants, and utilizing the already existing infrastructure and gird connections new investment cost can be saved. The electromechanical machines like generators are not required which reduce the amount of steel structures in the plant. Therefore, such plants are comparatively more eco-friendly.

By building electrical plants over water bodies, government and energy companies can save valuable real estate and they become source of income by better usage of the reservoir surface, which is anyway lying idle.

Disadvantages with FSPV plants: It costs more to introduce an on-water-type solar power generation system than a ground/roof top-based PV system. The construction period is longer with on-water system because a series of processes to set up the solar panels, such as mounting solar panels onto floats and connecting the floats together. Operation and maintenance is still uncertain in some respects. Although it requires no weeding, the algae that grow on the floats and panels need to be removed. There will be an impact on ecosystem, such as marine aqua culture due to reduced sun light. There will be an obstruction for the visitors for boating in the pond. From fall to winter, various kinds of birds including migrating birds and waterfowl rest on the floating mounts of the solar panels.

Installation and deployment: In general, installation of a typical FSPV plant is simpler and easy as compared to land-mounted solar PV plants. This is because, (a) the civil works like soil levelling and grading are not required to prepare the site; (b) floating platform used to mount solar modules are made in form of a modular individual floats which are prefabricated and are interconnected to form a large section; and (c) floating platforms are assembled on land by adding rows of these modular interconnecting floats. Each of these rows is pushed into water as the next row being added to form a large platform. Once completed, the entire platform is towed to the exact location on the water body with the help of boats. In general, the cost for setting megawatt scale FSPV plants depends upon the following: Size of the plant, project location, depth of water body, water-level variation, site conditions such as wind speed and its direction, solar irradiations, ambient temperature, humidity levels, etc.

Components of FSPV system: All the components used for FSPV system are similar to the ground-based systems except the floating platform, anchoring, and mooring systems. The solar modules see very high humidity levels. So, cables and connectors frequently touch the water surface. So more care has to be taken in selecting the solar modules. The back sheet performance at higher moisture level and higher humidity conditions has to be checked. The suitability of the back sheet, adhesive, junction box, and connectors has to be checked for this environment by conducting extra tests. The edge sealing, junction box bonding, junction box, cables and connectors, and performance of frame for this humid conditions are to be checked by qualifying them by conducting severe tests. The test methodology is under development. Frameless glass-glass modules with POE encapsulant with proper edge sealing can be used for FSPV system. Bifacial solar modules may not be advantageous for FSPV applications, as the albedo factor is very less, because of less gap between the water surface and the module.

String inverters can be used for FSPV. The system voltage of 1500 V cause arcing issues and accelerate the PID issue, so 1000 V can be maintained as system voltage. The efficiency of inverter will be higher due to lower temperature of operation. DC to AC ratio can be optimized by estimating the energy yield corresponding to the lower temperature of the module. Depending upon scale and distance from shore, the inverters can be mounted either on the banks of the lake or on a separate floating platform. Generally, for smaller capacity FSPV, inverter may be located on land near to PV arrays, otherwise for large capacity plants it is advisable to place inverter on a floating platform to avoid excessive resistive losses.

Cabling: To accommodate the movement of floating platform due to wind load and water-level variations, there is a requirement of extra length of cable in form of slackness. If sufficient cable length and slackness is not provided, it may result snapping and rupturing of cables due to the tension. The cable has to be properly sized for minimum voltage drop and power losses. The cables can be routed in either via floating platform or via submarine cables. To keep the cables on the water surface via floating platform, cable trays, cable conduits, cable ties, or clamps, and cable clip holders are required to be used to avoid cables coming into contact with water. Care must be exercised in the selecting the cables for FSPV systems. Cables used must be UV-resistant, and uses of wiring trunks are recommended to protect them from direct sunlight. Similarly AC cables can be routed either via separate dedicated floats or via use of submarine cables for the connection to main electrical infrastructure onshore in case of FSPV plants. Plot size, distance to shore, placement of inverters, transformers, quality of water, and variations in water level, play vital role while sizing the cable and its routing.

Module mounting structure (floating platform)

This component is the main difference with the ground mounted systems. MMS is a floating platform, supports all necessary components like solar PV modules and keep them in the floating condition. The floating platform should be anchored by mooring system to avoid its displacement during water-level variations.

The modules will be arranged in rows of certain number of modules each. The inter connection will be through pins so that slight rotational movements caused by wind or ripples in water surface are permitted which will relieve stresses in the platform members. HDPE is the most popular material being used for the floating platform in a majority of the FSPV power plants across the globe. Other materials like FRP, medium density polyethylene (MDPE), and ferro-cement are also been utilized as materials for the floating platform. There are floating platforms with different type of designs.

Pure-floats design: It uses a specially designed float that can hold PV panels directly. The entire system is made in a modular fashion and has a provision to join with pins or bolts to make a large platform. Every single unit of such a system typically consists of the main and secondary floats. The main purpose of the secondary float is to provide a walkway for maintenance and additional buoyancy. The design is being used by a few manufacturers, some of them are as follows: The Ciel & Terre is a French technology provider, provides Hydrelio floats made of HDPE and connection pins are made of polypropylene combined with fiberglass. The floats are manufactured by blow molding and pins are made from injection molding. Sumitomo Mitsui Construction Co. Ltd provides some additional features for HDPE like more regularly shaped float for denser packing and easy transportation. Additionally, a float is filled with polystyrene foam reducing the risk of sinking even when damaged and there is a usage of binding bands in the connecting part, which reduces the risk of structural failure for floats. The Yellow Tropus is an India-based design and engineering company focused and specialized in the development of FSPV power plants. The company is offering Seahorse technology—One single unit of this consists of one walkway float and two solar PV panel floats for supporting two solar PV panels.

Pontoon with metal structures design: The other common design which is used by some project developers uses a metal structure similar to land-based system and pontoons to provide buoyancy, hence eliminating the need for specially designed floats.

Anchoring and mooring system: The variations in water levels induced by monsoon, wind velocity, or increase/decrease in water quantity could be problematic for the plants. To avoid this situation, FSPV plants are anchored through mooring systems. The placement of a mooring system must take into account the location, bathymetry, soil conditions, and water-level variations. Mooring systems include quays, wharfs, jetties, piers, anchor buoys, and mooring buoys. Mooring system for a floating platform is generally attached with nylon polyester or nylon nautical ropes that are further tied to bollards on the bank and lashed at each corner. Mooring can be done by anchoring to near by bank or bottom of the lake or reservoir or to the pilings formed on to the subsoil of water body.

Design: FSPV design is similar to a conventional solar PV system except it requires a special arrangement to float on the water surface. The typical floating structure supports the PV arrays, inverters, combiner boxes, lighting arresters, etc. on a floating bed, which is made of fiber-reinforced plastic (FRP) or high-density poly ethylene (HDPE) or metal structures. The whole floating bed is buoyed with the help of anchoring and mooring systems. The initial site assessments of the proposed waterbody should be carried out to understand the feasibility of the site for the FSPV project development. The site assessments can be executed in the two following steps:

GIS-based assessment: The GIS tool provides information on water surface variations due to seasonal changes in the past years. The outcome of the analysis gives an approximate surface area of the waterbody which remains permanent irrespective of seasonal water-level deviations.

*Meteorological data***:** Solar irradiance, wind direction, wind velocity, wave height and amplitude, water-level variations, humidity, ambient temperature, rainfall, etc. will be collected.

Field assessment: An initial site visit is essential to understand the ground reality of the site conditions such as site access and pathways, distance from substations, nearby buildings infrastuctures, shadow-free area, type of water body and its current purpose, and coverage surface area. The visit mainly covers the assessment of type of waterbody and its purpose, accessibility to the site, location of nearest substations, and information on restricted area. In case, the waterbody is used for some industrial purposes or in power plants or water treatment plants, etc., details like fire hazard buffer zone, nearby building infrastructure for calculating the shadow-free area, availability of space while installation, space available near shore while placing the floating platform into waterbody, etc., are some of the important points that need to be looked into. Eventually, the initial site assessments assist in estimating the suitable area for FSPV installation and water quality of reservoir/pond/lagoon.

Bathymetry: The bathymetric and/or hydrographic surveys are useful in understanding the depth of a waterbody and topography of water bed. It will provide shape of the boundaries, average depth and depth distribution, structure of the water bed, locating any bed-rock outcrops, obstacles at bed level if any, for finding the optimum locations for placing anchors and mooring for the floating platform. This step is vital to judge the overall techno-commercial viability of the projects, and failure in conducting this properly could cause extensive damage to the FSPV plant. Soil testing is also an important step for deciding the type and design of anchoring that would be required. It provides information on soil composition of the banks and waterbody beds, etc. [14].

Challenges: The plant has to be designed to withstand maximum wind speed up to 200 km/h. The water current could be maximum speed of 2 m/s. As solar panel structure is surrounded by the water and the moisture due to which there occurs corrosion phenomenon which will affect the strength of the floating solar structure. The system problems arising due to high moisture content, corrosion as a result of adverse environmental conditions as well as stability in times of change in weather. Besides, there are also challenges in safely transporting the power from the floating objects. The module technology has to be developed to block the moisture that would penetrate from the backside and side surfaces of a solar modules. The technology has to be developed to prevent short circuits caused by moisture penetrating the electric joints. The technology has to be developed to prevent electricity loss due to lower insulation resistance in the humid environment. Even though the plastic mounting systems are floating on the water, they still have to prevent the solar panels from moving excessively and the solar panels from tilting due to changes in the water level. The conductors and cables should be of marine type quality having water proof. Figs. 9.31 and 9.32 show the images of solar PV plants on water bodies.

9.8 Bench mark cost of ground mounted solar PV systems

The cost of a ground-mounted solar PV system varies from region to region in the world.

The cost of the components, equipment, duties, and transportation charges are different in different countries. The cost of installation activity of solar PV system is different in different countries due to variation of labor charges for technicians and workers in different countries. NREL has done a study for a utility-scale 100 MWp solar PV project considering fixed tilt and a single-axis tracker [15]. Fig. 9.33 gives the variation of the benchmark cost of the utility system with and without a single-axis tracker from 2010 to 2018. The system cost is falling.

FIG. 9.31

Solar PV floating system with pontoons. The HDPE floating platforms form walk way. The floating platforms are anchored at different locations.

FIG. 9.32

Solar PV floating system.

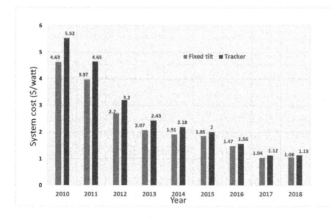

FIG. 9.33

Variation of the benchmark cost of a utility system from 2010 to 2018. The system cost for fixed tilt PV system is getting reduced from \$4.63/W in 2010 to \$1.06/W in 2018. This is due to reduction in the module prices and installation costs. In 2018, the PV system cost with horizontal single axis tracker is \$1.13/W.

References

[1] Wikipedia (https://en.wikipedia.org/wiki/Growth_of_photovoltaics).

[2] E. Bellini, PV Magazine Report, https://www.pv-magazine.com/2019/05/14/global-cumulative-pv-capacity-may-reach-1-3-tw-in-2023-solarpower-europe-says/, May 14, 2019.

[3] DNV GL, 2018 PV module reliability scorecard, https://www.dnvgl.com/publications/2018-pv-module-reliability-scorecard-117982, 2018.

[4] FIMER SPA, Inverters are manufactured by Fimer Group under ABB trademark licence. https://www.fimer.com/pvs980/pvs980-58-2mva

[5] S. Pandey, Evaluation of various PV module cable connectors and analysis of their compatibility, Int, J. Curr. Eng. Technol. 7 (5) (2017) 1721–1727.

[6] S. Smith, PV trackers, Solar Pro 4 (4) (2011) 28–54 June–July.

[7] V. Sheldon, Optimizing array-to-inverter power ratio, Solar Pro 7 (6) (2014) 14–20 October–November.

[8] White paper on solar DC cables, Renewable Watch Magazine; 2018, (2018)July. https://renewablewatch.in/2018/07/09/white-paper-solar-dc-cables/.

[9] LaveM., et al., Performance results for the prism solar installation at the New Mexico Regional Test Center: Field Data from February 15 to August 15, Sandia National Laboratories, 2016 SAND2016-9253.

[10] J. Libal, et al., bifi PV workshop, October 26; 2017, (2017).

[11] Longi Solar, Hi-MO3: reducing PV LCOE by the mono PERC Bifacial Half-cell Technology. White paper. https://en.longisolar.com/uploads/attach/20190726/5d3a8ea5e987e.pdf.

[12] REC, Riding the wave of solar energy: why floating solar installations are a positive step for energy generation, https://www.recgroup.com/sites/default/files/documents/wp_-_floating_pv_rev_d_web.pdf.

[13] M. Acharya, S. Devraj, Floating solar photovoltaic (FSPV): a third pillar to solar PV sector, Output of the ETC India Project. The Energy and Resources Institute, New Delhi, TERI Discussion Paper, 2019.

[14] World Bank Group, ESMAP and SERIS, Where Sun Meets Water: Floating Solar Handbook for Practitioners, World Bank, Washington, DC, 2019.

[15] F. Ran, D. Feldman, R. Margolis, U.S. Solar Photovoltaic System Cost Benchmark: Q1 2018. National Renewable Energy Laboratory; 2018, (2018).

Grid integration, performance, and maintenance of solar PV power systems

10

10.1 Introduction

Solar PV power plants have DC side power generation using a solar array. The inverter converts the DC power to three-phase AC power output at 415 V AC or a higher AC voltage. To export this AC power to the grid, it has to be stepped up to the required high AC voltage and evacuated to the grid. The components and equipment from the inverter output circuit(s) to the grid interconnection point, such as transformers, switchgears, circuit breakers, wiring, and switchyard elements, are considered as the AC interconnecting system. Fig. 10.1 shows the typical schematic of the grid-connected solar PV power system.

Solar modules are connected in series to form a string and the strings are combined in a string combiner box. The outputs of string combiner boxes are connected to the inverter. The inverter converts the solar array-generated DC power to AC power. The outputs of the one or two inverters are connected to the transformer for stepping up to a higher voltage. This transformer is called an inverter duty transformer. The outputs of transformers are combined in the HT switchgear and connected to the switchyard. From there, the power is carried through a transmission line to the substation and the power is evacuated. If the power evacuation is more than 33 kV, the HT panel output is stepped up to the required voltage such as 110 or 132 kV using a power transformer. The power is transmitted through the transmission line to the substation. At the substation end, there will be a bay extension and a control panel for the solar plant.

10.2 AC system requirements for solar PV power system

The considerations for the required medium-voltage or high-voltage AC system for a solar PV plant are equivalent to any typical conventional AC power distribution system. The best designed AC power system is to be cost-effective as well as safe. Also, it must address the present and future probable loads. The difference in the AC system for a grid-connected solar PV power plant is that it doesn't supply power to the loads directly, but is designed to supply power to interconnected transmission lines.

The electrical distribution design goal addresses the following aspects [1]:

- Maximize safety of equipment and personnel.
- Minimize investment.

Solar PV Power. https://doi.org/10.1016/B978-0-12-817626-9.00010-1

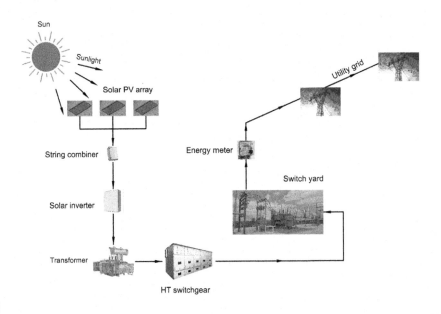

FIG. 10.1

Schematic of a grid-connected solar PV power plant.

- Maximize continuity of service.
- Maximize expandability for future requirements with flexibility.
- Maximize operational electrical efficiency.
- Minimize operational and maintenance costs.
- Maximize quality of power.

Different components of an AC collection system are briefly discussed.

10.2.1 ACDB: An interface for a string inverter with a power transformer

An AC distribution board is required in the case of a grid-connected solar PV power plant using string inverters. This is used to combine the output of several string inverters and connect to the LV windings of the inverter duty transformer through various pieces of input protection equipment. More details are presented in Chapter 9.

10.2.2 Transformer requirement

A transformer is a static device that transforms AC electrical power from one voltage to another voltage by the principle of electromagnetic induction, keeping the frequency the same. There are different types of transformers by application. They are power transformers, inverter duty transformers, current transformers (CTs), and potential transformers, which are being used in grid-connected solar PV systems.

10.2.2.1 Power transformer

A grid-connected solar PV power system makes use of a transformer to provide galvanic isolation while stepping up the voltage and transferring energy to the utility grid. The power transformers step up voltages from, say, 11/22/33 to 66/110/132 kV for a medium-voltage system. Power transformers, which are part of the AC electrical system, help to transfer energy from the solar PV array to the grid. The combined output of different inverter duty transformers such as 33 kV can be stepped up to 110/132 kV and transmitted through transmission lines.

10.2.2.2 Inverter duty transformer

Principle of operation of a multiple windings transformer is same as that of normal two winding transformer. The advantage of multiple windings is, the same transformer can be connected with more Inverters at the same time so that we can consume less space. If there is any problem with one of the inverters, transformer can work with other inverters without any problem. There will be less operating cost and Maintenance cost and the cost for oil filtration required for transformer oil is minimized.

This is used to supply the solar power to grid. And it's user friendly—daily switch ON and Daily switch OFF of the transformer is possible. This is a transformer that is used to step up the inverter AC voltage of 415 or 600 V AC to higher voltages such as 11/22/33 kV. This has to be designed taking into account the specifications of inverter harmonics, impedance, etc. Fig. 10.2 shows the transformer and some of its parts.

Solar Inverters are rated for unity power factor. So Sizing of Transformer should be based on unity power factor and No load losses shall be low and load losses shall have bit higher compared to distribution transformer. While manufacturing the Inverter Duty Transformers the following parameters should be taken care

- Use of high grade, good quality core material to see that no load losses shall be low.
- Electrostatic Shield should be provided in between LV and HV windings.
- Increased margin for core over excitation, withstanding high magnetizing inrush current.
- Possibility to have two, three, or four different windings on the low voltage side.
- Arrangement of windings
- Vector group connections
- Tank Fabrication

The parts of a transformer include, primary and secondary windings, laminated core, Insulating materials, tank, transformer oil, Bucholtz relay, magnetic oil gauge, winding and oil temperature gauges, conservator, breather, bushings, radiators, air vent plugs, drain valve, cooling tubes, and an onload or offload tap changer.

10.2.2.3 Windings

Coils are necessary to induce electromagnetic force in one coil due to a change in voltage in another coil. The coil in which the voltage is given is the primary coil and the coil in which the voltage is induced is the secondary coil. Thus windings are classified as primary and secondary windings based on the applied input and induced output voltage. They can be classified based on voltage range as high-voltage and low-voltage windings. The conductors made from electrolytic-grade copper free from scales and burrs are used for windings due to the high conductivity and ductility of copper. The high

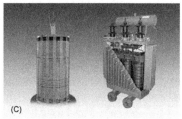

FIG. 10.2

The basic parts of a transformer. (A) A view of transformer with all parts, (B) Five copper winding coils mounted on core. There are three shoulders of core considered as three phases and each phase is equipped with four coils and each coil is having one LV and one HV winding. For R-phase of coil 1, LV winding goes to R-phase of the inverter 1. Similarly LV windings of coil 1 of Y,B phases will be going to Y,B phases of inverter 1. The fourth coil LV windings go to inverter 4. The individual HV windings of four coils fitted to R-phase of core are paralleled and the output goes to R-phase of HV side of transformer, and (C) Two winding coil and its fitting in a transformer.

Courtesy: Esennar Transformers Pvt. Ltd.

conductivity of copper minimizes losses. The high ductility of copper facilitates easy bending of wires into tight winding around the core.

Depending on the requirement, two, three, or five windings are used in the inverter duty transformer. In a two-winding transformer, one coil is used to connect LV and the other coil is used for connecting HV. LV and HV windings are done as single coil as one over other. For each phase one coil containing LV and HV windings will be fitted on one shoulder of core. In a three-winding transformer, there are two coils for each phase: two windings are used for connecting inverters independently in the LV side and one the two windings of HV coils are paralleled by brazing and connected to the phase for the HV side connection. For a five-winding transformer, there will be four coils per phase and each coil having LV-HV windings. From four coils of the phase, the independent LV connections go to four individual inverters. The HV windings of four coils of each phase are paralleled and considered as single HV winding.

10.2.2.4 Transformer oil

It serves as an insulating material as well as cooling media for the core and its windings. It is a hydrocarbon mineral oil in which the core and windings of the transformer must be completely immersed. It should be protected from moisture. In recent days, synthetic vegetable based oil (KNAN) is also used in the market for better cooling and insulation.

10.2.2.5 Core

To support the windings in a transformer, a core material is required. It is usually made of nonaging, high-grade, low-loss, cold-rolled grain-oriented (CRGO) silicon steel. The CRGO core reduces losses arising due to eddy currents and hysteresis.

10.2.2.6 Insulating materials

The primary and secondary windings are to be electrically isolated from each other as well as the transformer core. For this purpose, compressed papers and cardboard are used as insulating materials.

10.2.2.7 Conservator

The conservator is a metallic tank that provides adequate space for the expansion of oil, whenever the transformer is loaded or during temperature variations. It protects the oil from the deterioration of its insulating property.

Separator inside conservator: A separator installed inside the conservator to separate the transformer insulating oil from the atmosphere prevents gas or moisture contamination of the oil. It is also called an air cell and maintains a constant atmospheric pressure on the transformer oil. Under normal operation, the air cell is completely surrounded by oil and floats as high as it can in the conservator. As the transformer oil volume changes, the air cell inflates or deflates by an equivalent volume. The float of the magnetic oil level gauge makes contact with the underside of the air cell and follows the motion up and down, thereby giving an indication of the oil level inside the conservator.

10.2.2.8 Breather

The breather is filled with silica gel and has the shape of a cylindrical container. The air that enters the tank is kept moisture-free inside the breather. With the ingress of moisture, the insulation properties of the transformer oil get affected. So, the area above the surface of the oil in the tank should be kept free from moisture. So, the breather, which is a cylindrical container filled with silica gel, is used to absorb the moisture contained in the air entering the tank.

10.2.2.9 Radiators

The radiator units are made from very thin steel sheets of metal for good conduction of heat from oil to air and are used to take out the heat from the tank. Each radiator will have top and bottom shut-off valves, a top filling plug, a bottom drain plug, lifting lugs, thermometer pockets at the inlet and outlet pipes, air release devices, earthing provisions, filter valves, and all other necessary accessories.

10.2.2.10 Tank

The tank provides rigidity to the transformer to withstand pressure due to short circuit current. It is capable of bearing all stresses during transportation and operation without any deformation.

10.2.2.11 Oil gauge and WTI

One oil gauge of magnetic type with provision of alarm for low levels of oil is mounted on the conservator to indicate the minimum, normal, and maximum oil level. Temperature-sensing equipment is connected through the capillary tube. A winding temperature indicator (WTI) is an oil temperature thermometer. Its bulb is associated with a heater coil that carries a current proportional to the load on the transformer. The terminals of the heating coil are connected to the secondary of a ring-type

CT, which is fitted on the bushing lead under the transformer tank cover. The heater coil introduces an increment of temperature above that of the oil to correspond to the gradient between the winding and oil temperatures. The instrument thus indicates a figure that, although a reasonable analogue of the temperature of the windings, is not a direct measurement thereof.

Oil and winding temperature indicators are provided for oil and winding, respectively. Remote monitoring provisions will be there for both the oil and winding temperatures.

10.2.2.12 Buchholtz relay

This is a very sensitive gas and oil operated instrument that detects the formation of gas or the development of sudden pressure inside the oil transformer. Any electrical fault occurring inside a transformer is accompanied by an evolution of gas. Appreciable quantities of gas may be produced before the fault develops to such an extent that it can be detected by the normal electromagnetic protection equipment. The Buchholtz relay is connected between the transformer tank and the conservator.

10.2.2.13 Tap changer

To keep the secondary voltages reasonably constant at the user's end when incoming voltage and/or load on the transformer changes, it is necessary to adjust the voltage ratio (i.e., turns ratio of the windings) of the transformer. This is achieved by operating the tap changing switch. There are two types of tap changers. One is the online tap changer (OLTC), in which tapping of the transformer can be changed automatically without disconnecting the transformer from the grid. The other is the off circuit tap changer (OCTC). In OCTC tap changers, the transformer has to be disconnected and the tap change has to be done. Depending on the voltage fluctuations of the grid, the requirements of the OLTC for the inverter duty transformer are decided. For power transformers, OLTC is followed in grid-connected solar PV systems depending on the voltage fluctuations of the grid voltage of the substation.

10.2.2.14 Connection and vector group

The star, delta, or zigzag connection (of a set of phase-windings of a three-phase transformer or windings of the same voltage of single-phase transformers associated in a three-phase bank) shall be indicated by the letters Y, D, or Z for the HV winding and y, d, or z for the intermediate and LV windings. If the neutral point of star or a zigzag connected windings is brought out, the indication shall be YN or ZN and yn or zn, respectively.

The vector relating to the HV winding shall be taken as the vector of origin. For multiwinding transformers, the vector for HV winding remains the reference vector and the symbol for this winding shall be given first. Other symbols shall follow in a diminishing sequence of rated voltages of the other windings.

The inverter duty transformers are copper wound on both the HV and LV sides as well as being three-phase, natural cooled, core-type construction. They are oil immersed and suitable for outdoor installation with three-phase 50/60 Hz in which the neutral is effectively earthed. They should also be suitable for service under fluctuations in supply voltage from −15% to 10%.

Presently, five winding transformers are used. Four inverters can be connected to the LV side of four windings independently and one winding will be at the HV side.

The transformer is designed to take into account the harmonics that the inverter provides and the impedance the inverter is offering. The system fault level current at the power evacuation voltage also can be considered. The transformers meant for solar PV plants are working only during the day.

These are designed with cooling by oil natural air natural, but power transformers are cooled by oil natural air forced (with fans).

The hotspot temperature rise over a maximum yearly weighted temperature of 32°C will be 98°C. The transformer should have lower no-load losses and total load losses for better energy yield.

As part of the basic insulation level, the impulse withstand voltage on the HV side is 170 kV peak. The power frequency voltage on the HV side is 70 kV rms for a 33 kV transformer. It varies according to the HV voltage level. The voltage withstand level of the LV winding is 3 kV, and inverter output voltage varies from 400 to 800 V depending on the system voltage of 1000 or 1500 V.

Voltage rise time withstand level of LV winding is a min. 1000 V/μs against ground.

On the HV side, the delta connection is followed usually while the LV side star is considered. The short circuit thermal withstand time is 2 s.

The impedance at a 75°C rated current and frequency should be <6% between the HV-LV windings as well as between the LV1-LV2 windings. This is dictated by the inverter manufacturer. Static shields are provided between the low and high voltage windings and the shields are grounded according to relevant standards.

The transformer conforms to IEC 60076 and IEC 61378-1 and IEE-STD-C57 standards.

The following routine tests are to be conducted on the transformer:

- Measurement of winding resistance.
- Measurement of voltage ratio and check of voltage vector relationship.
- Measurement of impedance voltage/short circuit impedance and load-loss.
- Measurement of no-load loss and current.
- Measurement of insulation resistance.
- Induced overvoltage withstand test.
- Separate-source voltage withstand test.
- Noise measurement test of transformer.

In addition to the routine tests, the following tests are also conducted:

Lightning impulse-test, temperature rise test, short-circuit test, air pressure test, and an unbalanced current test (the value of unbalance current shall not be more than 2% of the full load current). The functional tests such as transformer tap changers, cooling equipment, emergency stop, door interlocks, temperature relays, their controls, and all other auxiliary equipment are operated to prove that they are functioning satisfactorily before the transformers are put into service. The insulation resistance between low-voltage windings is tested with a Megger test after installation. The measurement of the voltage ratio at every tap position is also measured.

10.2.3 Instrument transformers

Potential transformer and Current transformers are called Instrument transformers. The potential transformer is also known as a voltage transformer. It is used in an AC power system network to step down from a higher to a lower voltage level. It is used to monitor three-phase voltage and is also used for electrical metering. The potential transformer is used to convert the actual voltage suitable to rated metering equipment or for protection relays. The voltage can be measured with voltmeter, watt meter, or watt hour meter.

The current transformer converts actual current to rated metering equipment or other equipment such as protection relays.

10.2.4 Switchgear

It is critical to provide various switchgears on the DC and AC side of the PV power plant for protection and isolation purposes while complying with grid connection standards. Switchgear is the combination of electrical disconnect switches, fuse, or circuit breaker used to control, protect and isolate the electrical equipment. It is used to both to deenergize the equipment to allow work to be done and clear the faults down stream. The switchgear must be capable of interrupting both the normal current as well as fault currents. MV Switchgear should be capable of normal on/off switching operation. It should interrupt the short-circuit current and switch of capacitive and inductive currents. It carries out all the above operations with high degree of safety and reliability.

The AC switchgear combines the output of all the transformers, provides protection to the transformer, and supplies the power to the switchyard.

It carries the load current. It makes, breaks, and carries the bus charging current. It makes and breaks the load current. It makes and breaks the fault current. It carries the short-term fault current. The devices and components of switchgear are: Current Transformer, Potential Transformer, fuses, circuit breaker, isolator, relays, indicating instrument, Lightning Arrestor, Multifunctional Meter, and control panel.

It consists of control devices to check and/or regulate the flow of power. It has metering devices to measure the flow of electric power. Switching and interrupting devices are used to turn power on or off. Protective devices are used to protect power service from interruption, and to prevent or limit damage to equipment. It has different components that are described below.

A circuit breaker is used to turn the power on/off. A CT is used for converting the actual current suitable to the rated metering equipment or protection relays. Potential transformer is used for converting the actual voltage suitable to the rated metering equipment or protection relays. Metering equipment is used for metering of energy, power, load, etc., during operation.

Control elements are used for various modes of operation such as local/remote, auto/manual, and indications regarding status. Control circuit is used for controlling breaker operation with logic and using protection relay commands. Control protection is used to protect the control circuit and avoid damage in the case of external fault. Auxiliary contactor is used for logic development and input/output control of equipment operation. Power contactor is used for turning power on/off. Isolator is used for positive power isolation. Fuse switch unit is used for positive power isolation and protection. Power fuse is used for the protection of equipment at a high level fault.

The switchboard is usually the metal-enclosed, fully drawn out, freestanding, dust- and vermin proof, totally enclosed, fully compartmentalized, floor-mounted type. The circuit breaker panels are drawn out, multicompartmental units suitable for indoor as well as outdoor installation. The unit withstands the stresses encountered in the event of an electrical fault. It consists of AC isolators, circuit breaker, protection relays, CT and PT, and busbars.

10.2.4.1 Circuit breaker

A circuit breaker is an automatically operated electrical switch designed to protect an electrical circuit from the damage caused by the excess current from an overload or short circuit. Unlike fuse, which operates once and then must be replaced, a CB can be reset to resume normal operation. The function of a circuit breaker is to interrupt or close all currents, including the fault current. It provides protection to electrical equipment and personnel during faults. The breakers used in high-voltage switchgear applications must be capable of being operated safely from terminal faults, transformer magnetization current, charging capacitor, and switching of out of phase sequence.

The circuit breakers are different types and designated based on the arc interrupting media. They are called oil, air blast, SF6, vacuum, and air circuit breakers. The bulk oil and air blast types are used for the voltage level of 275 kV and above. The SF6-based circuit breaker is used for 66, 132, 275, and 400 kV voltage levels. Vacuum circuit breakers (VCB) are used for 11, 22, and 33 kV voltage levels. Air circuit breakers are used up to 11 kV voltage level.

The solar PV systems mainly use vacuum and SF6-based circuit breakers.

The VCB is a triple-pole arranged for motor/manual-operated spring-charged, independent closing and shunt tripping from a suitable voltage from the battery. The close/trip control switch will be interlocked to the trip before close. The closing and tripping circuits are self-opening upon completion of their respective functions irrespective of the position of the control switch. In VCB, arc quenching takes place in vacuum. Service life is more longer than the other types of CBs. There is no chance of fire hazard like in case of oil CB. It is environmental friendly compared to SF6-based CB.

Standard specification of circuit breaker are as follows:

Rated voltage: Highest RMS voltage for which the circuit breaker is designed and is the upper limit for continuous operation.

Rated current: The maximum RMS current the breaker is capable of carrying continuously without exceeding the given temperature rise at the given ambient temperature.

Rated frequency: Frequency at which the breaker is designed to operate.

Rated interrupting current: Current at the instant of contact operation. The interrupting current rating can be given as one of the following values.

Symmetrical interrupting current: RMS value of the AC component of the short circuit current the breaker is capable of interrupting.

Asymmetrical interrupting current: The RMS value of the total short circuit current the breaker can interrupt.

Rated making current: RMS value of the short circuit current on which the breaker can safely close at the rated voltage.

Rated short time current: RMS value of the current that the circuit breaker can carry in a fully closed position without damage for a specified short time interval. Normal considered values are 1 or 3 s. These ratings are based on thermal limitations.

Rated interrupting time: Maximum interval from the time the trip coil is energized until the arc is extinguished.

Rated impulse withstand voltage (basic insulation level): Maximum short duration impulse voltage that the breaker can withstand.

K factor (voltage range factor): For most circuit breakers, the rated interrupting current is independent of the operating voltage. For some breakers, mostly oil breakers, the rated interrupting current increases if the operating voltage is lowered to a certain limit that is given by the K factor.

This adjusted rated interrupting current is called the current interrupting capability (CIC). CIC = rated interrupting current × (rated voltage/system voltage).

Parts of a Vacuum circuit breaker

(a) Breaker spring charging motor: Drives to store energy in closing the spring for breaker closing.

(b) Closing spring and tripping spring: Delivers energy for closing and tripping of breakers.

(c) Main isolating contacts: Disengages breaker from main power.

(d) Closing coil and tripping coil: Actuates levers for closing and tripping action of breakers.

(e) Breaker auxiliary contacts: Provides breaker status inputs to use in control logic.

(f) Manual spring charging handle: Standby arrangement for spring charging in emergency.

(g) Fixed and moving contacts: Crucial parts of breaker and current flows through them.

(h) Arc chutes: Helps to quench arc during breaker opening on load.

Table 10.1 gives specifications of a typical 33 kV vacuum circuit breaker suitable for 5 MW solar PV plant. Fig. 10.3 shows an 11 kV Vacuum circuit breaker.

Table 10.1 Specifications of 33kV VCB suitable for 5 MW solar PV plant.	
Type of circuit breaker	**VCB-based**
Rated frequency	50 Hz
Number of poles	Three
Rated operating duty cycle	O—0.3 s CO—3 min—CO
Reclosing	Three-phase high-speed autoreclosing
Total closing time	Not more than 150 ms
Noise level	Maximum 140 dB at 50 m distance from base of circuit breaker
Rated terminal load	Adequate to withstand 100 kg static load as well as wind, seismic, and short circuit forces without impairing reliability or current carrying
Type of operating mechanism	Spring-operated
Minimum creepage distance	25 mm/kV
Rated ambient temperature	50°C
System neutral earthing	Effectively earthed
Rated voltage	36 kV
Operating voltage	33 kV
Rated continuous current at an ambient temperature of 50°C	630 A
Symmetrical interrupting capability	630 A
Impulse withstand voltage	170 kV
Short time current carrying capability for 3 s	26.5 kA, rms
Capability for 1 s out of phase breaking current capacity	6.25 kA, rms

FIG. 10.3

11 kV Roller type Vacuum circuit breaker. Left: Inside view with horizontal and vertical contacts, Right: Front side view of circuit breaker.

10.2.4.2 Protection relays

The switchgear is equipped with microprocessor-based numerical protection relays with communication facility. Numerical multifunctional relays can be used for three Phase Time Over-current Protection (50/50N), Three Phase Instantaneous Protection, Earth Time Over-current and Earth Instantaneous Over Current protection (51/51N), Circuit Breaker Failure Detection. Auxiliary Relays for Transformer Protection (63A,63B), Standby earth fault relay (51G), master trip relay (86), trip circuit supervision relay (95), Under/Over Voltage Relay (27/59), PT Fuse Failure Relay (60) and DC supervision relay (80) also will be integrated in the switchgear.

10.2.4.3 Measuring instruments and analogue meters

A digital metering unit consists of Multifunctional meters (MFM) will be there for incoming and outgoing feeders to measure energy. The digital metering unit will have provisions to display the parameters such as: phase and line voltage; phase currents; kVA, kW, kVAR, kVARH, kVAH, and kWH; power factor; frequency; total harmonics; and alarm output relay. Digital Ammeter and Digital Voltmeter for measuring current and voltage will be provided in the switchgear.

The Breaker Control Switch, Trip-Neutral-Close, Spring return to Neutral, Pistol Grip 2Pole 2Way Local Remote Switch, Alarm Relay (8/12 Window Annunciator), Hooter for alarm, Indicating lamps, power pack to provide power to tripping and closing coils are also integrated with the switchgear.

10.2.4.4 Current transformers

The magnitude of the current and voltage in power circuits is usually high and becomes too high in fault condition. To handle it safely for the secondary measuring equipment or relays, current and voltage transformers are used. They normally scale down a replica of the primary inputs within the required accuracy.

10.2.4.5 Potential transformers

The PTs will be made of epoxy cast resin. The burden, ratio, and class of accuracy will be specified in the single line diagram of the AC system as per the design. Generally, the PT shall have a specified accuracy class from 10% to 120% of the normal voltage. However, potential transformers shall have sufficient capacity to operate with the burden imposed by the devices shown on the drawing with their accuracy classification.

10.2.5 Substation for solar PV system

A substation is an interface between the AC output from the inverter and the grid while the transformer helps to step up the inverter AC voltage to grid voltage such as 11/33/66 kV or higher voltage as may be designed. Fig. 10.4 shows a 132 kV substation.

It will have electrical power transformers, current and potential transformers, conductors and insulators, isolators, busbars, lightning arresters (LAs), circuit breakers, control and relay panels.

The following are the specifications of the individual equipment.

10.2.5.1 Lightning arrester

LA is the protection device which gives protection against any overvoltage caused due to lightning surges. It is kept in the substation in a suitable place. It is connected between the ground and the line conductor. There are many types of LAs. They are: rod gap, sphere gap, horn gap, multigap, electrolytic type, and metal oxide type. Metal oxide type surge arresters are used in the substations, switchyards of solar PV system. Each surge arrester (SA) shall be a hermetically sealed single-phase unit.

The LA will have accessories such as a surge monitor, an insulated base, and bushings.

Fig. 10.5 shows an LA of a substation of 132 kV. Table 10.2 gives specifications of a 33 kV LA.

FIG. 10.4

Substation of 132 kV.

FIG. 10.5

Lightning arrester of a substation of 132 kV.

Table 10.2 Specifications of a 33 kV lightning arrester.	
Type	**Station class, gapless ZnO**
Nominal voltage rating	33 kV
Rated voltage	30 kV
Nominal discharge current	5 kA
Impulse withstand voltage	170 kV peak
System neutral earthing	Solidly earthed
Installation	Outdoor
Discharge current at which insulation is done	20 kA of 8/20 μs
Rated frequency	50 Hz
Long duration discharge class	Class III
Current for pressure relief test	40 kA rms
Prospective symmetrical fault current low current long duration test value (2000 μs)	40 kA rms for 1 s 1000 A
Pressure relief class	Class A of Table VII of IS:3070 or equivalent IEC. Not more than 50 p.C.

Continued

Table 10.2 Specifications of a 33kV lightning arrester—cont'd	
Type	**Station class, gapless ZnO**
Partial discharge at 1.05 MCOV (continuous operating voltage)	
Minimum discharge capability	5 kJ/kV (referred to rated arrestor voltage corresponding to minimum discharge characteristics)
Maximum continuous operating max. residual voltage (1 kA)	24 kV rms
Max. residual voltage at 10kA nominal discharge current wave)	70 kVp
Max. switching impulse residual voltage at 500 A peak	85 kVp
Max. steep current residual voltage	70 kVp
High-current short-duration test value (4/10 μs-wave)	93 kVp at 10kA 100 kAp
One minute power frequency withstand voltage of arrestor housing (dry and wet)	70 kV (rms)
Creepage distance	25 mm/kV
Radio interference voltage at 156 kV	Not more than 1000 μV
Seismic acceleration	0.3 g horizontal
Reference ambient temp.	50°C

10.2.5.2 Isolator

Isolator is a manually operated mechanical switch that isolates the faulty section of substation. It is used to separate faulty section for repair from a healthy section in order to avoid the occurrance of severe faults. It is also called disconnector or disconnecting switch. There are different types of isolators used for different applications. They are: single break, double break, bus isolator, and line isolator. The isolator will be a horizontal double break central rotating type with an earth switch. Isolators and earth switches can be hand operated. Earth switches and Isolators (in closed position) are designed to withstand thermal effects and other conditions due to short circuit current. Table 10.3 gives specifications of an isolator for 33 kV.

10.2.5.3 Current transformer (outdoor)

CT reduces power system current to lower value suitable for measurement. It insulates the secondary current from the primary. It allows to use standard current rating for secondary equipment. There are two types of CT and they are bar type and wound type. The CT will have different type of cores, oil-filled, self-cooled hermetically sealed, and dead tank. Table 10.4 gives specifications of a 33 kV CT.

10.2.5.4 Potential transformer (outdoor)

The potential transformer is single-phase, oil-filled, self-cooled, hermetically sealed dead-tank. Table 10.5 gives specifications of a 33 kV potential transformer. Fig. 10.6A shows an outdoor potential transformer of 132 kV.

Table 10.3 Specifications of 33 kV isolator.

Type of isolator	Outdoor type
Rated frequency	50 Hz
Number of poles	Three
Operating time	Not more than 12 s
Control voltage	110 V DC
Auxiliary contacts on Isolator	As required plus 8 NO and 8 NC contacts per pole/isolator as spare
Auxiliary contacts on earth switch	Total 6 NO and 6 NC
Rated mechanical terminal load	As per table III of IEC 60129
Temperature rise over ambient	As per IEC:60129
Minimum creepage distance	25 mm/kV
Rated ambient temperature	50°C
System neutral earthing	Effectively earthed
Seismic acceleration	0.3 g horizontal
Normal system voltage	33 kV
Highest system voltage	36 kV
Rated current at 50°C ambient temperature	630 A
Rated short time current of isolator and earth switch	25 kA (rms) for 3 s
Rated dynamic short time withstand current of isolator and earth switch	62.5 kA (peak)
Impulse withstand voltage with 1.2/50 μs wave	170 kVp to earth 195 kVp across isolating distance
One minute power frequency withstand voltage	70 kV (rms) to earth and 80 kV (rms) across isolating distance
Phase to phase spacing	1500 mm

Table 10.4 Specifications of a 33 kV current transformer used for 5 MW solar PV plant.

Highest system voltage	36 kV
Rated frequency	50 Hz
System neutral earthing	Effectively earthed
Installation	Outdoor
Rated short time thermal current	25 kA for 3 s
Rated dynamic current	63 kA (peak)
Rated min power frequency withstand voltage (rms value)	70 kV
Rated lightning impulse withstand voltage (peak value)	170 kV
Partial discharge level	10 pico Coulombs max.
Minimum creepage distance voltage	25 mm/kV of highest system
Temperature rise	As per IEC 60044
Type of insulation	Class A
Number of cores	Two with one protection core and one metering core
Burden	10 VA
Accuracy class	0.2 s

Table 10.5 Specifications of a 33 kV potential transformer used for 5 MW solar PV plant.	
Highest system voltage	36 kV
System neutral earthing	Effectively earthed
Installation	Outdoor
System fault level	25 kA
Rated min power frequency withstand voltage (rms value)	70 kV
Rated lightning impulse withstand voltage (peak value)	170 kV
Standard reference range of frequencies for which the accuracy is valid	96%–102% for protection and 99%–101% for measurement
Rated voltage factor	1.2 continuous and 1.5 for 30 s
One minute power frequency withstand voltage for secondary winding	2 kV rms
Partial discharge level	10 pico Coulombs max.
Rated insulation level	36 kV
Rated lightning (class of insulation)	B
Rated voltage factor	1.2 Cont. and 1.9 for 8 h
Applicable standard	IEC 60044-1/IS: 3156
Rating	$33,000/\sqrt{3}/110/\sqrt{3}/110/\sqrt{3}$ V
Number of cores	2
Purpose	Metering and protection
Burden	25 VA
Class of accuracy	0.2

FIG. 10.6

(A) Potential transformer for 132 kV substation. Every phase has one Potential transformer and is used for metering and protection purpose. (B) SF6 Circuit breaker for 132 kV substation.

10.2.5.5 SF6 Circuit breaker (outdoor)

Fig. 10.6B shows an SF6 circuit breaker. Here, SF6 gas is used in arc suppression. This can be used for switchgears of 33 kV and above voltage level. Pressurized sulfur hexafluoride (SF6) gas is used to extinguish the arc in a circuit breaker is called SF6 circuit breaker. SF6 gas has excellent dielectric, arc quenching, chemical, and other physical properties, which have proved its superiority over other arc quenching mediums such as oil or air. SF6 gas is extremely stable and inert, and its density is five times that of air. It has a unique property of fast recombination after the source energizing spark is removed. It is 100 times more effective as compared to arc quenching medium. Its dielectric strength is 2.5 times than that of air and 30% less than that of the dielectric oil. At high pressure, the dielectric strength of the gas increases.

Principle of SF6 circuit breaker: In the normal operating conditions, the contacts of the breaker are closed. When the fault occurs in the system, the contacts are pulled apart, and an arc is struck between them. The displacement of the moving contacts is synchronized with the valve which enters the high pressure SF6 gas in the arc interrupting chamber at a pressure of about 16 kg/cm^2. The SF6 gas absorbs the free electrons in the arc path and forms ions which do not act as a charge carrier. These ions increase the dielectric strength of the gas and hence the arc is extinguished. This process reduces the pressure of the SF6 gas up to 3 kg/cm^2 thus; it is stored in the low-pressure reservoir. This low-pressure gas is pulled back to the high-pressure reservoir for reuse.

10.2.5.6 Energy meter

It is required to know the amount of the energy generation the solar plant is providing.

Energy meters are used to measure the energy generation, and they are calibrated and sealed. At the point of power evacuation, main and check meters are provided. In the solar plant premises, energy meters are also installed. The LT or HT switchgears are equipped with multifunctional meters (MFM). From the inverter level to power evacuation, energy meters are installed. With the meter reading, loss of energy from the inverter output to the substation, the loss of generation due to the transmission line can be estimated. The following are the requirements of an energy meter:

- Energy meter is microprocessor-based and conforms to IEC 60687/IEC 62052-11/IEC 62053-22/ IS 14697.
- It carries out the measurement of active energy (both import and export) and reactive energy (both import and export) by three-phase. It is based on the four-wire principle suitable for balanced/ unbalanced three-phase load.
- It will have an accuracy of energy measurement of at least Class 0.2 for active energy and at least Class 0.5 for reactive energy according to IEC 60687, and is connected to Class 0.2 CT cores and Class 0.2 PT windings.
- The active and reactive energy are directly computed in the CT and PT primary ratings.
- The reactive energy shall be recorded for each metering interval in four different registers as MVARh (lag) when active export, MVARh (lag) when active import, MVARh (lead) when active export, and MVARh (lead) when active import.
- It will compute the net MWh and MVARh during each successive 15-min block metering interval along with a plus/minus sign, instantaneous net MWh, instantaneous net MVARh, average

frequency of each 15 min, net active energy at midnight, and net reactive energy for voltage low and high conditions at each midnight.

- Each energy meter will have a display unit with a seven-digit display. It displays the net MWh and MVARh with a plus/minus sign and average frequency during the previous metering interval; the peak MW demand since the last demand reset; the accumulated total (instantaneous) MWh and MVARh with a plus/minus sign, date, and time; and instantaneous current and voltage on each phase, depending on the requirement.
- There are registers that are stored in nonvolatile memory. Meter registers for each metering interval as well as accumulated totals are downloadable. All the net active/reactive energy values are displayed or stored with a plus/minus sign for export/import.
- It will have a built-in clock and calendar with an accuracy of less than 15 s/month drift without the assistance of an external time-synchronizing pulse.
- Date/time are displayed. The is synchronized with the SCADA time synchronization equipment at the station.
- The meter will have the means to test MWh and MVARh accuracy and calibration at site.

Technical details of meter (typical)
- IEC 62053 accuracy Class 0.2%, 0.5%, and 1.0%.
- UV-protected, polycarbonate IP54 enclosure.
- Wide operating voltage range: 46–528 V.
- Wide operating range of current: 1 mA to 10 A (CT connected).
- Temperature of operation: −40°C to 85°C (inside meter).
- Energy plus demand for kWh, kVARh, and kVAh.
- Four-quadrant metering for export import.
- Data stored in nonvolatile memory.
- Easily replaceable battery located under the terminal cover.
- Precision internal clock with backup timekeeping provided by supercapacitor and long-life battery.
- 16-Segment character LCD with optional backlight.

10.2.5.7 Two/four/six pole structure
A two/four/six pole structure is of suitable height. The material used for the pole is galvanized structural steel. A joist will be there to accommodate the HV equipment along with the hardware.

10.2.5.8 Auxiliary power supply
To meet the power for the control room, HT panels, street lights, office room, and SCADA systems, a power supply is required. The power is taken from an auxiliary transformer, which is a step down to 415 V from the power evacuation voltage. For a 33 kV level 33/0.415 kV, ONAN-type or dry transformer with OCTC with tapping at the rate of 2.5%, an impedance of 4.5% can be used. For the requirement of power for the inverter and SCADA system in the inverter room, the inverter output is tapped and stepped down to 415 V using a suitable size transformer.

10.3 **Transmission line requirement**

In order to evacuate the generated power from the solar PV power plant to grid, a transmission line is required. The transmission line transfers the solar-generated power to the nearest grid substation.

The solar plant-generated electrical power is transmitted to the substation by carrying it with the help of a transmission line and is evacuated at the substation. The transmission line is long conductors designed to carry bulk amount of generated power over a large distance from one substation to other substation with minimum loss and distortion. The inductance, capacitance, resistance, and conductance are the parameters associated with the transmission line. These electrical parameters are uniformly distributed along the length of the transmission line. A sound knowledge of these parameters is required for good electrical design of the transmission line.

Type and size of the conductor, efficiency of transmission, corona loss, power flow capability, stability, and economic aspects are the factors to be considered while deciding the transmission line. The successful design of the transmission line depends on the mechanical design. Overhead transmission lines and their supporting towers are subjected to extreme weather conditions such as storms, winds, rain, and high and low temperature climates. For tower design, minimum ground clearance, length of insulating string, minimum clearance between conductor to conductor and conductor and tower, location of ground wire with respect to outer most conductor, and mid span clearance are the parameters to be considered. Based on the above considerations, there are different configurations of towers and poles to support the hanging conductors depending on the transmitting voltage.

Transmission lines can be classified as overhead transmission lines or underground cables. The overhead transmission lines are further classified as short transmission lines, medium transmission lines, and long transmission lines depending on the line voltage and the length of the transmission line. For overhead transmission line, bare conductors are hanging from supports. Porcelain insulators insulate the conductors from the support. This is less costly, but prone to occurrence of faults. Overhead transmission line has got the components such as: conductors, supports, insulators, cross arms, dampers, spacers, and jumpers.

Copper as a conductor has higher conductivity and better tensile strength, but is costly. In a steel-core copper conductor, the steel is used in the center as the core surrounded by a copper conductor. This gives high tensile strength and an increase in flexibility. All aluminum conductors have low conductivity and low strength, but are cheaper. If Al is reinforced with steel, the tensile strength will be increased. The Al conductor alloyed with magnesium and silicon provides better tensile strength and less conductivity while being suitable for TL. ACSR is used for larger spans keeping in view of minimum sag. It may consists of 7 or 19 strands of steel surrounded by Al strands concentrically. These strands provide flexibility, prevent breakage, and minimize skin effect. The number of strands depends on the application and may be 7, 19, 37, 61, and more.

The overhead conductors of TL are generally supported by towers and poles. The towers and poles are properly grounded, so there must be an insulator between them. The insulator electrically isolates the conductor with the structure. There are different type of insulators based on the application. They are pin type, suspension type, strain type, shackle type, or egg/stay type insulators. Porcelain, glass, and synthetic resin-based materials are used as insulators. Porcelain is a ceramic material, with very high insulation resistance that is free from cracks, holes, etc. The dielectric strength of porcelain is about 60 kV/cm. Glass has higher compressive strength than porcelain but less cheaper than porcelain.

The synthetic resin-based insulators are becoming popular due to a high tensile strength and low weight while being comparatively cheaper.

In transmission line design, voltage drop, transmission line efficiency, and line loss are the important factors to considered. These are affected by the resistance, inductance, and capacitance associated with conductors of TL.

The voltage regulation which is the ratio of difference between the sending voltage and TL receiving end voltage, between the conductors with and without load. Power factor has an effect on voltage regulation of TL. The suspended TL conductors possess sag and this has to be considered for deciding the mid span of the towers.

For under ground cable-based TL, the conductors are insulated and are costly and less likelihood of developing faults. In case of faults, it takes more time to repair.

10.4 Protection requirements for solar PV power system

In line with various standards such as IEC 929/UL 1751 and others, adequate protection is to be provided on both the DC and AC sides to protect the solar PV array as well as the grid due to fault conditions or conditions arising out of the malfunctioning of any components.

The objective of protection scheme is to keep the power system stable by isolating the components that are under fault. The devices used to protect the power system from fault are called protective devices. A good electric power system should ensure the availability of electric power without interruption to every load connected to it. Protection relays and systems detect abnormal conditions with faults in the electrical circuit and automatically operate the switchgear to isolate the faulty equipment from the system as quickly as possible. The protection system limits the damage at the fault location and prevents the effect of the fault spreading into the system. The protection should be able to recognize the fault or an abnormal condition in the electrical circuit and take suitable steps to isolate the fault without affecting the normal operation of the system.

The main fault types and disturbance conditions in solar PV systems are classified as: phase faults, ground faults, abnormal voltage, unbalanced currents, abnormal frequencies, breaker failures, and system faults. In addition to the above faults and disturbance conditions, loss of synchronism is also protected.

The faults normally occur due to reduction in the insulation resistance between the phase conductors, between phase conductors and earth and the earth screen surrounded by the conductor. Breakdown at normal voltages may occur due to deterioration of insulation resistance. Surges due to switches and lightning may also cause abnormal conditions. The over voltage may result in deterioration of insulation resistance, failure of insulation, and cause fire also. Under voltage may damage some of the systems. Over frequency occurs due to unstable power system and sudden decrease of huge electric load. Under frequency occurs with sudden increase of load. So, the system has to be protected for over voltage, under voltage, over frequency, and under frequency conditions. The commonly used components for the protection system are CT, VT, CB, tripping, and auxiliary supplies and relays.

There are different ways to discriminate the type faults such as: discrimination by time; by current magnitude and time; by time and direction; by distance measurement; by time + current magnitude or distance; by current balance; by comparison of power direction; and by phase comparison. For this purpose, different types of relays are used.

Relay: Relay is a device by which electrical circuit is indirectly controlled during a fault condition. The purpose of relay is to operate the correct circuit breaker, so as to disconnect only the faulty equipment from the system as quickly as possible, thus minimizing the trouble and damage caused by faults when they do occur. The essential qualities of protection relay are:

Reliability: Protection scheme must operate, when the system condition calls upon to do so. Failure in the trip and control circuit of the breaker can be determined by continuous supervision arrangement.

Selectivity: Protective system must be such that it should correctly select the faulty section and cut off the same from the system without disturbing other healthy sections.

Speed: To avoid unnecessary damage to plant, protection must operate quickly.

Stability: The protection system should be stable and it must actuate from the concerned signal only and not from any other similar signal.

Back up relaying: If due to some reason the primary relaying system fails to operate, the backup relays must operate and isolate the faulty equipment.

Auto reclosing relays: These relays are used to reconnect the circuit so that if the fault is of transient nature, the system is returned to normal operation. This system is used mostly on overhead lines where 80% to 90% faults are of transient nature such as Lightning, birds passing near or through lines, tree branches, etc.

There are different types of relays:

Distance relay: This Relay is directional type and works on the principal of impedance rather than current/voltage. This relay functions upon the distance of fault in the line. More specifically, the relay operates depending upon the impedance between the point of fault and the point where relay is installed. These relays are known as distance relays or impedance relays.

Differential relay: This relay compares the currents in the windings of the transformer through CTs whose ratios are such as to make their currents normally equal. The polarities of the CTs are such as to make the current circulate without going through the relay during load conditions and external faults. During internal faults, the balance condition is disturbed and relay operates.

Over current and earth fault relay: These relays work on IDMT characteristic and are made directional and nondirectional as per requirement. It is a current actuated relay.

Auxiliary relay: Relay that operates in response to the opening or closing of its operating circuit to assist another relay or device in performing a function. The relays are classified by the ANSI numbers.

Instantaneous over current and earth fault [50/51]: This protective relay initiates a breaker trip based on current sensed by the CT if it exceeds a preprogrammed "pickup" value for length of time. Ideally, the relay operates as soon as the current in the coil gets higher than pick up value of setting current. There is no intentional time delay applied. A few order of milliseconds is operating time required for this relay.

IDMT over current and earth fault [50N/51N]: IDMT is an acronym for inverse definite minimum time lag. This relay will have higher resolution compared to the instantaneous over current relay. This operates when CT gets saturated, due to the fault current presence. This relay will operate very fast to trip the circuit against over current and earth fault condition.

Over voltage [59]: Overvoltage may occur during a load rejection. The overvoltage relay (59) is used to protect the PV generator from this condition. Three limits can be set which are in the range of

110%–135% with time range 0.05–2 s. If the circuit breaker experiences over voltage against preprogrammed set value, the over voltage activates and trips, so that the transformer gets isolated from the grid. Otherwise, the high voltage will damage the transformer coil.

Under voltage [27]: Under voltage may occur due to sudden reactive power demand which can be taken care of by under voltage relay (27). Under voltage range is 50%–90%. If the circuit breaker experiences under voltage against preprogrammed set value, the breaker immediately gives proper trip signal to the master trip relay.

Trip circuit supervision [95]: Trip circuit plays a very important role in protection system. A trip circuit supervision relay is the watch dog to the trip circuit of any electrical system. If by some means the trip circuit is not functioning, it will send out alarm or whatever means is assigned to.

Master trip relay [86]: This relay has a unique name called as master trip, because of its function. All relays will give signal to this relay, to act immediate against system faults like over current, over voltage, earth fault, and many on. Transformer auxiliary relays also having communication with this relay, to trip the breaker if any internal faults happen in the transformer.

Antipumping relay [94]: The antipumping relay is a device in circuit breaker whose function is to prevent multiple breaker closures. For instance, if the operator gives the closing command to the breaker by pressing the close button and the breaker closes. However, a fault in the system causes the breaker to trip. Since the close command is still in the pressed condition, there is a chance of the breaker closing again and being tripped by the relay multiple times. This can damage the closing mechanism of the breaker. The antipumping relay prevents this by ensuring that the breaker closes only once for one close command from the control panel.

Transformer auxiliary relays [49 OTI and 49 WTI]: Generally, transformer marshalling box having oil temperature indicator (OTI) and winding temperature indicator (WTI) in a closed enclosure. Transformer auxiliary relays are connected with OTI and WTI. If OTI experiences higher temperature against set value, auxiliary relay will give trip signal to the master trip. The same action similar to the WTI also.

Phase fault protection: Phase faults in an incoming feeder can be detected and protected by a pack of three number of over current relay (50) and one number of IDMT earth fault relay (51N). One IDMT nondirectional over current relay (51) is also provided. Phase faults in a transformer winding can cause thermal damage to insulation, windings and the core. Primary protection for transformer phase-to-phase faults is best provided by a differential relay (87T) used for unit protection. Differential relaying will detect phase-to-phase faults, three-phase faults, and double-phase-to-ground faults. The voltage-restrained/controlled over current function (51V) can also be used for this backup function. The voltage-restrained/controlled over current relay will restrain operation under emergency overload conditions and still provide adequate sensitivity for fault detection.

Ground-fault protection: One of the main causes of ground faults is insulation failure. Depending on the location of the fault, separate ground-fault protection is usually provided by instantaneous over current relay (50N). Backup protection for ground faults can be provided by an inverse definite time over current relay (device 51N) in conjunction with an instantaneous over current relay (device 50N) applied at the generator neutral to detect zero sequence unbalance current which flows during ground faults. Backup for system ground faults can be accomplished with a time over current (51) relay connected in the neutral of the step-up transformer primary. A master trip relay (86) connects to all the relays and connects to the tripping devices such that its command is used for tripping the circuit breaker.

Over/under frequency protection: Frequency variation in the grid requires a response from the PV system for safety of the equipments at point of common coupling. The PV system should operate in synchronism with the grid with $\pm1\%$ and for exceeding range must trip with in 0.2 s. Again on restoration of frequency with in this range needs resynchronization. The multifunction relays provide a two-set point over frequency relay (81O) that can be set to alarm or trip on an over frequency condition. Overloading of a generator, perhaps due to loss of system generation and insufficient load shedding, can lead to prolonged operation of the generator at reduced frequencies. While load-shedding is the primary protection against generator overloading, under frequency relays (81U) should be used to provide additional protection. If breaker experiences over frequency or under frequency, it will give trip signal to the master trip relay, to shut down the system. In addition to relays, fuses are provided and fuse failure is also protected by fuse failure protection (97).

Breaker failure protection: Backup protection must be provided for the case where a breaker fails to operate when required to trip. This protection consists of a current detector, in conjunction with a timer initiated by any of the protective relays in the generator zone. The breaker failure relay (95) in association with trip coil supervision will initiate tripping of the backup breakers.

10.5 Grid integration of solar PV power system

The grid integration of a solar PV power system is of paramount importance.

The following steps describe the processes involved in the grid integration of a solar PV power system.

Step 1: The solar array is interconnected to generate the required DC system voltage (within the DC input range of the inverter).

Step 2: The proper solar PV array-to-inverter having grid management features such as LVRT/HVRT, Reactive power support, antiislanding etc. has to be considered.

Step 3: Grid parameters (AC voltage and frequency of, say, 50/60 Hz) are well within the range of inverter operations.

Step 4: Switching of DC and AC breakers and comparison of voltage and frequency with the grid parameters. Fig. 10.7 shows the connection scheme of a solar plant's power evacuation to the grid.

The power in the stepped voltage is carried from the plant through a transmission line to the substation, where it has to be evacuated to the grid. The point of interconnection has to be identified in the substation. A bay extension of the voltage level has to be created in the substation. Fig. 10.8 shows the bay extension between the transmission line entry to the grid connection point in a 110 kV substation.

It consists of an isolator with an earth switch to isolate the bus. It also consists of a circuit breaker with CT and PT for protection. CT, PT, and metering equipment along with an LA also will be there. There will be an isolator on the line side. There will be a control panel in the substation at the point of power evacuation.

The following are the grid connection requirements, and the grid code varies from country to country.

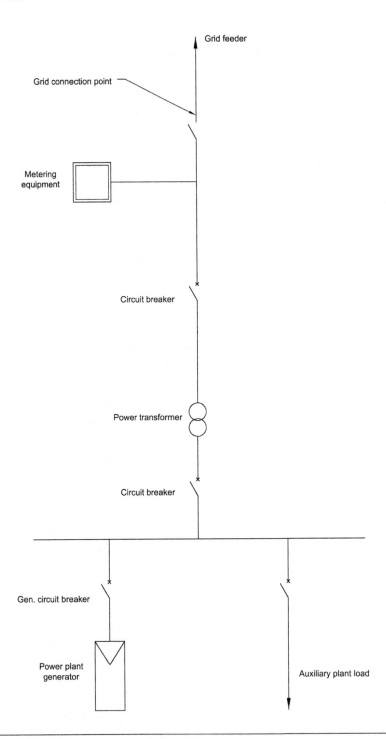

FIG. 10.7

Schematic of a solar PV plant's connection with the grid.

FIG. 10.8

110 kV bay extension for power evacuation at a 110 kV substation.

Range of grid voltage and range of grid frequency

The solar PV plant will not be connected to the grid if the voltage and frequencies of the grid are out of the range set in the inverter. The inverter monitors the voltage and frequency by grid synchronization and disconnects if it is out of the set limits.

Harmonics on AC side

The nonlinear loads such as drives, rectifiers, and furnaces connected to the system as loads cause harmonic distortion. The limits of harmonics are stipulated in the grid code of connectivity.

In India, the harmonic distortion values for total voltage, individual voltage, and current are 5%, 3%, and 8%, respectively.

The voltage unbalance

It is defined as the deviation between the highest and lowest line voltages divided by the average line voltage of three-phases. Solar PV plants should withstand a voltage unbalance not exceeding 2% for at least 30 s.

The fluctuations in voltage can occur because of switching operations of capacitor banks, and collection circuit transformers within the solar plant due to inrush currents. These will affect the operation and performance of the plant. The maximum permissible limit for the occasional fluctuations in voltage is 3%, where the limit is 1.5% for the voltage fluctuations that occur repetitively.

Power quality requirements are to be taken care of by the power condition unit. As per IEC 61727, the DC injection into the grid should be within 1% of the rated current of the inverter. DC injection causes the saturation of the transformer, which reduces the efficiency and life of the system.

As per IEC 61000-3-7, the flicker caused by the solar plant at the grid connection point must be ≤ 0.35 for a long-term flicker factor over time periods of 2 h. The short-term flicker factor over a time period of 10 min should be ≤ 0.25.

Fault ride through [2]

A solar PV plant has to ride through the grid fault without disconnection from the grid for temporary voltage drops, when the positive sequence voltage is above the curve, as shown in Fig. 10.9.

The solar plant shall trip if all phase-to-phase voltages are below the curve in Fig. 10.9.

The inverter will be able to take care of all voltage ride through requirements.

10.5.1 Reactive power control

The solar plant must be able to control the reactive power at the grid connection point in a range of 0.95 lagging to 0.95 leading at maximum active power and, according to Fig. 10.10, for Medium and Large scale solar PV plants. The solar plant must be able to control reactive power as follows:

(a) Set-point control of reactive power Q.
(b) Set-point control of power factor ($\cos\varphi$).

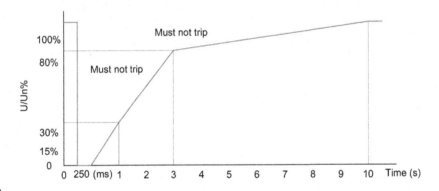

FIG. 10.9

Fault ride through profile for a solar plant.

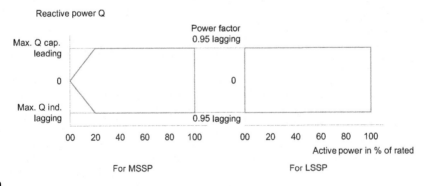

FIG. 10.10

P-Q Diagram for a solar plant. MSSP is Medium size solar plant whose capacity is <20 MW and LSSP is Large size solar plant whose capacity is >20 MW.

(c) Fixed power factor ($\cos\varphi$).
(d) Characteristic: power factor as a function of the active power output of the solar plant, $\cos\varphi(P)$
(e) Characteristic: reactive power as a function of voltage, $Q(U)$.

For LSSP, even at zero active power output (e.g., during the night), reactive power can be injected to the Grid Connection Point, fully corresponding to the P-Q Diagram taking the auxiliary service power, the losses of the transformers and the Solar Plant cabling into account. The inverter has to be selected such that it will take care of the reactive power requirements.

10.6 Methodology for performance analysis of solar PV power plant

Developers and system designers use various methods to check the performance of an installed and commissioned solar PV power plant. In this process, theoretical design parameters and output details of the PV system are checked against actual performance data using various measurement techniques.

10.6.1 Performance ratio

To check the performance of the solar PV plant, metrics such as performance ratio (PR) and plant load factor (PLF) are used. Once the plant is constructed and commissioned by an EPC contractor, the metric PR is commonly used for acceptance and operations testing of the plant. IEC 61724 [3] defines the PR as the ratio of the actual and theoretical possible energy from the plant. It tells how effectively the plant converts sunlight collected by the PV panels into AC energy. PR is [4]:

$$\text{Performance Ratio} = \frac{\text{Actual reading of plant output in kWh per annum}}{\text{Theoretical nominal plant output in kWh per annum}}$$

The actual meter reading gives the total generation of the plant that is exported to the grid.

The theoretical nominal plant output can be obtained by multiplying the total area of the solar module with the number of modules and the module efficiency. This is nothing but the name plate capacity of the solar array installed in the plant. This total array capacity in wattage multiplied by irradiance falling on the plane of array gives the theoretical energy available from the plant. In another way, theoretical energy is the expected total power generation in standard test conditions based on the PV plant total installed power capacity.

Instead of a complex nature, the performance ratio in % can be defined as

$$PR = \frac{Y_f}{Y_r} \times 100$$

where Y_f = specific yield over a year expressed in (kWh/kWp).

The specific yield is the measure of total power generation output for one kWp. This can be calculated as $\dfrac{\text{Total energy yield in kWh}}{\text{Total name plate solar array capacity of the plant in kWp}}$.

Y_r is the total insolation measured in the plane of array at the PV plant through the irradiance measurement equipment installed at the site in kWh/m^2 during the test period.

The Y_r defines the solar irradiation resource for the PV system. It is related to the site location, the direction and tilt angle of the solar PV array, and the variations in weather conditions over various months of a year. The specific yield can be expressed as shown below:

$$\frac{\int P \cdot dt}{P_{stc}} = \frac{E}{P_{stc}}$$

P: The instantaneous power output of the solar PV array in kWp.
P_{stc}: The STC-rated capacity of the array in kWp.
E: The energy output of the solar PV array in kWh.

The irradiance in the plane of array can be expressed as shown below:

$$\frac{\int G \cdot dt}{G_{stc}} = \frac{I}{G_{stc}}$$

G: The instantaneous irradiance measured in the plane of the array (POA) in W/m². G_{stc}: STC irradiance, that is, 1000 W/m².
I: The irradiance falling on the solar PV modules in the plane of array expressed in Wh/m².

The performance ratio is a metric used for the comparative performance of grid-connected PV systems irrespective of their orientation, technology, and location. It is unitless and tells how the plant performs per unit kWh incidence of energy on the panels.

The PR depends on the inclination angle, the temperature of the modules, and the technology of the modules used. The PV system electrical output changes with temperature (typically ~0.4%/°C), irradiance (typically can vary by as much as 5%–10%, especially for modules with high shunting or series resistance), and spectrum (typically varies by up to ~3%, depending on the difference in responses of the irradiance sensor and the PV module). So, a weather-corrected PR is required [5]. Usually, the temperature correction factor will be incorporated in the PR. It is called the temperature-corrected PR, and it is obtained by multiplying the PR with Y_w. Y_w is an adjustment value based on the difference between the average cell temperature at the PV plant and the average temperature taken from the PVSYST or any other model and Y_w can be expressed as

$$Y_w = \frac{1}{1 + \gamma(T_{module} - T_{ref})}$$

where γ is the temperature coefficient of the power of the PV module as recommended by the manufacturer, T_{module} is the irradiance weighted average module temperature based on measurements over the *period*, and T_{ref} is the irradiance weighted average module temperature estimated from a model such as PVSYST.

The PR calculation test is carried out for a period of 7–10 days after commissioning and stabilization of the plant. During the PR testing period, the PV plant should experience irradiance (POA) of at least 600 W/m² for at least 2 h/day.

Further, the medium-voltage panel's export meter is utilized to measure the power generation output of the solar PV plant and the calculation of the PV plant performance ratio.

The most significant and direct impacts on PV performance are: in-plane irradiance received by the PV array, the temperature of the solar cell, shading losses due to soiling. Secondary factors that may enter the assessment are clipping of the inverter and power curtailment, in which the network may not accept the available power. Due to higher DC to AC inverter ratio, the inverter cannot output more than a certain amount of power to grid is called clipping. So, all the solar modules are to be cleaned and grass and bush cutting under the panels has to done before taking up the performance ratio test and the measurement equipments such as pyranometer, temperature sensors are to be calibrated.

Measurement errors in instruments (meter, pyranometer, and thermocouple) and energy estimation errors (from PYSYST or any other yield estimation tool) should be taken into account while making a decision on the test results.

So, the PR can be used to compare the performance of plants in different locations. This is an attractive feature of the metric because it helps to understand which locations will provide the most productive plants. A site location with a colder temperature will yield a higher PR, which implies additional electricity generation assuming that all other parameters remain equal.

Hence, PR is directly dependent on the ambient temperature of the site location and varies over the season.

10.6.2 PLF calculation

PLF is plant load factor, which is called the capacity utilization factor. It is common practice to calculate the capacity utilization or plant loading for 24-h operated plants such as coal or gas thermal power plants. Thermal power plants operate 24 h a day even on cloudy or rainy days. However, a solar PV plant can operate only during sunny times of the day, that is, in the day time during sunshine hours.

PLF or CUF is calculated as

$$PLF = \frac{\text{Energy exported to grid by generation of the plant}}{\text{Total Capacity of the plant} \times 365 \times 24}$$

If the total name plate capacity of solar array is considered for the total capacity of the plant, the obtained PLF is called DC PLF. The sum of the total capacities of inverters of the solar plant is called AC capacity of the plant, the obtained PLF is called AC PLF, if AC capacity of the plant is used. Some government bodies fix the tariff rates and plant capacities based on AC PLF. For example, if a government body asks for a 20 MW solar PV plant with a PLF factor in the operational life of the plant between 12% and 19%, the plant should be designed such that the total capacity of the inverters should be 20 MW. Based on the requirement of the minimum and maximum required values of PLF, the DC-to-AC ratio of the inverter can be adjusted, and the array capacity has to be estimated.

In the solar PV Industry, DC capacity is synonymously referred to as solar PV array capacity.

10.7 Performance analysis of solar PV systems and calculations—Typical example

Example 1

The following example demonstrates how theoretical design parameters have been verified and complied based on various measurements based on actual system performance data.

The estimated versus actual energy generation figures with irradiance values are shown for a 1 MW solar PV plant in Fig. 10.11. The plane of irradiance values are estimated based on meteonorm. The

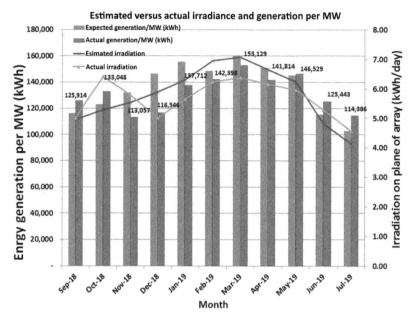

FIG. 10.11

The energy yield of a 1 MW solar PV plant.

actual observed plane of irradiance values are lower in the most of the months compared to estimation. So, the energy generation figures are changing accordingly and lower than the estimated values.

Table 10.6 shows the actual versus estimated energy yields per kWp per day are compared along with the estimated versus actual plane of array solar irradiance values. Fig. 10.11 data is tabulated in different form.

Example 2: Grand Solar, Chennai, India, has done an EPC for a 11 MWp solar PV plant, the details of which are given in Table 10.7.

Photos of the 11 MWp solar PV plant are given in Figs. 10.12 and 10.13.

The monthwise energy yield is given in Fig. 10.14 and it performs more than the expectations.

10.8 Reliability of grid-connected solar PV power systems

In view of the huge investment, developers expect that solar PV systems are to be designed in such a manner that they have built-in reliability for the total system. The reliability aspect is of paramount importance considering calculations such as LCOE that lead to decision making for solar PV power plant installations. The reliability of the total system is dependent upon the reliability of individual components

Overall Reliability = sum (Individual reliability)
= reliability of solar pv module × reliability of cables × reliability of inverter
× reliability of transformer × reliability of switchgears × reliability of grid substations
× reliability of transmission line

Table 10.6 Average day wise actual versus estimated energy yields per kWp capacity in a solar PV power plant for different months of a year.

Month	Estimated irradiation	Actual irradiation	Expected generation (kWh/KWp/day)	Actual generation (kWh/KWp/day)
Sep-18	4.94	5	3.88	4.20
Oct-18	5.28	3.31	3.96	4.29
Nov-18	5.53	5.78	4.40	3.77
Dec-18	5.89	5.01	4.73	3.76
Jan-19	6.30	5.63	5.01	4.44
Feb-19	6.94	6.24	5.31	5.09
Mar-19	7.07	6.37	5.16	4.94
Apr-19	6.66	6.16	5.03	4.73
May-19	6.23	5.96	4.69	4.73
Jun-19	4.83	5.28	3.85	4.18
Jul-19	4.14	4.56	3.32	3.69

The observed and estimated irradiance values are also compared.

Table 10.7 Features of 11 MWp solar PV plant.

Location	Madurai, India
EPC done by	Grand Solar, Chennai, India
Commissioned on	March 2019
Installed SPV capacity 11,000 kWp	11 MWp
Solar PV panel capacity	325 and 330 Wp
Number of modules used for the plant	33,579
System voltage	1000 V
Power evacuation voltage	22 kV
Transmission line	Single circuit 6 km
Solar PV panel make	Trina Solar
Inverter capacity	ABB 1732 kW × 6
Type of MMS	MS to HDG
String length	30 modules
String monitoring box	24 × 1
HT panel 22 kV	Individual VCBs × 2 22 kV group control VCB × 1
Transformer	4 Winding 550/22 kV, 6.3 MVA × 2
Earthing system—pits	36
Monitoring system	Dataglen
Average number of units generated per day	55,000

FIG. 10.12

A photo of an 11 MW solar PV plant. The modules are mounted in fixed tilt angle and each MMS with 30 modules in 2 × 15 portrait configuration.

Courtesy Grand Solar.

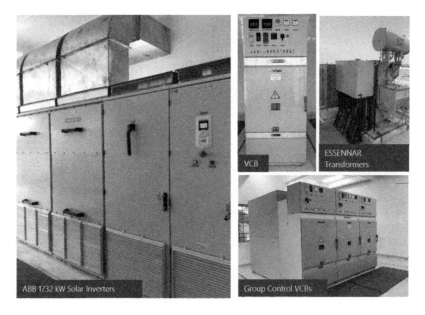

FIG. 10.13

Inverter, VCB, transformer, and HT panel used for an 11 MW solar PV plant.

Courtesy Grand Solar.

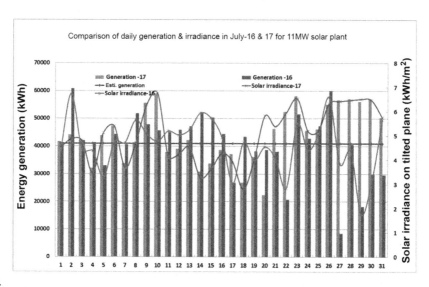

FIG. 10.14

The expected versus actual generation of 11 MW plant. Average Generation per day in July 2017 is 42458 units, whereas 39933 units is the average Generation per day in July 2016. Average Irradiance in July 2017 is 4.86 KWh/m2/day, whereas 4.45 KWh/m2/day is the average Irradiance 2016. Estimated Irradiance is 4.61 KWh/m2/day and Estimated generation per day is in July as per pvsyst is 40936 units. The energy yield depends on the radiation and irradiance fluctuates year to year.

Courtesy Grand Solar.

The system reliability is defined as the probability that a system, including all hardware, firmware, and software, will satisfactorily perform the task for which it was designed or intended for a specified time and in a specified environment. A PV system is a compilation of different components, such as solar modules, inverters, cables, transformer, HT panels, etc. A system reliability assessment is required in order to optimize decisions in design, engineering, procurement, construction, and service.

To have reliable long-term performance of the plant, a solar PV system should have better reliability. The solar modules, inverters, and AC systems are critical components deciding the reliability of solar PV systems. The reliability aspects of solar PV modules and the selection criteria are presented in Section 8.2.5 of Chapter 8 and Section 9.3.1 of Chapter 9.

Sun Edison has presented a paper [6] based on the analysis of more than 3500 issues of the plant tackled for 350 systems during 27 months without considering modules. The plants are designed and operated by Sun Edison. As per their study, they found that early stage inverter failures are common problems with the plants and another one is the failure or outage of AC systems. Fig. 10.15 shows the failures observed with different components of the systems.

The inverter problems are attributed to the failure of different components in the inverter and their percentage is shown in Fig. 10.16.

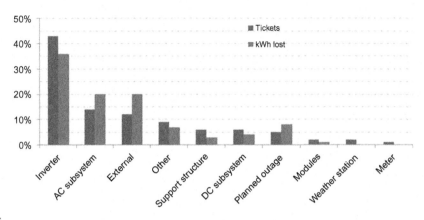

FIG. 10.15

Failures and issues with different components in solar PV systems.

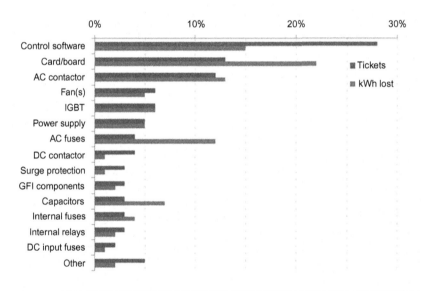

FIG. 10.16

Components affecting the failures of inverters.

So, the inverters should be maintained properly.

As shown in Fig. 10.15, the AC subsystem comes second in terms of frequency of issues in the operation of the plant. So, care must be exercised in the design and selection of the AC systems of solar PV plants. Fig. 10.17 shows the different type of failure modes of the solar module during the operation period in the field.

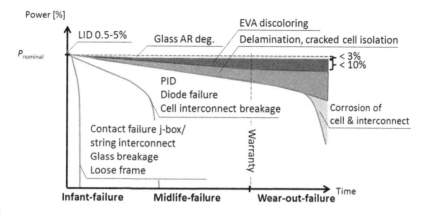

FIG. 10.17

Type of failures in solar modules in the PV plant [7].

Courtesy: International Energy Agency Photovoltaic Power Systems Programme (IEA PVPS).

10.9 O&M aspect of grid-connected solar PV power system

To meet the designed and guaranteed generation from a solar PV power plant, it is necessary that the operational and maintenance aspect is taken care of in a scheduled manner—say, periodic checks with an operation and maintenance schedule. Better operation and maintenance services help to address potential risks, while improving the LCOE and impacting the return on investment (ROI). The following are the international standards that need to be followed in the operation and maintenance practice of a solar PV power plant.

IEC 62446-1: 2016—Photovoltaic (PV) systems—Requirements for testing, documentation, and maintenance—Part 1: Grid-connected systems—Documentation, commissioning tests, and inspection.
IEC 62446-2—Photovoltaic (PV) systems—Requirements for testing, documentation, and maintenance—Part 2: Grid-connected systems—Maintenance of PV systems.
IEC YS 63049: 2017—Terrestrial photovoltaic (PV) systems—Guidelines for effective quality assurance in PV system installation, operation, and maintenance.
IEC 60364-7-712: 2017—Low-voltage electrical installations—Part 7-712: Requirements for special installations or locations—Solar photovoltaic (PV) power supply systems.
IEC 62548—Photovoltaic (PV) arrays—Design requirements.
IEC TS 62446-3:2017—Photovoltaic (PV) systems—Requirements for testing, documentation, and maintenance—Part 3: Photovoltaic modules and plants—Outdoor infrared thermography.

Operational aspect of solar PV power plants
– Operating the plant efficiently.
– Spare parts management.

– Warranty management.
– Document management.
– Interacting with stakeholders.

Performance analysis and improvement includes Solar PV power Plant, Inverter and string level monitoring. The analysis should also include the option for having custom alarms-based on specific thresholds.

The following list shows typically controlled parameters in a PV plant:

– Absolute active power control.
– Power factor control.
– Ramp control (active and reactive power if needed).
– Frequency control.
– Reactive power control.
– Voltage control.

10.9.1 Maintenance aspect of solar PV power plants

Predictive maintenance

Actions performed to help assess the condition of a PV system and its components, predict/forecast, and recommend when maintenance should be performed. Predictive maintenance has several advantages, including:

- Optimizing the safety management of equipment and systems during their entire lifetime.
- Anticipating maintenance activities (both corrective and preventive).
- Delaying, eliminating, and optimizing some maintenance activities.
- Reduce time to repair and optimize maintenance and spare part management costs.
- Reduce spare part replacement costs.
- Increase availability, energy production, and performance of equipment and systems.
- Reduce emergency and nonplanned work.
- Improve predictability.

Raw data such as GHI, irradiance on POA, active energy produced measurements, active energy consumed can be obtained.
Key Performance Indexes of PV plants such as reference yield, specific yield, and performance ratio will be higher.

10.9.2 Preventive and corrective maintenance

Preventive maintenance comprises regular visual and physical inspections as well as verification activities on all the key components of the solar PV power plant.

Corrective maintenance includes three activities:

Fault diagnosis, also called troubleshooting, to identify the fault cause and localization.
Temporary repair to restore the required function of a faulty item for a limited time until a repair is carried out.
Repair to restore the required function permanently.

Actions need to be initiated for failure correction, system breakdowns, component malfunctioning, system anomalies, or component damage that have been observed during regular inspections and also through system monitoring, system alarms, operation and maintenance personnel fault, or other reporting.

10.9.3 Performance and operational aspects

Measurements, calculations, trends, comparisons, inspections, etc., performed in order to evaluate the PV plant, segments and/or single-component performance, site conditions, equipment behavior, etc., and to provide reports and assessment studies to interested parties.

Operation and maintenance aspects of grid-connected solar PV power systems:

- Monitoring, controlling, troubleshooting, maintaining logs and records, registers.
- Supply of all spares and consumables and fixing applications as required.
- Conducting periodical checking, testing, overhauling, and preventive and corrective action.
- Upkeep of all equipment, building, roads, solar PV modules, inverters, etc.
- Periodic cleaning of solar modules as per the recommendations of the existing site conditions.
- Repair and replacement of components of the solar power plant, including all other associated infrastructure developed as a part of the EPC works that have gone faulty or worn-out components, including those that have become inefficient.
- Continuously monitoring the performance of the solar power plant and regular maintenance of the whole system including modules, PCUs, transformers, overhead HT cables as applicable, outdoor/indoor panels/kiosks as required, and other infrastructure developed as a part of EPC works in order to extract and maintain the maximum energy output from the solar power plant and serviceability from the associated infrastructure.
- Preventive and corrective maintenance of the complete solar power plant and associated infrastructure.
- Comprehensive repair, operation, and maintenance all other facilities such as roads, drainage, water supply system, CCTV network, streetlight network, and air conditioning.
- System fire detection and protection system and other civil, mechanical, electrical, and plumbing systems developed during the project as part of the solar power plant.
- PV site maintenance—Module cleaning, vegetation management, snow or sand removal, waste disposal, pest control, road management, pending repairs of plant facilities.

10.9.4 Solar PV module care

The dust deposited on the solar PV module has to be cleaned regularly, that is, twice a month. The modules that get more soiling due to nearby roads are to be frequently cleaned to avoid energy generation loss. For regular dust accumulation, simply hose with water and a brush can be used for module cleaning. If there are significant accumulations of tree sap or bird droppings, it may need a sponge or squeegee with a mild soap and water solution. The tightening of the module mounting bolts and inspection of the cables shall have to be checked frequently.

The water quality has to be tested for total dissolved salts, PH, for mineral contents and hardness. If the available water is having low mineral content and a total hardness that is less than 75 mg/L can be used for module cleaning. Reverse osmosis (RO) water is the ideal option to clean the solar PV modules. The hard water causes stains or scaling on the glass that reduce the light transmission characteristics. The water pressure is the force with which the water will exit and the water will hit your solar panels. Water pressure should not exceed 40 bar at the nozzle end. The temperature of the water being used for cleaning should ideally be the same as that of solar module temperature. Solar panels should be cleaned only during low light conditions. The surface of the solar panel should not be scratched or scrubbed to remove the stains. To remove the stub-born stains, the stain has to be soaked in water and with the usage of brush made up of soft sponge, soft bristles, nonconductive type, and nonabrasive type has to be used.

Automatic dry cleaning of the modules can be done with robot-based cleaning system. The robot is equipped with spiral bristle brushes. The brushes rotate as the robot moves over the panel, causing the dust to fall onto the ground, without requiring a single drop of water. Some of the key features of the robot are listed below. Extremely lightweight and portable type or permanently fixed types are available. The cleaning systems are fitted with lithium iron phosphate-based lithium-ion batteries having better cycle life. It uses spiral bristles type brushes for dry cleaning and there is no requirement of water. It uses BLDC motors and the cleaning speed is 4–8 m/min features like scheduled cleaning and weather-based cleaning allow the system to run completely autonomous. Some robots are equipped with on board solar panel and batteries that helps in performing night-time cleaning operation. The alignment of the structure has to be maintained properly and MMS has to be designed with the aim that modules are cleaned with robot cleaning system. The gap between the adjacent structures is fitted with a bridge and the robot continuously moves in a single row of MMS.

10.9.5 Inverter care

The best thing you can do for your inverter is to keep it cool, clean the fans, and fill the coolant, if the inverter is the cooling type.

10.9.6 Transformer care

- The oil and WTI are to be checked through a monitoring system.
- The breather has to be checked for silica gel color.
- Test the transformer oil for BDV and moisture content on yearly basis.
- Measure IR values with a suitable Megger according with rating.
- Physical examination of the diaphragm of vent pipe for any crack.
- Test OLTC oil for BDV and moisture.
- Test oil for dissolved gas analysis, acidity, tan delta, and interface tension.
- Measure DC winding resistance.
- Turns ratio test at all taps.
- Overhaul tap change and mechanism.
- Change the gaskets at all locations when leakage is found or gasket is damaged.

10.9.7 Potential transformer care

Daily, the PT should be checked for any abnormal sound/oil leakages. If any oil leakage has been observed, it should be arrested immediately. Periodically, inspections should be done for any damage to insulators. Yearly, all power cable connections should be tightened to avoid unnecessary arcing.

Thermal imaging of the primary terminals and the top dome of an operational PT needs to be carried out once a year.

Using an insulation resistance (IR) tester, an IR test can be done.

(i) Primary to earth by 5 kV.
(ii) Secondary each core to earth by 500 V.
(iii) Primary to secondary by 5 kV.
(iv) Secondary core to core by 500 V.

The turns ratio test can be done by injecting variable AC voltage in the primary winding and measuring the induced secondary voltage at different voltages and verifying the same with the transformer.

10.9.8 Circuit breakers (outdoor)

- Tightness of power connections and control wiring connections.
- Cleaning of insulators.
- Lubrication of moving parts.
- Checking of insulation resistance.
- Checking of gas pressure for SF6 circuit breaker (leakages if any).
- Checking of air pressure for pneumatic operated breaker (leakages if any).
- Checking of controls, interlocks, and protections (as per control schematic) like checking of pole discrepancy system, that is, whether all three poles are getting ON-OFF at the same time.
- Cleaning of auxiliary switches by CTC or CRC spray and checking its operation.
- Check the resistance of trip and close coils (refer to manual/test reports for resistance values).

An IR test can be done by keeping the breaker condition open. The voltage of 5 kV has to be applied between the terminals R-R′,Y-Y′,B-B′. In breaker close condition, 5 kV can be applied between the R, Y, and B phases and earth separately and insulation resistance can be measured. The following can be done: air pressure check, air cleaning with blower, auxiliary contact cleaning, motor control check, tightening of nuts and bolts, IR values of power and control circuits, and checking and adjustment of track alignment and interlocking mechanism.

10.9.9 Lightning arrestor

Insulator cleaning should be done during yearly maintenance. The tightness of the connections should be maintained to prevent arcing. On a daily basis, the leakage current should be noted and if any abnormality is found, the respective LA should be replaced.

IR testing is done between stack to stack and between each stack to earth by a suitable IR tester.

Surge counter test—By applying a 230 V AC power supply across the counter, the pointer movement should be checked for movement in the clockwise direction.

The following checks such as cleaning of porcelain insulator, checking the reading of the surge arrestor, monitoring the total leakage current and resistive current, and testing of counters can be done.

10.9.10 Isolator

Check both the male/female contacts for rigid connection and condition.

Check for alignment of male and female contacts and in the event of any misalignment, rectification needs to be carried out.

Cleaning of insulators is to be done on regular intervals.

Lubrication of all moving parts on a regular basis.

Tightness of all earthing connections.

In case of an isolator with an earth switch, check the electrical and mechanical interlock, that is, the isolator can be closed only when the E/switch is in open condition and vice versa.

As isolators are operated on no load, check the interlock with the circuit breaker, if provided. Isolators can be operated when the breaker is in the OFF condition.

The motor operating mechanism box, in case of motor operated isolators, should be checked for inside wiring, terminal connectors, etc.

Check the panel indications, that is, the semaphore and bulbs if provided (isolator and earth switch—close and open condition) and rectify if required.

IR Testing—Phase to phase and phase to earth by a 5 kV Megger.

Contact resistance check—Measure the contact resistance by a suitable microohmmeter.

10.9.11 Current transformer (outdoor)

Daily, the CT should be checked for any abnormal sound/oil leakages. If any oil leakage has been observed, it should be arrested immediately. Periodically, an inspection should be done for any damage to the insulators. Yearly, all power cable connections should be tightened to avoid unnecessary arcing. All secondary connections should be kept earthed of any spare cores and care should be taken to avoid the open circuits on the secondary sides. Scanning of primary terminals, the top dome of an operational, and the active CT should be performed once a year using an infrared thermal vision camera. This scanning can be done with help of an infrared thermal vision camera.

IR testing with IR tester (Megger)

(i) Primary to earth by 5 kV.
(ii) Secondary each core to earth by 500 V.
(iii) Primary to secondary by 5 kV IR.
(iv) Secondary core to core by 500 V IR tester.

Ratio test—Inject current in the primary winding and measure the induced secondary current for different readings and verify with the CT ratio.

Polarity test: The Polarity of the CTs is extremely important. Just like a battery, a CT has a polarity. The polarity determines the direction of the secondary current in relation to the primary current.

A wrong connection of the CTs can cause false operation of the protection relays. Hence, it is vital to ensure that the CTs are connected with the correct polarity.

Winding resistance test—Measure secondary winding resistance by a microohmmeter.

10.9.12 Switchyard, control panel and relays

- Check switchyard at regular intervals for any abnormal observations, visual defects, live sparks, contact loosening, red hotspots, and loosening of bolts.

- Check the earth resistance of earthing.
- Check the protection and control circuits of each piece of equipment monthly.
- Check operation and interlock of all equipment monthly.

Periodical tightness of all terminals should be checked in the control panel.

All relays should be calibrated periodically. Periodical cleaning of the panel should be done. Incoming and outgoing cables should be checked for any fissures of insulation. Panel body earthing should be checked and tightened if necessary.

L T panel can be checked for the following
- Visual inspection of panel.
- Checking and sealing of cable entry holes.
- Checking of DC supply and control switchgear.
- Checking of indication lamps, replacement if required.
- Checking of indication meter and rectification/replacement if required.
- Checking/replacement of fuses if required.
- Checking of busbar connection, tightening of nut bolts, cleaning of busbar if required.
- Cleaning and tightening of busbar in the busbar chamber.
- Tightening of all earthing connections.
- Cleaning of the inside and outside panels using blowers and a vacuum cleaner.

Earthing system
- Checking all earthing connections and joints, and cleaning and tightening thereof.
- Putting an adequate quantity of water in the earth pits.
- Checking and recording earth resistance of all points and pits and taking corrective action to improve if required.

10.9.13 Infrared thermography

Infrared (IR) thermographic data provide clear and concise indications about the status of the PV modules and arrays, and they are used in both predictive and corrective maintenance.

In order for the thermographic data to be usable, a number of minimum requirements have to be met. The irradiance shall equal a minimum of $600\,W/m^2$ and shall be continuously measured onsite, ideally orthogonally to the module surface. Infrared cameras need to possess a thermal resolution of at least 320×240 pixels and a thermal sensitivity of at least $0.1\,K$.

IR thermography can also be used to inspect other important electrical components of a PV plant, such as cables, contacts, fuses, switches, inverters, and batteries. The utilization of IR thermography alone is sometimes not enough to reach a conclusive diagnosis on the cause and the impact of certain PV module failures. Therefore, it is usually combined with the following complementary field tests.

IR is a great tool, which enables to assess entire sites at a reasonable cost. However, the IR method is limited in its effectiveness of providing conclusive detail on module defects or deterioration phenomena. IR will find issues such as strings out, modules out, activated diodes, and interpretation of hot spots. The identified defective modules are checked for the *I-V* curve by a hand-held array *I-V* tester. The inaccuracy in *I-V* curve tracing with this equipment is very high, up to 30%.

The solar modules will be tested using a mobile van-based flash tester. The EL test also can be performed to identify the defects in the solar PV modules.

At present, thermal imaging cameras mounted on drones are being used for thermal scanning of solar modules by an aerial survey. The software will interpret the post thermal imaging data. This technique is becoming very useful for solar PV plants. Aerial IR survey finds the PID affected modules on sites, where electroluminescence (EL) and *I-V* testing is required to verify the PID. IR thermography is the best way to spot activated, or faulty bypass diodes, and cheaper and quicker than string testing at site. Drone IR enables to see strings out, modules out and indicate where the issues causing hot spots. The diode failures can be recognized with a certainty of 99.5%. Hot spots are symptoms to a variety of causes. Defective diodes themselves can cause hot spots on the module. But defective diodes—like other failure causes—show distinctive characteristics. Diode failures, for example, result in a sudden drop of voltage, which is around 33% for each burnt diode. Figs. 10.18 and 10.19 show the infrared thermal images of loose connection in a string combiner box. The loose contact is seen with high spot temperature. If the defect is not rectified, the burning of string combiner takes place due to dc arc faults. So, it is essential to tight all the connections of SCB once in 6 months.

Fig. 10.20 shows the thermal image of the solar module with broken glass. A temperature difference of 11–17°C has been observed in the image. Fig. 10.21 shows the thermal image of 230 W solar module having failed bypass diode. One complete column in the image is showing bright spot and the temperature are higher. Due to failure of the bypass in short mode, the corresponding string does not contribute power and generated power is dissipated in the circuit causing of temperature rise. This confirmed in *I-V* curve as shown in Fig. 10.22. The *I-V* curve shows a kink at 31 V and gives two-thirds of V_{oc} of the good module and the produced power is 88.6 W at 56°C.

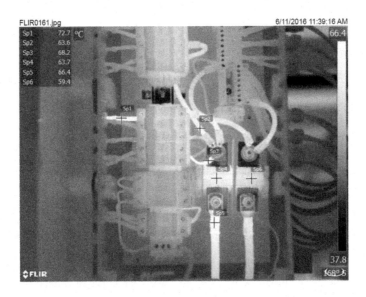

FIG. 10.18

Thermal image of a loose connection in a string combiner box.

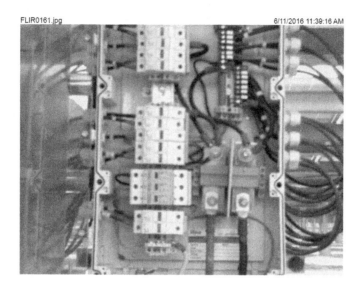

FIG. 10.19

Image of a string combiner box.

FIG. 10.20

Thermal image of a broken solar PV module.

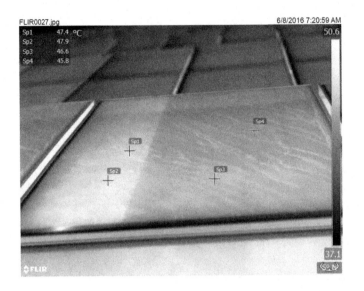

FIG. 10.21

Thermal image of a bypass diode failure in the solar module.

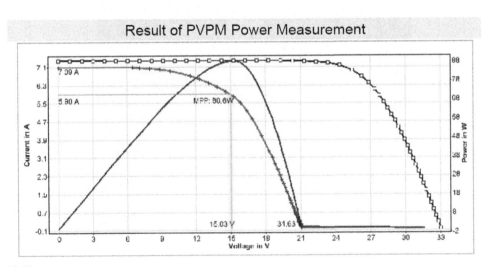

FIG. 10.22

I-V curve of a solar module whose bypass diode failed.

Fig. 10.23 shows the IR images of module having hot spot. The bright hot spots are seen in the images. The hot spot arises due to localized soiling, shadowing due grass, and bushes are any defect causing less current in the modules. These hotspots will have developed into perfect defects in the module later, if corrective steps are not taken. Figs. 10.24 and 10.25 show the front and rear side IR images

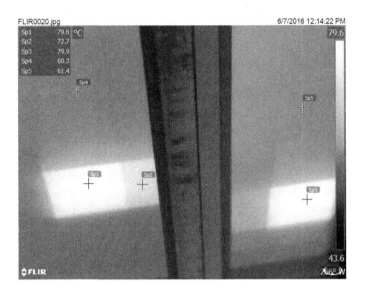

FIG. 10.23

Thermal image of a solar module with a hotspot.

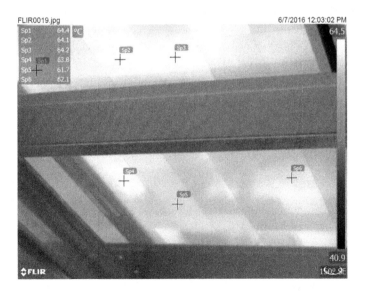

FIG. 10.24

Thermal image of a solar module with a PID problem.

FIG. 10.25

Rear side of a solar module with a PID problem.

of module affected with PID problem. The observed chess pattern is the signature for PID problem. To confirm the PID effect, the same module has been tested by portable *I-V* tester and the obtained *I-V* curve is shown in Fig. 10.26. In *I-V* curve, the module is exhibiting very low fill factor producing 126 W at 55°C. The shunt resistance is drastically reduced and series resistance increased causing the reduction of fill factor, which confirms the module degradation is due to PID effect. Thus thermal imaging combined with *I-V* curve testing and EL testing of modules in the site are useful for the operation and maintenance of the solar PV plant.

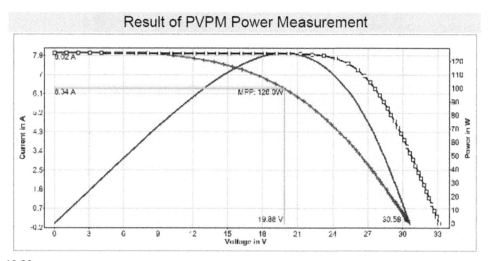

FIG. 10.26

I-V curve of a solar module with a PID problem.

References

[1] D. Simpson, Basics of medium-voltage wiring for PV power plant AC collection systems, Solarpro Magazine (6.1) (2012–2013) 58–80 (December-January).

[2] O.H. Abdalla, A.A.A. Mostafa, Technical Requirements for Connecting Solar Power Plants to Electricity Networks; 2019. (2019). https://doi.org/10.5772/intechopen.88439.

[3] IEC 61724, Photovoltaic System Performance Monitoring—Guidelines for Measurement, Data Exchange and Analysis; 1998, (1998).

[4] Technical Note on 'Performance Ratio—Quality Factor for the PV Plant', SMA, http://files.sma.de/dl/7680/Perfratio-TI-en-11.pdf.

[5] T. Dierauf, A. Growitz, S. Kurtz, J.L.B. Cruz, E. Riley, C. Hansen, Weather-Corrected Performance Ratio, (Technical Report-NREL/TP-5200-57991)(April 2013).

[6] A. Golnas, PV system reliability: an operator's perspective. IEEE J. Photovoltaics (2013). https://doi.org/10.1109/JPHOTOV.2012.2215015.

[7] IEA PVPS, Review of Failures of Photovoltaic Modules Report IEA-PVPS T13–01:2014, Programme. International Energy Agency Photovoltaic Power Systems (2014).

Index

Note: Page numbers followed by *f* indicate figures and *t* indicate tables.

Printed in the United States
By Bookmasters